Yearbook on Space Policy

More information about this series at http://www.springer.com/series/8166

Yearbook on Space Policy

Edited by the
European Space Policy Institute
Director: Jean-Jacques Tortora

Cenan Al-Ekabi • Stefano Ferretti
Editors

Yearbook on Space Policy 2016

Space for Sustainable Development

Springer

Editors
Cenan Al-Ekabi
European Space Policy Institute
Vienna, Austria

Stefano Ferretti
European Space Policy Institute
Vienna, Austria

ISSN 1866-8305 ISSN 2197-9405 (electronic)
Yearbook on Space Policy
ISBN 978-3-030-10205-0 ISBN 978-3-319-72465-2 (eBook)
https://doi.org/10.1007/978-3-319-72465-2

Softcover re-print of the Hardcover 1st edition 2018

Printed on acid-free paper

This Springer imprint is published by Springer Nature
The registered company is Springer International Publishing AG
The registered company address is: Gewerbestrasse 11, 6330 Cham, Switzerland

Preface

The United Nations recently defined the 2030 Agenda which consists of 17 Sustainable Development Goals to be achieved by all countries by 2030. In this period, the space sector is also on the verge of a new revolution, which is linked to the increased digitalization of the industrial and service sectors and the increasing availability of large amounts of free and open data on planet Earth. This context opens up new opportunities for overcoming the many challenges ahead by working together, pooling resources and information from a variety of key actors, and integrating them in a holistic approach toward the full implementation of the 2030 Agenda. In order to clarify and shed additional light on the expected impacts of these trends, the European Space Policy Institute (ESPI) has decided to focus on the topic of space for sustainable development for its *Yearbook on Space Policy 2016*.

Traditionally, the first part of the yearbook sets out a comprehensive overview of the economic, political, technological, and institutional trends that affected space activities in 2016. It is prepared in-house in ESPI, and while its perspective is European, it also provides a comparative analysis of space developments around the world.

The second part of the ESPI yearbook approaches the overall theme from an analytical perspective. This year, 13 contributions are included, bringing together the views of professionals from space agencies, the wider development community, academia, and industry and new private actors, as well as European and international institutions aiming at a stronger coordination among space agencies, IGOs, NGOs, private sector, academia, and sustainable development actors. Several key prerequisites for a successful contribution of space activities to the Sustainable Development Goals were identified. For instance, space actors are urged to adopt an end-to-end approach on identifying user needs; to that end, a greater inter-sectoral, interinstitutional (e.g., state and NGO), and international cooperation and information sharing should be sought. Moreover, there is a need to improve technology awareness among all actors to strengthen capacity building, beckoning a formalized cross-sectoral dialogue platform. And there is a need to define basic common requirements serving the Sustainable Development Goals for the next generation of space infrastructure, in order to improve access to space services and enhance international cooperation. The contributions in the second part of the yearbook help to put forward concrete proposals for improved dialogue and cooperation.

The third part of the yearbook continues its character as an archive of space activities. Again prepared in-house by ESPI, a chronology, data about institutions, and a bibliography are provided where readers of the now ten volumes of the yearbook can identify statistical developments and evolutions.

In closing, I would like to thank the contributors of the articles in the second part for their engagement in this publication, as well as the ESPI staff that have been instrumental for its production.

Vienna, Austria

Jean-Jacques Tortora
Director of ESPI

Contents

List of Acronyms: Acronym Explanation

A

A3R	Arkyd 3 Reflight spacecraft
AAD	Advanced Air Defence
ABS	Asia Broadcast Satellite
ADPC	Asian Disaster Preparedness Center
AG	Aktiengesellschaft
AGRHYMET	Agriculture, Hydrology and Meteorology Regional Center
AIA	Atmospheric Imaging Assembly
Airbus D&S	Airbus Defence and Space
AIM	Asteroid Impact Mission
AIS	Automatic Identification Satellites
ALR	Austrian Aeronautics and Space Agency
AMESD	African Monitoring of the Environment for Sustainable Development
AMS	Alpha Magnetic Spectrometer
APAC	China and other Asia Pacific
ARISE	Agriculture Resource Inventory and Survey Experiment
ASAP	Austrian Space Application Programme
ASAT	Anti-Satellite
ASEAN	Association of Southeast Asian Nations
ASI	Agenzia Spaziale Italiana (Italian Space Agency)
ASL	Airbus Safran Launchers
ASPERA-3	Mars Express' Analyzer for Space Plasmas and Energetic Atoms
ATK	Alliant Techsystems Inc.
ATV	Automated Transfer Vehicle
AWE	AWE Management Limited
AWS	Automatic Weather Stations

B

BDS	BeiDou Navigation Satellite Systems
BELSPO	Belgian Federal Science Policy Office
BHRS	Belgian High Representation for Space Policy
BIS	Business, Innovation and Skills
BMD	Ballistic Missile Defence
BMVIT	Austrian Federal Ministry for Transport, Innovation and Technology

C

CAD	Computer Aided Design
CALET	CALorimetric Electron Telescope
CAPE	Crop Acreage and Production Estimation
CASC	China Aerospace Science and Technology Corporation
CAST	China Aerospace Science and Technology Corp.
CATHALAC	Water Center for the Humid Tropics for Latin America and the Caribbean
CBERS	China-Brazil Earth Resource Satellite
CD	Conference on Disarmament
CDOP 3	Third Continuous Development and Operations Phase
CDRA	Carbon Dioxide Removal Assembly
CDTI	Centre for the Development of Industrial Technology
CEC	Consortium for Educational Communication
CELAC	Community of Latin American and Caribbean States
CENI	Commission Électorale Nationale Indépendante
CEOS	Committee on Earth Observation Satellites
CERSGIS	Centre for Remote Sensing and Geographic Information Services
CET	Centre for Education Technology
CFAS	Federal Commission for Space Affairs
CGWIC	China Great Wall Industry Corporation
CHF	Swiss franc
CIET	Central Institute of Educational Technology
CILSS	Comité permanent Inter-Etats de Lutte contre la Sécheresse dans le Sahel (Ghana)
CIS	Communications, Intelligence & Security
CLARREO	Climate Absolute Radiance and Refractivity Observatory
CMA	Governing Body of the Paris Agreement
CME	Coronal Mass Ejection
CMSA	China Manned Space Agency
CNES	Centre National d'Études Spatiales (French Space Agency)
CONAE	Argentinian Space Agency
CONCORDi	European Commission's biennial Conferences on Corporate R&D and Innovation

COP	Conference of the Parties
COPUOS	United Nations Committee on the Peaceful Uses of Outer Space
COSTIND	Commission for Science, Technology and Industry for National Defense
CRESDA	Centre for Resources Satellite Data and Applications
CRISM	Compact Reconnaissance Imaging Spectrometer for Mars
CRS	Commercial Resupply Services
CSA	Canadian Space Agency
CSE	Centre de Suivi Ecologique (Senegal)
CSES	China Seismo-Electromagnetic Satellite
CubeSats	Cube Satellites

D

DAMPE	Dark Matter Particle Explorer
DARS	Digital Audio Radio Service
DBS	Direct Broadcast Services
DECU	Development and Educational Communications Unit
DJEI	Department of Jobs, Enterprise & Innovation
DLR	Deutsches Zentrum für Luft- und Raumfahrt (German Aerospace Center)
DoD	Department of Defence
DRDO	Defence Research and Development Organisation
DSC	Decision Support Center
DSCOVR	Deep Space Climate Observatory
DSI	Deep Space Industries
DTH	Direct To Home

E

EBIT	Earnings Before Interest and Taxes
EBITDA	Earnings Before Interest, Taxes, Depreciation and Amortization
ECA	Evolution Cryotechnique type A
EDM	ExoMars Entry, Descent and Landing Demonstrator Module
EDT	Electrodynamic Tether
EELV	U.S. Evolved Expendable Launch Vehicle Program
EIB	European Investment Bank
EIF	European Investment Fund
EIT	Extreme ultraviolet Imaging Telescope
ELIRG	Extremely Luminous Infrared Galaxies
ELV	European Launch Vehicle
EM	Exploration Mission

EMEA	Europe, the Middle East and Africa
EMMRCs	Educational Multimedia Research Centres
EO	Earth Observation
EON-MW	Earth Observing Nanosatellite-Microwave
EPS-SG	European Polar System Second Generation
ERG	Exploration of energization and Radiation in Geospace
ESA	European Space Agency
ESA DG	ESA Director General
ESPI	European Space Policy Institute
ESSO	Earth Systems Science Organization
ETC	Emergency Telecommunications Cluster
EU	European Union
EUMETSAT	European Organisation for the Exploitation of Meteorological Satellites
EUTELSAT	European Telecommunications Satellite Organisation
EVE	EUV Variability Experiment

F

FAA	Federal Aviation Administration
FASAL	Forecasting Agricultural output using Space, Agrometeorology and Land based observations
FCT	Foundation for Science and Technology
FFG	Austrian Research Promotion Agency
FFL	Fondation Follereau Luxembourg
FOCAC	Forum on China-Africa Cooperation
FSS	Fixed Satellite Services
FY	Fiscal Year

G

GDP	Gross Domestic Product
GEO	Geostationary Earth Orbit
GEO	Group on Earth Observation
GEOSS	Global Earth Observation System of Systems
GERD	Gross Domestic Expenditure on R&D
GFDRR	Global Facility for Disaster Reduction and Recovery
GGIM	Global Geospatial Information Management
GmbH	Gesellschaft mit beschränkter Haftung
GMT	Greenwich Mean Time
GNI	Gross National Income
GNSS	Global Navigation Satellite Systems

GOES-R	Geostationary Operational Environmental Satellite R
GPS	Global Positioning System
GRaND	Gamma Ray and Neutron Detector
GSA	European GNSS Agency
GSLV	Geosynchronous Satellite Launch Vehicle
GSRT	General Secretariat for Research and Technology
GSSAP	Geosynchronous Space Situational Awareness Program
GTO	Geosynchronous Transfer Orbits

H

HAT	Human African trypanosomiasis/Sleeping sickness
HDTV-EF2	High Definition TV Camera-Exposed Facility 2
HFA	Hyogo Framework for Action
HMI	Helioseismic and Magnetic Imager
Hot DOG	Hot, Dust-Obscured Galaxy
HTV	H-2 Transfer Vehicle
HR	High-Resolution
HSO	Hungarian Space Office
HSTI	Human Space Technology Initiative

I

I&B	Information and Broadcasting
IAA	International Academy of Astronautics
IAC	International Astronautical Congress
IADC	Inter-Agency Space Debris Coordination Committee
IAEG-SDGs	UN Statistical Commissions' Interagency Expert Group
IARI	Indian Agriculture Research Institute
IASC	Inter-Agency Standing Committee
ICG	International Committee on Global Navigation Satellite Systems
ICIMOD	International Centre for Integrated Mountain Development
ICoC	Draft International Code of Conduct for Outer Space Activities
ICS	Information and Communication Systems
ICRC	International Committee of the Red Cross
ICT	Information and Communications Technology
IEV	Intermediate Experimental Vehicle
IFIs	International Financial Institutions
IKAR	Interdepartmental Committee for Space Affairs
IGS	International GNSS Service
ILS	International Launch Services
IMF	International Monetary Fund

IMU	Inertial Measurement Unit
INCOIS	Indian National Centre for Ocean Information Services
INTA	National Institute of Aerospace Technology
IODC	Indian Ocean Data Coverage
IOs	Regional Organizations and International Organizations
IoT	Internet of Things
IPP	International Partnership Programme
IR	Intermediate Result
IRIS	Interface Region Imaging Spectrograph
IRNSS	India Regional Navigation Satellite System
ISC	International Satellite Company Limited
ISED	Innovation, Science and Economic Development
ISIS	Islamic State
ISRO	Indian Space Research Organization
ISS	International Space Station
ITAR	International Traffic in Arms Regulations
ITU	International Telecommunication Union
IUCAA	Inter-University Centre for Astronomy and Astrophysics
IUVS	Imaging UltraViolet Spectrograph

J

J-PAL	Abdul Latif Jameel Poverty Action Lab
J-SSOD	JEM Small Satellite Orbital Deployer
JAXA	Japan Aerospace Exploration Agency
JIRAM	Jovian Infrared Auroral Mapper
JPSS	Joint Polar Satellite System
JUICE	Jupiter Icy moon Explorer

K

K2	Kepler 2
KARI	Korea Aerospace Research Institute (Korean Space Agency)
KITE	Kounotori Integrated Tether Experiment

L

L2	Earth-Moon Lagrange
LAXPC	Large Area X-ray Proportional Counter
LEO	Low Earth Orbit
LRO	Lunar Reconnaissance Orbiter
LULC	Land Use and Land Cover

M

MARSIS	Mars Advanced Radar for Sub-Surface and Ionospheric Sounding
MAVEN	Mars Atmosphere and Volatile Evolution
MDA Corp.	MacDonald, Dettwiler and Associates Ltd.
MDGs	Millennium Development Goals
MDI	Michelson Doppler Imager
Melco	Mitsubishi Electric Co.
MEO	Medium Earth Orbit
MERLIN	Methane Remote Sensing LIDAR Mission
MESA	Monitoring the Environment and Security
MESSENGER	Mercury Surface, Space Environment, Geochemistry and Ranging
Metop	Meteorological Operational Satellite
Metop-SG	Metop Second Generation
MEXT	Ministry of Education, Culture, Sports, Science and Technology
MFG	Meteosat First Generation
MIT	Massachusetts Institute of Technology
MIUR	Ministry of Education, University and Research
MMO	Mercury Magnetospheric Orbiter
MOD	Ministry of National Defense
MOKV	Multi-Object Kill Vehicle
MOM	Mars Orbiter Mission
MOSDAC	Meteorological and Oceanographic Satellite Data Archival Centre
MoU	Memorandum of Understanding
MPO	Mercury Planetary Orbiter
MRO	Mars Reconnaissance Orbiter
MSF	Médecins Sans Frontières
MSL	Mars Science Laboratory
MSM	Methane Sensor for Mars
MSG	Meteosat Second Generation
MSS	Mobile Satellite Service
MTG	Meteosat Third Generation
MTM	Mercury Transfer Module

N

NASA	National Aeronautics and Space Administration
NATO	North Atlantic Treaty Organization
NCERT	National Council for Educational Research and Training
NCSTE	China's National Centre for Science and Technology Evaluations
NDAA	National Defense Authorization Act
NDCs	Nationally Determined Contributions
NEC	Nippon Electric Company
NEO	Near-Earth Orbit

NGA National Geospatial-Intelligence Agency
NGCV Next Generation Crew Vehicle
NGO Non-Governmental Organization
NNRMS National Natural Resources Management System
NOAA National Oceanic and Atmospheric Administration
NRO National Reconnaissance Office
NRSC National Remote Sensing Centre
NSC National Space Council
NSC Norwegian Space Centre
NSO Netherlands Space Office
NOW Netherlands Organisation for Scientific Research

O

OBIA Object-Based Image Analysis
OCO Orbiting Carbon Observatory
ODA Official Development Assistance
OECD Organisation for Economic Co-operation and Development
OHB Orbitale Hochtechnologie Bremen
OOF Other Official Flows
OPEC Organization of the Petroleum Exporting Countries
ORU Orbital Replacement Units
OST Outer Space Treaty

P

PACE Plankton, Aerosol, Cloud, ocean Ecosystem
PAD Prithvi Air Defense
PAROS Prevention of an Arms Race in Outer Space
PAS Polish Academy of Sciences
PES Payment for Ecosystem Services
PHA Potentially Hazardous Asteroids
PLA People's Liberation Army
PND Portable Navigation Devices
PNTAB Position, Navigation and Timing Advisory Board
POLSA Polish Space Agency
PPP Public-Private Partnership
PPWT Draft Treaty on the Prevention of the Placement of Weapons in Outer Space and of the Threat or Use of Force against Outer Space Objects
PRM Period Reduction Maneuver
PRS Public Regulated Service
PS-TEPC Position Sensitive Tissue Equivalent Proportional Chamber

PSA	Programme on Space Applications
PSLV	Polar Satellite Launch Vehicle
PROBA	Project for Onboard Autonomy
PUMA	Preparation for the Use of MSG in Africa programme

Q

QZSS	Quasi-Zenith Satellite System

R

RCM	RADARSAT Constellation Mission
RCMRD	Regional Centre for Mapping of Resources for Development
REDD+	Reducing Emissions from Deforestation and Degradation
RHESSI	Reuven Ramaty High Energy Solar Spectroscopic Imager
RKV	Redesigned Kill Vehicle
RLV	Reusable Launch Vehicle
ROSA	Romanian Space Agency
Roscosmos	Roscosmos State Corporation
RRS	Regional Radiocommunication Seminars
RSCC	Russian Satellite Communications Company

S

SAARC	South Asian Association for Regional Development
SAB	Space Advisory Board
SAC	Space Applications Center
SAF	Satellite Application Facilities
SAHEL	Sub-Saharan initiative for Telemedicine
SDGs	Sustainable Development Goals
SDO	Solar Dynamics Observatory
SDP	Space for Development Profile
SEI	Stockholm Environment Institute
SEPs	Solar Energetic Particles
SERI	State Secretariat for Education, Research and Innovation
SES	Société Européenne des Satellites
SIETs	State Institutes of Educational Technology
SHARAD	Shallow Subsurface Radar
SIA	Satellite Industry Association
SIG	Spatial Informatics Group
SITE	Satellite Instructional Television Experiment

SLS	Space Launch System
SMPAG	Space Mission Planning Advisory Group
SNC	Sierra Nevada Corporation
SNSB	Swedish National Space Board
SOHO	Solar and Heliospheric Observatory
SpaceX	Space Exploration Technologies
SRC	Space Research Centre
SRON	Netherlands Institute for Space Research
SS/L	Space Systems/Loral
SSO	Sun-synchronous orbit
SST	Space Surveillance and Tracking
STEREO	Solar Terrestrial Relations Observatory
STSC	Scientific and Technical Subcommittee

T

TCBM	Transparency and Confidence-Building Measures
TDP	Technology Demonstration Programme
TEU	Treaty on European Union
TGO	ExoMars Trace Gas Orbiter
THAAD	Terminal High Altitude Area Defense system
TRAI	Telecom Regulatory Authority of India

U

UAE	United Arab Emirates
UAV	Unmanned Aerial Vehicle
UGC	University Grants Commission
UK	United Kingdom
ULA	United Launch Alliance
UN	United Nations
UNCOPUOS	United Nations Committee of Peaceful Uses of Outer Space
UNCTAD	United Nations Conference on Trade and Development
UNDP	United Nations Development Programme
UNDSS	United Nations Department of Safety and Security
UNFCCC	United Nations Framework Convention on Climate Change
UNGA	United Nations General Assembly
UNGIWG	United Nations Geographic Information Working Group
UNIDIR	United Nations Institute for Disarmament Research
UNOOSA	United Nations Office of Outer Space Affairs
UNSDI	United Nations Spatial Data Infrastructure
UNISPACE	United Nations Conference on the Exploration and Peaceful Uses of Outer Space

UN-SPIDER	United Nations Platform for Space-based Information for Disaster Management and Emergency Response
U.S.	United States of America
U.S. MDA	Missile Defense Agency
USAID	U.S. Agency for International Development
USAT	Ultra Small Aperture Terminals

V

VAST	Vietnamese Academy of Science and Technology
VHR	Very High Resolution
VIR	Visible and Infrared Mapping Spectrometer
VKO	Aerospace Defence Forces
VSAT	Very Small Aperture Terminals

W

WFP	World Food Programme
WG	Working Group
WGP	World Gross Product
WISE	Wide-field Infrared Survey Explorer
WRS	World Radiocommunication Seminar

Part I
The Year in Space 2016

Chapter 1
European Space Activities in the Global Context

Cenan Al-Ekabi

1.1 Global Political and Economic Trends

1.1.1 Global Economic Outlook

The "World Economic Situation and Prospects" report is the United Nation's lead publication in the annual discussion of current economic trends and prospects. In 2016, the global economy appeared stuck in a prolonged period of slow economic growth and dwindling international trade growth, with both rates at their lowest since the 2009 recession that followed the financial crisis. World gross product (WGP) had dropped to 2.2% in 2016, below the average rate of 2.5% since 2012, and well below the 3.4% growth rate observed in the decade before the crisis, with the sluggishness characterised by diminished productivity growth, increased levels of debt, low commodity prices, and continued conflict and geopolitical tensions.[1]

WGP growth in developed economies dropped to 1.5% in 2016 from 2.1% growth in 2015; moreover, growth in output was expected to remain below 2% for 2017 and 2018. In the eurozone, new EU members showed the most growth at 3.0%, while Western European economies continued with 1.7% for 2016; overall, growth in the European Union had decreased to 1.8% in 2016 from 2.2% in 2015, and it was expected to remain steady for the upcoming years. US growth in global output dropped to 1.5% in 2016 from 2.6% in 2015 but was expected to increase to 1.9% in 2017 and 2.0% in 2018. Japan's global output also increased by 0.5% in 2016,

[1] "World Economic Situation and Prospects 2017." 17 Jan. 2017. United Nations 16 Mar. 2017 https://www.un.org/development/desa/dpad/wp-content/uploads/sites/45/publication/2017wesp_full_en.pdf

C. Al-Ekabi (✉)
European Space Policy Institute (ESPI), Vienna, Austria
e-mail: Cenan.Al-Ekabi@espi.or.at

C. Al-Ekabi, S. Ferretti (eds.), *Yearbook on Space Policy 2016*,
Yearbook on Space Policy, https://doi.org/10.1007/978-3-319-72465-2_1

lowering by 0.1 from the 0.6% growth in output in 2015; its output growth was expected to increase to 0.9% for 2017 and 2018.[2]

Growth in transition economies declined for the second consecutive year, contracting by 0.2% in 2016 after a previous contraction of 2.8% in 2015, but was expected to increase by 1.4% in 2017 and 2.0% in 2018, driven mainly by increased performance in South-Eastern Europe. Developing economies showed the most growth, increasing by 3.6% in 2016 and 3.8% in 2015; growth in output was expected to increase by 4.4% in 2017 and 4.7% in 2018. Developing economies remained the fastest growing, driven mainly by India, China, and other East and South Asian economies; African economies also continued to show positive growth, while South American economies continued to show weak performance for 2016.[3]

WGP was expected to increase by 2.7% in 2017 and 2.9% in 2018, due mainly to the stabilisation from some short-term shocks that had restrained growth in 2016, such as the nonfarm inventory destocking cycle and contractions in oil-related sector investment in the United States and sharp terms-of-trade shock experienced by commodity exporters. Rather than signalling a revival of the economy, as the factors underpinning sluggish economic growth tend to be self-reinforcing and will likely prolong the slowdown, this relatively low rate of growth risks hampering progress towards achieving the Sustainable Development Goals (SDGs) of the United Nations 2030 Agenda for Sustainable Development, which aims to eradicate extreme poverty and create decent work for all.[4]

1.1.2 Political Developments

1.1.2.1 Geopolitics

Several significant world events in 2016 are likely to continue in 2017.

On 23 June 2016, UK citizens voted to end the UK's membership of the European Union (EU). Despite a narrow split where 51.9% of voters (mainly in rural parts of England and Wales) chose to leave and 48.1% of voters (mainly Scotland and North Ireland) chose to remain[5] and a November 2016 High Court ruling that the British government must get Parliament's approval before the "Brexit" process could begin,[6] the UK's separation from the EU seems imminent. In order for the United Kingdom to withdraw from the EU, it must trigger Article 50 of the Treaty on European Union (TEU). Article 50 is triggered when an EU member state has noti-

---

[2] Ibid., 3.

[3] Ibid.

[4] Ibid. 1–38.

[5] "EU Referendum – Results." BBC News 17 Aug. 2017 http://www.bbc.com/news/politics/eu_referendum/results

[6] Castle, Stephen, and Steven Erlanger. "'Brexit' Will Require a Vote in Parliament, U.K. Court Rules." 3 Nov. 2016. The New York Times 17 Aug. 2017 https://www.nytimes.com/2016/11/04/world/europe/uk-brexit-vote-parliament.html

fied the European Council of its intent to leave, opening a 2-year period in which a leaving agreement is negotiated setting out the arrangements for the withdrawal and outlining the UK's future relationship with the EU.[7] On 2 October 2016, Theresa May – who replaced David Cameron as prime minister when he stepped down the day following the Brexit vote – announced that she would trigger Article 50 by the end of March 2017.[8]

Donald Trump won the US presidential election to become the 45th president of the United States. In an election race where it appeared inevitable that Hillary Rodham Clinton would easily sweep both the popular and electoral vote from the vitriolic Trump campaign, the turmoil that followed the hacking of the Democratic National Committee's email systems by Russian intelligence groups in the late 2015 and their release in mid-2016 via WikiLeaks mortally wounded the front-running candidate's campaign. The FBI's announcement that it was reopening and once again closing its investigation into Hillary Clinton's poor handling of emails just days before the ballot served to reignite mistrust in the candidate. On 8 November 2016, Donald Trump won the electoral vote with 306 of the 538 votes available (270 votes are needed to win); Hillary Clinton had won the popular vote with a 48.5% share of the votes cast to Donald Trump's 46.4% share of votes.[9] In his campaign, Donald Trump promised to build a wall on the US-Mexican border, to pull out of major US trade agreements, to review the benefit of the NATO alliance, and to take a tougher line with China and a softer line with Russia. Incidentally, the US CIA and the FBI have concluded with "high confidence" that Russian President Vladimir Putin had personally authorised the Kremlin operation to help elect Trump.[10]

In Syria, Bashar al-Assad's regime, backed by Russian air support, Lebanese Hezbollah, and Iranian militia, began launching an offensive operation against the rebel-held parts of the city of Aleppo in June 2016.[11] The rebels are supported by the United States, Turkey, Saudi Arabia, and other Gulf states. Despite a short-lived ceasefire attempt in September 2016, brokered by Russia and the United States, the assault by Syrian and Russian forces continued, developing into a humanitarian crisis as humanitarian convoys could not deliver aid because of the danger and the inability to obtain simultaneous security guarantees from all sides.[12] An evacuation

[7] Article 50. Consolidated version of the Treaty on European Union. OJ C 326, 26.10.2012, pp. 13–390 http://eur-lex.europa.eu/legal-content/EN/TXT/HTML/?uri=CELEX:12012M/TXT&from=EN

[8] "Brexit: Theresa May to trigger Article 50 by end of March." 2 Oct. 2016. BBC News 17 Aug. 2017 http://www.bbc.com/news/uk-politics-37532364

[9] "Presidential results." CNN 17 Aug. 2017 http://edition.cnn.com/election/results/president

[10] Lindsay, James M. "Ten Most Significant World Events in 2016." 16 Dec. 2016. Council on Foreign Relations 17 Aug. 2017 https://www.cfr.org/blog/ten-most-significant-world-events-2016

[11] Darke, Diana. "Aleppo: Is besieged Syrian city facing last gasp?" 22 July 2016. BBC News 17 Aug. 2017 http://www.bbc.com/news/world-middle-east-36853689

[12] DeYoung, Karen. "United Nations: Aleppo faces widespread starvation without humanitarian aid." 10 Nov. 2016. The Washington Post 17 Aug. 2017 https://www.washingtonpost.com/world/national-security/united-nations-aleppo-faces-widespread-starvation-without-humanitarian-aid/2016/11/10/883b2d28-a77a-11e6-ba59-a7d93165c6d4_story.html

deal was reached between Russia and Turkey by mid-December 2016, to remove the last remaining residents of the rebel-held parts of the city.[13] Just days later, in what appeared to be a backlash against Russian military involvement in the Syrian civil war and to disrupt the normalisation of Russian-Turkish relations, Russia's ambassador to Turkey was assassinated by an off-duty Turkish police officer.[14]

North Korea conducted its fourth underground nuclear test on 5 January 2016, claiming to have detonated its first hydrogen bomb[15]; its fifth underground nuclear test took place on 8 September 2016.[16] In April 2016, it test-fired a ballistic missile from a Sinpo-class submarine and conducted three failed launches of its Musudan, which could be capable of reaching US military bases as far as Guam. The uptick in activities led US and South Korean intelligence officials to conclude that North Korea was now able to mount nuclear warheads on short- and medium-range missiles that would be capable of hitting Japan and South Korea.[17] In June 2016, North Korea successfully test-launched an intermediate-range ballistic missile into high altitude; it was followed by the successful test of a submarine-launched ballistic missile on 23 August 2016, just 2 days after the United States and South Korea began their annual joint military exercises. The threat posed to the region by North Korea's nuclear programme combined with its gradually increasing missile technology motivated the United States and South Korea to deploy the American-built Terminal High Altitude Area Defense (THAAD) system in South Korea by the end of 2017; while the move will likely be welcomed by Japan's strategic interests, it will be vigorously protested by China.[18]

---

[13] Hubbard, Ben and Hwaida Saad. "Aleppo Evacuation Effort Restarts, and Assad Calls It History in the Making." The New York Times 17 Aug. 2017 https://www.nytimes.com/2016/12/15/world/middleeast/aleppo-syria-evacuation-deal.html

[14] Walker, Shaun, Kareem Shaheen, Martin Chulov, and Patrick Wintour. "Russian ambassador to Turkey shot dead by police officer in Ankara gallery." 20 Dec. 2016. The Guardian 17 Aug. 2017 https://www.theguardian.com/world/2016/dec/19/russian-ambassador-to-turkey-wounded-in-ankara-shooting-attack

[15] Sanger, David E. and Choe Sang-Hun. "North Korea Says It Has Detonated Its First Hydrogen Bomb." 5 Jan. 2016. The New York Times 17 Aug. 2017 https://www.nytimes.com/2016/01/06/world/asia/north-korea-hydrogen-bomb-test.html

[16] Forsythe, Michael. "North Korea's Nuclear Blasts Keep Getting Stronger." 9 Sept. 2016. The New York Times 17 Aug. 2017 https://www.nytimes.com/2016/09/10/world/asia/north-korea-nuclear-weapons-tests.html

[17] Sanger, David E. and Choe Sang-Hun. "As North Korea's Nuclear Program Advances, U.S. Strategy Is Tested." 6 May 2016. The New York Times 17 Aug. 2017 https://www.nytimes.com/2016/05/07/world/asia/north-korea-nuclear-us-strategy.html

[18] Sang-Hun, Choe. "North Korea Test-Fires Missile From Submarine." 23 Aug. 2016. The New York Times 17 Aug. 2017 https://www.nytimes.com/2016/08/24/world/asia/north-korea-submarine-missile.html

1.1.2.2 Environment

The Paris Agreement aims to keep global average temperature increases to below 2 °C above pre-industrial levels and to make more ambitious efforts to limit temperature increases even further to 1.5 °C and eliminate the increase of greenhouse gas emissions in the second half of the century.[19] Following its creation in the 21st UN Framework Convention on Climate Change Conference of Parties (UN FCCC/COP), it rapidly entered into force amid uncertainties brought on by the US presidential election which threatened to undo the global initiative on combating climate change. The Paris Agreement entered into force on 4 November 2016, triggered by the ratification of the European Union on 5 October 2016, which met the threshold that at least 55 parties, accounting for at least an estimated 55% of total global greenhouse emissions, ratify the instrument. China and the United States, representing nearly 40% of global greenhouse gas emissions, ratified the Paris Agreement in September of 2016, followed by India at the beginning of October 2016.[20] And while Russia has yet to ratify the Paris Agreement, 121 parties to the UN FCCC/COP representing more than 79% of global emissions had ratified the Paris Agreement by the end of 2016.[21]

The 22nd UN FCCC/COP took place in Marrakech, Morocco, from 7 to 18 November 2016.[22] The event also served as the first meeting of the governing body of the Paris Agreement (CMA) and marked the beginning of the Paris Agreement's implementation phase, following years of negotiation. Despite its rapid entry into force, in order for the Paris Agreement to be fully operational, its parties first need to elaborate and adopt decisions on a wide range of topics including mitigation (e.g. nationally determined contributions (NDCs)), adaptation communications, finance, transparency, "global stocktake", and market and non-market mechanisms; they aim to do so by 2018, ahead of the 2020 timeline from which the agreement was intended to begin. Developed countries also released a roadmap for obtaining $100 billion per year in climate funding by 2020, with estimates by the UN FCCC reaching $741 billion for 2014. The 23rd UN FCCC/COP will be held from 6 to 17 November 2017 in Bonn, Germany.[23]

[19] "The Paris agreement marks an unprecedented political recognition of the risks of climate change." 12 Dec. 2015. The Economist 24 June 2016 http://www.economist.com/node/21683990/

[20] "Paris Agreement – Status of Ratification." United Nations Framework Convention on Climate Change 28 Apr. 2017 http://unfccc.int/paris_agreement/items/9444.php

[21] "COP 22 Summary Report." 20 Nov. 2016. IETA 28 Apr. 2017 http://www.ieta.org/resources/UNFCCC/COP22/COP22WRAP_FINAL.pdf

[22] "OUTCOMES OF THE U.N. CLIMATE CHANGE CONFERENCE IN MARRAKECH | 22nd Session of the Conference of the Parties to the United Nations Framework Convention on Climate Change (COP 22) November 7–18, 2016." 18 Nov. 2016. Centre for Climate and Energy Solutions 28 Apr. 2017 https://www.c2es.org/international/negotiations/cop22-marrakech/summary

[23] Ibid.

1.1.2.3 Energy

With the Paris Agreement on climate change having entered into force in November 2016, renewable energy is expected to lead the transformation of the power sector to the promise of low-carbon energy production. The energy sector currently accounts for at least two-thirds of greenhouse gas emissions; however, by 2015, growth in energy-related carbon dioxide emissions had stalled completely, thanks to gains in energy efficiency and the expanded use of renewables and other clean energy sources worldwide. Energy demand is expected to grow by 30% in 2040, with demand declining in OECD countries and shifting towards increased growth in India, Southeast Asia, China, and other non-OECD countries in Africa, Latin America, and the Middle East. To meet the growing demand, an estimated $67 trillion will need to be invested in the energy sector by 2040, with $44 trillion needed for the global energy supply and $23 trillion for improvements in energy efficiency. It should be noted that 60% of the $44 trillion would go towards oil, gas, and coal extraction (down from 70% of total supply investment between 2000 and 2015), while nearly 20% will go towards renewable energies, and the rest likely towards nuclear energy. Yet even with intensive efforts to meet growing energy demand, more than 500 million people, mainly in rural areas of sub-Saharan Africa, are expected to remain without basic energy services in 2040.[24]

Whereas the consumption of all modern fuels is expected to increase in 2040, oil and natural gas will remain the bedrock of the global energy system for many decades to come. While oil markets are currently in a downturn due to greater global production, the recent drop in upstream spending on new crude oil resources could affect the rhythm of the oil market in the early 2020s, as the lead times from investment to first oil are typically between 3 and 6 years. Over the long term, global oil demand will be concentrated in freight, aviation, and petrochemical products, which have few alternatives, with supply increasingly concentrated in the Middle East, while the largest source of future demand growth will likely come from India, followed by China. The demand for natural gas is expected to increase by 1.5% annually from 2015 to 2040, with consumption increasing almost everywhere, except for Japan where it is expected to decrease with the reintroduction of nuclear power. The largest sources of growth in demand for natural gas are China and the Middle East, but there remains a question of how the global market for gas will rebalance in light of the current surplus and as new capacity is developed in the United States, East Africa, and Australia which could bring tighter markets in the 2020s.[25]

Growth in coal demand essentially plateaued amid environmental concerns in 2015, and its use is expected to decline in 2040. In this light, to return to market equilibrium, cuts to the supply capacity will be needed for coal, particularly in China and the United States. While nearly 75% of China's power generation comes from coal, almost all growth in its power demand has come from other sources; its

[24] "International Energy Agency. World Energy Outlook 2016 – Executive Summary" 16 Nov. 2016. IAE 30 Mar. 2017 http://www.iea.org/Textbase/npsum/WEO2016SUM.pdf

[25] Ibid.

use of coal will drop to less than 45% by 2040. India is the second-largest coal producer in the world, after China; currently 75% of its power generation comes from coal – which is expected to drop to 55% by 2040. And while the European Union and the United States together account for around one-sixth of current global coal use, coal demand is expected to fall by over 60% and 40%, respectively, between 2015 and 2040, with mostly flat or declining overall energy needs as large strides are made in displacing coal with low-carbon alternatives.[26]

1.1.2.4 Resources

International trade continued to decelerate in 2015, lowering to a rate of around 1.5%, from about 2.3% in 2014. The situation worsened to 1.2% in 2016, as a result of a further slowdown in the first quarter of the year.[27] The slowed growth in 2015 was primarily due to lacklustre performance in merchandise trade, which dropped by 12.7% due to continuing commodity price declines in addition to currency fluctuations that favoured the US dollar. Moreover, global output growth in 2016 was also expected to decrease to about 2.3%, from 2.5% in 2015 and 2014, still well below the 4.0% growth posted in the years prior to the financial crisis.[28] The decelerated growth reflected a continued contraction in import volume demand, especially in transition economies and several developing economies in Asia and Latin America. Meanwhile, the appreciated US dollar muted positive values in export volume demand in Europe and Asia, which trade in their own currencies.

Developed economies showed an uptick in both the volumes of imports and exports, measuring 3.3% and 2.2%, respectively, in 2015. The EU showed growth in imports at 3.6%, owing to rising household consumption; moreover, its exports also increased to 3.2%, resulting from an acceleration of trade within the continent and other developed regions, while exports to transition economies and developing countries, including China, seemed subdued. The volume of imports to the United States increased by 4.8%, while its exports contracted by 0.2%, due to slow external growth and the increasing value of the dollar. However, Japan's imports and exports decreased by 2.8% and 1.0%, respectively, due mainly to domestic factors and to weak demand from its developing country neighbours. By contrast, import volumes in transition economies dropped once again by 19.4% in 2015, from the 7.6% decrease in 2014, due to steep currency depreciations, inflation, and recession, while export volumes again showed stunted growth at 0.9%, due to the drop in the global prices of oil, gas, and minerals. Developing economies also continued to slow down in aggregated growth of both imports and exports, except South Asia whose volume of imports departed from the downward trend; here, China's

[26] Ibid.

[27] World Economic Situation and Prospects. "Fact Sheet." 17 Jan. 2017. United Nations 16 Mar. 2017 https://www.un.org/development/desa/dpad/wp-content/uploads/sites/45/publication/2017wesp_factsheet.pdf

[28] According to the latest data available from UNCTAD.

contraction in both the volumes of imports and exports weighed heavily on trade flows within the region, as it is the largest export market for key manufacturing economies in Asia. For 2015, developing countries' growth in the volume of imports fell to 0.4% from 2.5% in 2014, while the volume of exports fell to 0.4% in 2015, from 3.1% in 2014.[29]

According to the United Nations Conference on Trade and Development (UNCTAD), while commodity prices experienced declines in all group categories, decreasing by 36.7% in 2015, the largest contraction was in crude oil which decreased by 47.2%; yet by the first half of 2016, there were signs that the downward trend had abated somewhat, decreasing by 14.5% and 23.6%, respectively. Crude oil prices bottomed at $30.8 per barrel at the start of 2016 but recovered in the following months reaching $50 per barrel by mid-year. The plunge in prices was attributed to greater global production stoked by several OPEC countries (including Iran which recently returned to the world oil markets eager to reach its pre-sanction production levels) in response to increased production by the United States. The price of oil began to increase in the first quarter of 2016, following financial difficulties and increased bankruptcies for many producers in the United States that reduced production levels, in addition to unplanned supply disruptions in Canada, Ghana, and Nigeria. While the Brent price of oil per barrel increased to $54 per barrel by the end of 2016, the average price for 2016 was $44.0 per barrel or $8 lower than the 2015 average.[30]

Low oil prices also had an effect on the prices of non-oil commodities, such as in reducing transportation and fertiliser prices, in addition to becoming more competitive with biofuels. Agriculture markets were still mainly determined by their own supply situation and weather conditions. The El Niño phenomenon affected some food commodities in Africa and Asia which heightened concerns over food prices and food security in those regions; however, a plentiful supply of agricultural products resulted in production exceeding consumption which prevented a significant impact on most agricultural prices in 2015. The price of metals, ores, and minerals also continued to fall, decreasing by 22.0% in 2015, after an 8.5% reduction in 2014; the trend looked set to continue in 2016, showing an 11.4% reduction by mid-year. Phosphate rock, a key component in the production of phosphate fertilisers for agriculture, was the only outlier to increase in price, rising by 6.5% in 2015; its price has decreased by 1.7% as at mid-2016.[31] Lastly, the price for gold and silver was on the upswing for 2016, increasing by 7.6% and 8.9%, respectively, while the price of platinum continued its decrease by 6.3% for the year; yet the price for each of these precious metals appeared to be contracting in the final quarter of 2016.[32]

---

[29] United Nations Conference on Trade and Development. Trade and Development Report, 2016. Geneva: UNCTAD, 2016. 1–26.

[30] "Crude oil prices increased in 2016, still below 2015 averages." 4 Jan. 2017. U.S. Energy Information Administration (EIA) 9 Jan. 2017 http://www.eia.gov/todayinenergy/detail.php?id=29412

[31] United Nations Conference on Trade and Development. Trade and Development Report, 2016. Geneva: UNCTAD, 2016. 1–26.

[32] World Bank Commodity Price Data (The Pink Sheet). 4 Jan. 2017. World Bank Group 9 Jan. 2017 http://pubdocs.worldbank.org/en/151321483568157298/CMO-Pink-Sheet-January-2017.pdf

1.1.2.5 Knowledge

By now the advantages of higher education should be seen as worth the effort as employment rates and earnings tend to increase as an adult's level of education and skills increases; moreover, the labour market still regards a diploma or degree as the primary indication of a worker's skills. For Europe, the expansion of its pool of highly skilled and specialised scientists and professionals should be a constant priority if it is to remain a leading actor in the field of space-related scientific and technological R&D. By 2016, the percentage of the European[33] working age population between 25 and 64 years of age with a higher education degree remained at 32%, nestled between the OECD average of 35% and the 30% average of G20 members; in contrast, 45% of the working age population in the United States and 55% in Canada have a higher education degree.[34] In Europe, Finland, Ireland, Norway, and the United Kingdom share the top spot, each with 43%, well above the OECD average, followed by Switzerland, Luxembourg, Sweden, Estonia, Belgium, and Denmark; the Netherlands and Spain meet the OECD average, while the remaining European countries under consideration were below that average. It should be noted that the percentage of Europeans between 25 and 34 years of age with a higher education degree in the Czech Republic, Germany, Hungary, Italy, Portugal, and Slovakia continues to fall behind the OECD average.[35]

According to the OECD, the enrolment rate of 20–24-year-olds in tertiary education has increased on average from 29% to 33% in the past decade across OECD countries. But while tertiary attainment is likely to continue rising, on average most students (69%) are taking longer than envisaged to graduate at least one tertiary degree before the age of 30.[36] This can be partly attributed to the fact that labour market demand for skills is changing much faster than education patterns, requiring youth to stay in school longer to acquire higher skills for complex jobs. According to the European Centre for the Development of Vocational Training, growth in employment in Europe will only reach pre-financial crisis levels by 2019; however, growth in employment from 2020 to 2030 is expected to be weaker than in the pre-crisis period, offset by the need to replace the ageing workforce. The strongest growth in employment near the end of this period is expected to be in Belgium, Cyprus, Iceland, and Ireland, while Germany, Estonia, Latvia, Poland, and Romania are expected to show a slight decrease in job growth.[37]

About 85% of all job openings over 2015–2025 will come from the need to replace workers leaving an occupation. Yet overall increases in both process complexity and the number of highly qualified students entering the labour force will

[33] That is, not including Bulgaria, Cyprus, Malta, Lithuania, Romania, and Croatia.

[34] OECD. Education at a Glance 2016: OECD Indicators, OECD Publishing, 2016: 42 http://download.ei-ie.org/Docs/WebDepot/EaG2016_EN.pdf

[35] Ibid. at 43.

[36] Ibid. at 30.

[37] "Future skill needs in Europe: critical labour force trends." 16 Dec. 2016. CEDEFOP 8 May 2017 http://www.cedefop.europa.eu/files/5559_en.pdf

lead to a decline in demand for low qualifications and the increased risk of skill mismatch. The employment of highly skilled workers in all occupations in Europe is expected to increase to 38% in 2025 from 32% in 2015. However, some highly skilled labour will have no alternative than to take up jobs that have typically not required such high formal qualifications in the past. For instance, taking elementary occupations as an example, by 2025 the share of employees with low qualifications will fall to 33% from 44% in 2015, whereas the share of high-skilled employees working in occupations that typically demand lower levels of skills will grow to 14% from 8% in 2015.[38]

1.1.2.6 Mobility

Maritime transport is the most commonly used form of transport for international trade, accounting for about 80% of global merchandise trade by volume; and in terms of value, observers estimate the share of maritime trade to be somewhere between 55% and just over 66% of total merchandise trade.[39] Maritime trade volumes expanded by 2.1%, exceeding 10 billion tons in 2015; however, while the recorded volume was unprecedented, the pace of growth was notably lower than the historical average of 3.0% since 2007. Growth in maritime trade is expected to increase marginally in 2016. Between 2015 and 2016, the *World Order* book continued to decline for most vessel types except for container ships, remaining far below the order peak of 2008–2009. During the 12 months to 1 January 2016, the global fleet of vessels increased by 3.48%, marking the lowest rate of growth since 2003; but with the supply of vessels increasing faster than demand, there is a continued state of overcapacity. As at 1 January 2016, the global commercial fleet consisted of 90,917 vessels, an increase of 1.6% from the 89,464 vessels at the beginning of 2015. Dry-bulk carriers accounted for 43.1% share of the world fleet capacity measured in terms of dead-weight tons, a decrease from its 43.6% share in 2015. The relative share of oil tankers decreased to 27.9% of the world fleet from 28.0% in 2015. Container vessels increased to 13.5% from 13.1% in 2015, while general cargo vessels remained at 4.2% in 2016, on par with the previous year's share. Moreover, the number of ships sold for demolition increased by 2.9% to 23,037 in 2015 from 22,394 in 2014. Once again, dry-bulk carriers accounted for the most tonnage sold for demolition, accounting for 73.0% in 2015 from 40.6% in 2014; meanwhile, the share of container ships and oil tankers halved and quartered, respectively, reaching 9.9% and 5.1% in 2015. Lastly, almost all of the known ship demolitions in 2015 took place in Asia, with over 93% occurring in India, Pakistan, and China.[40]

---

[38] Ibid.

[39] United Nations Conference on Trade and Development. Review of Maritime Transport 2016. Geneva: UNCTD, 2016. 6.

[40] United Nations Conference on Trade and Development. Review of Maritime Transport 2016. Geneva: UNCTD, 2016. 29–50.

Supply chain security is another challenge for the maritime industry, as there are heightened exposure and vulnerability to piracy, armed robbery, and other crimes. Between 1984 and the end of 2015, the number of incidents of piracy and armed robbery against ships totalled 7346 worldwide. While in the past, concern over supply chain security was localised to piracy incidents near East Africa, including Somalia's coastal line, the Gulf of Aden, and further in the Indian Ocean, the growth of piracy incidents in Asian waters has transformed the issue into a cross-sectoral global challenge capable of impacting regional economies and global trade. In 2015, there were 303 reports of piracy and armed robbery against ships, a modest increase of 4.1% from the 291 incidents reported in 2014. The narrow Straits of Malacca and Singapore were the most affected with 134 incidents (44.2%), followed by the South China Sea with 81 (26.7%), the Western Indian Ocean with 38 (12.5%), and West Africa with 35 (11.6%) and the remainder occurring near the Americas or in more temperate waters. And while the number of incidents by Somali-based pirates in the Arabian Sea increased to 15 from 12 in 2014, they were not successful in hijacking a ship in 2015. In total, 5 ships were hijacked in 2015, a notable decrease from the 21 ships hijacked in 2014. However, as ships increasingly rely on software, the Internet, and other technologies, they might become more vulnerable to current and emerging cyberattacks in the future.[41]

1.1.3 Main Science and Technology Indicators Relevant for Space Activities

1.1.3.1 Science and Technology Inputs

Science and technology inputs at the European level are constantly measured against the benefits they generate. In 2016, R&D investment continued to push for increased efficiency and effectiveness and stronger R&D integration in broader industrial and macroeconomic policies. The result in the reporting period is measured by gross domestic expenditure on R&D (GERD), a statistical tool showing nominal changes in those expenditures.

According to Eurostat, overall GERD as a percentage of GDP spent by the 28 EU countries was provisionally assessed to be 2.03% in 2015, a slight decrease from the 2.04% recorded in 2014.[42] The EU's GDP grew modestly to €14.820 trillion in 2016 at current prices, up a mere 0.7% from €14.714 trillion in 2015.[43] The EU's perfor-

[41] United Nations Conference on Trade and Development. Review of Maritime Transport 2016. Geneva: UNCTD, 2016. 93–94.

[42] "Gross domestic expenditure on R&D (GERD) – % of GDP." Eurostat 8 Mar. 2017 http://ec.europa.eu/eurostat/tgm/table.do?tab=table&init=1&language=en&pcode=t2020_20&plugin=1

[43] "Gross domestic product at market prices | At current prices." Eurostat 8 Mar. 2017 http://ec.europa.eu/eurostat/tgm/refreshTableAction.do?tab=table&plugin=1&pcode=tec00001&language=en; Total intramural R&D expenditure (GERD) by sectors of performance – per Million euro." 8 Mar. 2017. Eurostat 21 Apr. 2016 http://appsso.eurostat.ec.europa.eu/nui/submitViewTableAction.do

mance trailed Japan, South Korea, and now China whose data is limited to 2014, and the United States, whose data is limited to 2013. Japan's GERD grew 3.59% in 2014, amounting to €124.531 billion (¥18.035 trillion), from 3.47% or €128.645 billion (¥14.639 trillion) in 2013. Moreover, South Korea's GERD as a percentage of GDP has steadily increased in recent years, reaching 4.29% or €45.585 billion (₩66.422 trillion) in 2014. And China's GERD edged past the EU with 2.05% or €159.004 billion (¥1.340 trillion). And while US values are limited to 2013 on Eurostat, its GERD as a percentage of GDP grew to 2.73% or around €344.083 billion ($454.352 billion).

On the European national level, according to the latest available figures, the top five countries' GERD as a percentage of GDP in 2015 were Sweden (3.26%), Austria (3.07%), Denmark (3.03%), Finland (2.90), and Germany (2.87%). In terms of billion euro put towards GERD, Germany continued to spend the most in 2015, reaching €87.188 billion, once again followed by France (€48.463 billion) and the United Kingdom (€43.878 billion). Moreover, the number of central and eastern EU member states with expenditure below 1% decreased to 7 from 9 in 2015.[44]

A further breakdown of statistical data helps to underscore the impact of private R&D investment in Europe, when paired to publicly funded projects. In 2015, innovation investment in countries with higher GERD as a percentage of GDP tended to come from the business enterprise sector; the top five showed only slight variation in position, being Sweden (2.27%), Austria (2.18%), Germany (1.95%), Finland (1.94%), and Denmark (1.87%). And Denmark, Sweden, Austria, Finland, and also the Netherlands led in the amount put towards higher education research. Overall, Europe's private sector funded R&D accounted for 63.7% of the EU's total GERD in 2014, whereas private sector funded R&D in China, Japan, and South Korea accounted for 77.3%, 77.8%, and 78.2%, respectively, in 2014; US private sector funded R&D stood at 70.6% in 2013.[45]

1.1.3.2 Science and Technology Outputs

In contrast to R&D input, the outputs achieved from investments in innovation are harder to measure, due to the complex market dynamics within the private sector. It should be noted that statistics on the number of patents lodged by country, industrial sector, and individual companies are merely indicative of output – i.e. their effect on financial performance is also heavily dependent on external market competition and other factors. As observed in recent years, the recovery of R&D investment levels by companies based in the EU has been slow, whereas US companies appear more willing to return to the high R&D investment levels experienced prior to the financial crisis.

---

[44] "Gross domestic expenditure on R&D (GERD) – % of GDP." Eurostat 8 Mar. 2017 http://ec.europa.eu/eurostat/tgm/table.do?tab=table&init=1&language=en&pcode=t2020_20&plugin=1

[45] Ibid.

Based on the European Commission's 2016 EU Industrial R&D Investment Scoreboard, in 2015/2016 approximately 2500 companies worldwide (including 590 companies from the EU) each invested at least €21 million in R&D. Growth was mainly driven by high-tech industries (i.e. ICT, health, and auto); however, total expenditure decreased by 3.6% due to a decrease of sales in low-tech sectors (i.e. oil- and mining-related industries). R&D investments by the 2500 leading companies grew by 6.6% to €696.0 billion in 2015/2016, matching similar growth of 6.8% to €607.2 billion in 2014/2015. European companies accounted for 27.1% of that share (i.e. €188.3 billion, a 1.0 point decrease from the previous year mainly due to the depreciation of the euro with respect to other major currencies), while US companies accounted for 38.6% (€268.6 billion, up to 0.4 points), and Japanese companies remained steady at 14.4% (€99.9 billion, an increase of 0.1 points). Chinese companies accounted for a 7.2% share (€49.8 billion, up to 1.3 points), while companies in the rest of the world accounted for the remaining 12.8% share of R&D investments.

The increases in EU performance are largely attributed to German company investments in the Automobiles and Parts sector and also in ICT and health-related sectors; however, substantial decreases of revenue from oil and mining companies in the United Kingdom resulted in an overall decrease of 3.6% in net sales. R&D investment growth in US companies came mainly from Pharmaceutical and Biotechnology and Software and Computer Services sectors, but poor sales in oil-related sectors, Industrial Engineering, and Automobiles resulted in a decrease of 4.0% in sales. Next, Japanese company investment in R&D was modest in comparison to the EU and United States but showed slight positive growth in sales (i.e. 0.3%) coming mainly from its Automobiles and Parts sector. Companies in China reported the largest increase in R&D investment, particularly in ICT-related companies, but also experienced a significant decrease 6.2% in net sales. And lastly, while companies in the rest of the world showed slight R&D investment growth, they also experienced a significant decrease of 4.8% in net sales.[46]

The uneven benefits of corporate innovation on a European scale are further demonstrated in the European Commission's biennial Conferences on Corporate R&D and Innovation (CONCORDi). The 5th CONCORDi meeting, held on 1–2 October 2015, found that young innovative EU firms and start-ups face entry barriers that still need to be fully identified to ensure an effective shift towards a more knowledge-intensive and innovation-oriented European industry. Moreover, further research is needed to properly identify the type and intensity of innovation targeted by policy instruments to ensure their effectiveness. And more regular exchanges between science and policymakers are required to provide better information on the conditions for policy implementation.[47] The 6th CONCORDi meet-

[46] European Commission. "EU R&D Scoreboard | The 2016 EU Industrial R&D Investment Scoreboard." 14 Nov. 2016. EU JRC 9 Mar. 2017: 1–39 http://iri.jrc.ec.europa.eu/scoreboard16.html

[47] Dosso, Mafini, Petros Gkotsis, Fernando Hervás, and Pietro Moncada-Paternò-Castello. "Industrial Research and Innovation: Evidence for Policy." Nov. 2015. European Commission – JRC Policy Brief 22 Apr. 2016 http://iri.jrc.ec.europa.eu/documents/10180/12238/Industrial%20Research%20and%20Innovation.%20Evidence%20for%20Policy

ing will take place in September 2017 and will discuss issues related to, inter alia, the effects of firms' investment in R&D and innovation, entrepreneurship and firm demographics in knowledge-intensive activities, industrial modernisation and its dynamics, current technological frontiers (including "big data" and "artificial intelligence"), and the next generation of technological discoveries (identification, effects, and policy issues).[48]

1.2 Worldwide Space Policies and Strategies

1.2.1 The United Nations System

Various institutions within or associated with the United Nations are relevant for space policy. In this section, the UN General Assembly, its committees, and other UN bodies and organs that deal with space activities are discussed.

1.2.1.1 United Nations General Assembly

The United Nations General Assembly (UNGA) passed four resolutions in its 71st (2016–2017) session relating to the use and exploration of outer space. The first was on the "Prevention of an arms race in outer space" (Resolution A/RES/71/31 adopted on 5 December 2016).[49] The second was on "No first placement of weapons in outer space" (Resolution A/RES/71/32 adopted on 5 December 2016).[50] The third resolution was on "Transparency and confidence-building measures in outer space activities" (Resolution A/RES/71/42 also adopted on 5 December 2016).[51] The fourth resolution was on "International cooperation in the peaceful uses of outer space" (A/RES/71/90 adopted on 6 December 2016).[52]

[48] "CONCORDi 2017 – Innovation and Industrial Dynamics: Challenges for the next decade." 14 Nov. 2016. EU JRC 13 Mar. 2017 http://iri.jrc.ec.europa.eu/concord/2017/index.html

[49] United Nations General Assembly. Resolution adopted by the General Assembly on 5 December 2016 – 71/31. Prevention of an arms race – 71st Session. UN Doc. A/RES/71/31 of 9 Dec. 2016. Vienna: United Nations http://www.un.org/en/ga/search/view_doc.asp?symbol=A/RES/71/31

[50] United Nations General Assembly. Resolution adopted by the General Assembly on 5 December 2016 – 71/32. No first placement of weapons in outer space – 71st Session. UN Doc. A/RES/71/32 of 9 Dec. 2016. Vienna: United Nations http://www.un.org/en/ga/search/view_doc.asp?symbol=A/RES/71/32

[51] United Nations General Assembly. Resolution adopted by the General Assembly on 5 December 2016 – 71/42. Transparency and confidence-building measures in outer space activities – Seventieth Session. UN Doc. A/RES/71/42 of 9 Dec. 2016. Vienna: United Nations http://www.un.org/en/ga/search/view_doc.asp?symbol=A/RES/71/42

[52] United Nations General Assembly. Resolution adopted by the General Assembly on 6 December 2016 – International cooperation in the peaceful uses of outer space – 71st Session. UN Doc. A/RES/71/90 of 22 Dec. 2016. Vienna: United Nations http://www.un.org/en/ga/search/view_doc.asp?symbol=A/RES/71/90

First, Resolution A/RES/71/31 on the "Prevention of an arms race in outer space" repeated the positions of the General Assembly in previous years' resolutions. The instrument emphasised the paramount importance of strict compliance with existing arms limitation and disarmament agreements relevant to outer space, along with the mutually complementary nature of bilateral and multilateral efforts for the prevention of an arms race in outer space, and hoped that concrete results from those efforts would soon emerge. It also stressed that the growing use of outer space increases the need for greater transparency and better information on the part of the international community and recalled the importance of transparency and confidence-building measures (TCBM) to avoid an arms race in space. The resolution also called upon all states to contribute actively to the peaceful use of outer space and the prevention of an arms race in outer space and to refrain from actions contrary to that objective and reiterated that the Conference on Disarmament (CD) has the primary role in negotiating multilateral agreements on the prevention of an arms race in outer space in all its aspects. It also urged these states to keep the CD informed of the progress of bilateral and multilateral negotiations on the matter so as to facilitate the CD's work.[53] The resolution was adopted with 182 votes in favour to none against, with Israel, Palau, South Sudan, and the United States abstaining.

The second Resolution A/RES/71/32 on "No first placement of weapons in outer space" reiterated its previous years' Resolution A/RES/70/27, reaffirming that practical measures should be examined and taken in the search for agreements to prevent an arms race in outer space and that the legal regime applicable to outer space by itself does not guarantee prevention of an arms race in outer space, marking a need to consolidate and reinforce that regime. It once again welcomed the draft Treaty on the Prevention of the Placement of Weapons in Outer Space and of the Threat or Use of Force Against Outer Space Objects introduced by China and the Russian Federation at the CD in 2008, and updated in 2014, and noted the importance of the political statements made by a growing number of states (i.e. Argentina, Armenia, Belarus, Bolivia, Brazil, Cuba, Indonesia, Kazakhstan, Kyrgyzstan, Nicaragua, the Russian Federation, Sri Lanka, Tajikistan, and Venezuela) that they would not be the first to place weapons in outer space. It also urged an early start of substantive work based on the updated draft treaty and encouraged all states, especially spacefaring nations, to consider the possibility of making a political commitment not to be the first to place weapons in outer space.[54] The resolution was adopted with 130 votes in favour to 4 against (Georgia, Israel, Ukraine, and the United States) and 48 abstentions.

[53] United Nations General Assembly. Resolution adopted by the General Assembly on 5 December 2016 – 71/31. Prevention of an arms race – 71st Session. UN Doc. A/RES/71/31 of 9 Dec. 2016. Vienna: United Nations http://www.un.org/en/ga/search/view_doc.asp?symbol=A/RES/71/31

[54] United Nations General Assembly. Resolution adopted by the General Assembly on 5 December 2016 – 71/32. No first placement of weapons in outer space – 71st Session. UN Doc. A/RES/71/32 of 9 Dec. 2016. Vienna: United Nations http://www.un.org/en/ga/search/view_doc.asp?symbol=A/RES/71/32

Building on the previous years' resolution, Resolution A/RES/71/42 on "Transparency and confidence-building measures in outer space activities" reaffirmed the right of free exploration and use of outer space by all states in accordance with international law and that the prevention of an arms race in outer space was in the interest of maintaining international peace and security, in addition to fostering international cooperation in the exploration and use of outer space for peaceful purposes. In addition to noting that the UN COPUOS had a fundamental role to play in enhancing transparency and confidence-building among states and in ensuring that outer space was maintained for peaceful purposes, it noted the special report by the Inter-Agency Meeting on Outer Space Activities (UN-Space) on the implementation of the report of the Group of Governmental Experts on Transparency and Confidence-Building Measures in Outer Space Activities, submitted to the Committee at its 59th session in 2016. The resolution also noted the UN COPUOS endorsement to hold a joint half-day panel discussion in a plenary meeting of the Disarmament and International Security Committee (First Committee) and the Special Political and Decolonization Committee (Fourth Committee) during the 72nd session of the United Nations General Assembly in 2017 as a joint contribution to the 50th anniversary of the Outer Space Treaty. Moreover, it again welcomed Resolution 186 of 7 November 2014 on strengthening the role of the International Telecommunication Union with regard to transparency and confidence-building measures in outer space activities. It also called upon member states and the relevant entities and organisations of the United Nations system to support the implementation of the full range of conclusions and recommendations contained in the report.[55] The resolution was adopted without a vote.

Resolution A/RES/71/90 on "International cooperation in the peaceful uses of outer space" updated the resolution from 2015. Updates included an endorsement of most of the aspects of the report by UN COPUOS on the work of its 59th session, which will form a full compendium of guidelines to be adopted by it and referred to the UNGA in 2018. It once again noted with satisfaction the establishment of and work carried out by the International Asteroid Warning Network and the Space Mission Planning Advisory Group to implement the recommendations for an international response to the near-Earth object impact threat with the support of the UNOOSA, which serves as the Group's permanent secretariat. The resolution declared 30 June as International Asteroid Day, to observe annually at the international level the anniversary of the Tunguska impact over Siberia, Russian Federation, on 30 June 1908 and to raise public awareness about the asteroid impact hazard. It also noted with satisfaction that the UN COPUOS had agreed on seven thematic priorities of UNISPACE + 50, including their objectives and mechanisms. It also noted with satisfaction the tenth anniversary of UN-SPIDER and recognised the significant achievements made and the advisory support provided to 38 member

[55] United Nations General Assembly. Resolution adopted by the General Assembly on 5 December 2016 – 71/42. Transparency and confidence-building measures in outer space activities – 71st Session. UN Doc. A/RES/71/42 of 9 Dec. 2016. Vienna: United Nations http://www.un.org/en/ga/search/view_doc.asp?symbol=A/RES/71/42

states within the UN-SPIDER framework since its establishment in 2006 and reiterated the importance of the Sendai Framework for Disaster Risk Reduction 2015–2030.[56] This resolution was also adopted without a vote.

1.2.1.2 UNGA Committees

Several UNGA committees address space policy and associated matters. The main ones are discussed below.

The Disarmament and International Security Committee
The Disarmament and International Security Committee (First Committee) works in close cooperation with the United Nations Disarmament Commission and the Geneva-based Conference on Disarmament to deal with disarmament, global challenges, and threats to peace that affect the international community and seeks solutions to the challenges in the international security regime, including the prevention of an arms race in outer space.[57] On 2 November 2016, the First Committee forwarded 69 draft resolutions and decisions to the UNGA for adoption, including documents focussing on the recent proliferation of chemical weapons particularly in Syria and Iraq and the development of new and emerging threats. And on the topic of nuclear disarmament, the First Committee submitted a proposal to convene a United Nations conference in 2017 to move forward negotiations on a legally binding instrument banning nuclear weapons, the creation of a preparatory process for a fissile material cut-off treaty, and the establishment of a group of governmental experts to consider the role of verification in advancing nuclear disarmament.[58] On 5 December 2016, the UNGA adopted 68 documents, with continued divergence on the way forward in certain nuclear-related topics, including draft resolution III titled "Convention on the Prohibition of the Use of Nuclear Weapons", reiterating the UNGA's request to the Conference on Disarmament to commence negotiations in order to reach agreement on an international convention prohibiting the use or threat of use of nuclear weapons under any circumstances, which had a recorded vote of 128 in favour to 50 against, with 9 abstentions, and draft resolution V titled "Reducing nuclear danger", calling for a review of nuclear doctrines and immediate and urgent steps to reduce the risks of unintentional and accidental use of nuclear weapons, including through de-alerting and de-targeting nuclear weapons, which had a recorded vote of 126 in favour to 49 against, with 10 abstentions. Another

---

[56] United Nations General Assembly. Resolution adopted by the General Assembly on 6 December 2016 – International cooperation in the peaceful uses of outer space 71st Session. UN Doc. A/RES/71/90 of 22 Dec. 2016. Vienna: United Nations http://www.un.org/en/ga/search/view_doc.asp?symbol=A/RES/71/90

[57] "Disarmament and International Security." General Assembly of the United Nations 15 Mar. 2017 http://www.un.org/en/ga/first/index.shtml

[58] "First Committee Sends 69 Texts to General Assembly, Concluding Session by Approving Drafts on Chemical Weapons, Improvised Explosive Devices." 2 Nov. 2016. United Nations 15 Mar. 2017 https://www.un.org/press/en/2016/gadis3567.doc.htm

topic where nearly one-third of the UNGA abstained was in regard to the report "Conclusion of effective international arrangements to assure non-nuclear-weapon States against the use or threat of use of nuclear weapons", calling on all states, especially those possessing nuclear weapons, to work towards an early agreement on a common approach and formula that could be included in an international legally binding instrument, which had a recorded vote of 128 in favour to none against and 57 abstentions.[59]

The Committee on the Peaceful Uses of Outer Space (COPUOS)

The UN COPUOS's activities included its 59th plenary session from 8 to 17 June 2016, along with the 53rd session of its scientific and technical subcommittee held from 15 to 26 February 2016 and the 55th session of its legal subcommittee from 4 to 15 April 2016. The Committee added new topics to its plenary session agenda, including general exchanges of views on the legal aspects of space traffic management and on the application of international law to small satellite activities as part of its review of the report of the legal subcommittee on its 55th session. It also added a proposed strategic framework for the programme on the peaceful uses of outer space for the period 2018–2019 and considered the first set of guidelines for the long-term sustainability of outer space activities. The Committee also endorsed the findings and recommendations of its subcommittees from their preceding sessions and considered space and sustainable development; it reviewed the current status of the spin-off benefits of space technology, space and water, and space and climate change; the use of space technology in the United Nations system; the future role of the committee; and other related matters.[60]

At the 2016 scientific and technical subcommittee meeting, discussions and presentations expanded on a range of space-related issues, including space technology for socioeconomic development in the context of the United Nations Conference on Sustainable Development and the Post-2015 Development Agenda; matters relating to remote sensing of the Earth by satellite, including applications for developing countries and monitoring of the Earth's environment; space debris; space system-based disaster management; recent developments in global navigation satellite systems; space weather; near-Earth objects; the use of nuclear power sources in outer space; and the long-term sustainability of outer space activities. The subcommittee also examined the physical nature and technical attributes of the geostationary orbit and its utilisation and applications, including the field of space communications with particular note of the needs and interests of developing countries.[61]

---

[59] "Adopting 68 First Committee Texts, General Assembly Addresses New Threats, Use of Banned Weapons, Urges Drive to Curb Arms Proliferation." 5 Dec. 2016. United Nations 15 Mar. 2017 https://www.un.org/press/en/2016/ga11866.doc.htm

[60] United Nations General Assembly. Report of the Committee on the Peaceful Uses of Outer Space on its 59th session, Held in Vienna from 8 to 17 June 2016. UN Doc. A/71/20 of 28 June 2016. New York: United Nations https://cms.unov.org/dcpms2/api/finaldocuments?Language=en&Symbol=A/71/20

[61] United Nations General Assembly. Report of the Scientific and Technical Subcommittee on its 53rd session, held in Vienna from 15 to 26 February 2016. UN Doc. A/AC.105/1109 of 9 March

As in previous years, the proceedings of the 2016 legal subcommittee focussed on exchanging information on the activities of international governmental and non-governmental organisations related to space law, in addition to the status and application of the five United Nations treaties on outer space. It once again considered matters relating to the definition and delimitation of outer space and the utilisation of the geostationary orbit, including consideration of ways and means to ensure the rational and equitable use of the geostationary orbit without prejudice to the role of the International Telecommunication Union. National legislation relevant to the peaceful exploration and use of outer space was another agenda item, in addition to capacity building in space law and a review and possible revision of the Principles Relevant to the Use of Nuclear Power Sources in Outer Space. This year, in addition to the general exchange of information and views on legal mechanisms relating to space debris mitigation measures and on non-legally binding United Nations instruments on outer space, there was an exchange of views on the legal aspects of space traffic management and on the application of international law to small satellite activities. It also held a review of international mechanisms for cooperation in the peaceful exploration and use of outer space and welcomed proposals for new items to be considered by the legal subcommittee at its 56th session in 2017.[62]

1.2.1.3 Other UN Bodies and Organs Monitoring Outer Space Activities

Beyond the UN General Assembly and its committees, there are other UN bodies, programmes, and organs related to space activities. The following discusses the ITU (a specialised agency of the UN), the UN-SPIDER, the UN Programme on Space Applications, the International Committee on Global Navigation Satellite Systems (ICG), the United Nations Spatial Data Infrastructure (UNSDI), the Conference on Disarmament (CD), and the UNIDIR.

International Telecommunication Union (ITU)

The International Telecommunication Union (ITU) held a workshop focussing on non-geostationary satellite issues that were addressed by the director of the Radiocommunication Bureau in his report to WRC-15 in Geneva, Switzerland, on 21 April 2016.[63] The ITU also held the International Satellite Communication Symposium on "Interference-Free Satellite Frequency Spectrum: Myth or Reality in 2016" in Geneva, Switzerland, on 13 to 14 June 2016, which reflected on international cooperation to facilitate expanded access to satellite connectivity, and

---

2016. Vienna: United Nations https://cms.unov.org/dcpms2/api/finaldocuments?Language=en&Symbol=A/AC.105/1109

[62] United Nations General Assembly. Report of the Legal Subcommittee on its 55th session, held in Vienna from 4 to 15 April 2016. UN Doc. A/AC.105/1113 of 27 April 2016. Vienna: United Nations https://cms.unov.org/dcpms2/api/finaldocuments?Language=en&Symbol=A/AC.105/1113

[63] "Workshop on non-geostationary satellite issues." ITU 28 Mar. 2017 http://www.itu.int/en/ITU-R/space/workshops/2016-NGSO/Pages/default.aspx

continued on from a similar workshop in June 2013.[64] The ITU Workshop on the Efficient Use of the Spectrum/Orbit Resource took place on 6 September 2016, followed by the ITU International Satellite Symposium 2016 "Satellite Regulation, Market, Technology Trends and Industry Opportunities" on 7 to 8 September 2016, in Bali, Indonesia. The Workshop addressed the implementation of WRC-15 decisions, the impact of new technologies and future trends on satellite communications, and latest developments in non-GSO constellations and their evolution.[65] The Symposium looked at satellite international regulations: updates from WRC-15 outcomes; challenges and opportunities to space industry on the WRC-19 agenda; satellite policy and regulatory framework including country practices on satellite communications and broadcasting services; and satellite market and technology trends, as well as industry opportunities.[66] And lastly, the ITU Symposium and Workshop on small satellite regulation and communication systems took place in Santiago de Chile, Chile, on 7 to 9 November 2016 addressing sustainable development of small satellite systems, the outer space legal regime, the ITU Radio Regulations and the WRC-15 outcomes related to small satellites, authorisation of small satellites under national space legislation, small satellite projects in the region, and the advanced future small satellite systems.[67]

The ITU conducted its biennial World Radiocommunication Seminar 2016 (WRS-16), in Geneva, Switzerland, from 12 to 16 December 2016, dealing with the use of the radio-frequency spectrum, satellite orbits, and the application of the provisions of the ITU Radio Regulations. In addition to celebrating the 110th anniversary of the ITU Radio Regulations (1906–2016), the WRS-16 plenary sessions addressed the coexistence of terrestrial services (including its regulatory framework, frequency plans, notification of frequency assignments, harmful interference, and international monitoring) and space services (including the international regulatory framework for orbit/spectrum allocation, planned and non-planned satellite services, harmful interference, and space databases). It also held space service and terrestrial service workshops in the last 3 days of the WRS-16.[68] World Radiocommunication Seminars (WRS) are organised in complement to the cycle of Regional Radiocommunication Seminars (RRS) that take place to inform ITU member states on the development of general issues of spectrum management and international regulations for terrestrial and space services, including the relevant

[64] "International Satellite Communication Symposium (Geneva, 13–14 June 2016)." ITU 28 Mar. 2017 http://www.itu.int/en/ITU-R/space/workshops/SISS-2016/Pages/default.aspx

[65] "ITU Workshop on the Efficient Use of the Orbit/Spectrum Resource, Bali, Indonesia." ITU 28 Mar. 2017 http://www.itu.int/en/ITU-R/space/workshops/bali-2016/Pages/default.aspx

[66] "ITU International Satellite Symposium 2016." ITU 28 Mar. 2017 http://www.itu.int/en/ITU-D/Regional-Presence/AsiaPacific/Pages/Events/2016/Sep-ISS2016/home.aspx

[67] "ITU Symposium and Workshop on small satellite regulation and communication systems, Santiago de Chile, Chile, 7–9 November 2016." ITU 28 Mar. 2017 http://www.itu.int/en/ITU-R/space/workshops/2016-small-sat/Pages/default.aspx

[68] "ITU World Radiocommunication Seminar 2016 (WRS-16), 12 to 16 December 2016, Geneva, Switzerland." 27 Nov. 2015. ITU 18 Apr. 2016 http://www.itu.int/net/pressoffice/press_releases/2015/56.aspx

coordination, examination, and registration procedures. Two RRS meetings took place in 2016, with the RRS-16-Americas occurring in Port of Spain, Trinidad and Tobago, from 18 to 22 July 2016, followed by RRS-16-Asia-Pacific in Apia, Samoa, from 19 to 23 September 2016.[69]

UN-SPIDER

The United Nations Platform for Space-based Information for Disaster Management and Emergency Response (UN-SPIDER) organised several workshops and regional meetings in 2016. Set up by the United Nations General Assembly in 2006, the platform aims to provide universal access to all types of space-based information and services relevant to disaster management support. In this context, the International Charter on Space and Major Disasters was activated many times by the UN Office for Outer Space Affairs (UNOOSA) at the request of other UN entities and member states. Key geopolitical events that activated the Charter included floods in the United States, Argentina, Iran, Sri Lanka, China, Bangladesh, Sudan, India, Indonesia, Australia, Panama, and Costa Rica; hurricanes, tropical storms, and torrential rains in Fiji, Seychelles, Bangladesh, China, Haiti, Cuba, the Dominican Republic, the Bahamas, the United States, the Philippines, and India; and wildfires in Panama, Canada, the Russian Federation, and Israel, along with volcanic activities in Indonesia and earthquakes in Taiwan, Japan, Ecuador, and Indonesia. The Charter was activated a total of 36 times in 2016.[70]

UN Programme on Space Applications (PSA)

The UN Programme on Space Applications (PSA) is concerned with cooperation in space science and technology. Several activities were carried out under PSA's auspices in 2016, including workshops on space technology applications, on the use of Earth observation data, and on the use of global navigation satellite systems (GNSS). A symposium and a conference were also held addressing space technology applications for climate change and for wildlife management and protecting biodiversity.

The PSA conducted the United Nations/Costa Rica Workshop on "Human Space Technology" in San José, Costa Rica, from 7 to 11 March 2016, to exchange information on achievements in human space programmes and to consider ways to promote international cooperation by further facilitating the participation of developing countries and industries in human space exploration-related activities; it also focussed on raising awareness of the benefits of human space technology and its applications and building capacity in microgravity science education and research.[71]

[69] "Radiocommunication Seminars and Workshops." ITU 28 Mar. 2017 http://www.itu.int/en/ITU-R/seminars/Pages/default.aspx

[70] "Charter Activations." International Charter – Space & Major Disasters 17 Mar. 2017 https://www.disasterscharter.org/web/guest/activations/charter-activations

[71] "United Nations/Costa Rica Workshop on Human Space Technology." UNOOSA 20 Mar. 2017 http://www.unoosa.org/oosa/en/ourwork/psa/schedule/2016/workshop_costarica_human_space_technology.html

The United Nations/India Workshop on Use of Earth Observation Data in Disaster Management and Risk Reduction: "Sharing the Asian Experience" was held in Hyderabad, India, from 8 to 11 March 2016, to demonstrate operational programmes and tools to address disaster management (e.g. understanding disaster risks, responding to emergencies, assessing damage and loss, and providing inputs to mitigate disasters); it also synthesised the experiences and lessons learnt by Asian countries for the future effective use of Earth observation for disaster management.[72]

The United Nations/International Astronautical Federation Workshop on Space Technology for Socio-Economic Benefits: "Integrated Space Technologies and Applications for a Better Society" was held in Guadalajara, Mexico, from 23 to 25 September 2016, in conjunction with the 67th International Astronautical Congress (IAC). It discussed space technologies, applications, information, and services that contribute to sustainable economic and social development programmes and focussed on the use of satellite telecommunications for early warning systems and on increasing awareness among decision-makers and representatives of the research and academic community of space technology applications for addressing economic development.[73]

The United Nations/Islamic Republic of Iran Workshop on the Use of Space Technology for Dust Storm and Drought Monitoring in the Middle East Region took place in Tehran, Islamic Republic of Iran, from 5 to 9 November 2016, addressing the use of space technologies in drought monitoring and for dust storm tracking and monitoring, with a main focus on Central Asia, and including a session dedicated to considering the specific benefits of space tools in the wider environmental monitoring of the Caspian Sea basin.[74]

The last workshop was the United Nations/Nepal Workshop on the Applications of Global Navigation Satellite Systems, held in Kathmandu, Nepal, from 12 to 16 December 2016. It aimed to introduce GNSS and its applications to transportation and communications, aviation, surveying, mapping and Earth science, management of natural resources, the environment and disasters, precision agriculture, and high-precision mobile applications. It also looked at space weather effects on GNSS dual-frequency receivers, promoted a greater exchange of actual experiences with specific applications, and encouraged greater cooperation in developing partnerships and GNSS networks in the framework of the regional reference frames.[75]

[72] "United Nations/India Workshop on the Use of Earth Observation Data in Disaster Management and Risk Reduction: Sharing the Asian Experience." UNOOSA 20 Mar. 2017 http://www.unoosa.org/oosa/en/ourwork/psa/schedule/2016/workshop_india.html

[73] "United Nations/International Astronautical Federation Workshop on Space Technology for Socio-Economic Benefits: 'Integrated Space Technologies and Applications for a Better Society'." UNOOSA 20 Mar. 2017 http://www.unoosa.org/oosa/en/ourwork/psa/schedule/2016/workshop_mexico_uniaf.html

[74] "United Nations/Islamic Republic of Iran Workshop on the Use of Space Technology for Dust Storm and Drought Monitoring in the Middle East Region." UNOOSA 20 Mar. 2017 http://www.unoosa.org/oosa/en/ourwork/psa/schedule/2016/workshop_iran.html

[75] "United Nations/Nepal Workshop on the Applications of Global Navigation Satellite Systems." UNOOSA 20 Mar. 2017 http://www.unoosa.org/oosa/en/ourwork/psa/schedule/2016/2016-nepal-workshop-on-gnss.html

The PSA conducted the United Nations/Austria Symposium on "Integrated Space Technology Applications for Climate Change" in Graz, Austria, from 12 to 14 September 2016, to address, inter alia, how countries affected by climate change could make better use of space applications to assess their vulnerability, identify potential alternatives in the context of mitigation and adaption to climate change, and improve synergies among space agencies and organisations targeting efforts on climate change.[76] It also conducted the United Nations/Kenya Conference on Space Technology and Applications for Wildlife Management and Protecting Biodiversity from 27 to 30 June 2016, to address the growing demand for EO data for biodiversity monitoring and wildlife management, to present recent advancements in this domain, and to identify specific needs in Africa for further action in better applying benefits of space technologies in the biodiversity domain.[77]

International Committee on Global Navigation Satellite Systems (ICG)
The International Committee on Global Navigation Satellite Systems (ICG) was established in 2005 to promote voluntary cooperation on matters of mutual interest related to civil satellite-based positioning, navigation, timing, and value-added services. It aims to make positioning, navigation, and timing available globally for societal benefits at a lower cost through encouraging coordination among providers of GNSS, regional systems, and augmentation services to ensure greater compatibility, interoperability, and transparency. It also aims to promote the introduction and utilisation of navigation services and their future enhancements through providing assistance for integration into their infrastructures and by existing as a focal point for information exchange to assist GNSS users with their development plans and applications. The UNOOSA serves as the Executive Secretariat of the ICG and the associated Providers' Forum. The 11th meeting of the ICG was organised by the State Space Corporation "Roscosmos" on behalf of the government of the Russian Federation and took place in Sochi, Russian Federation, from 6 to 11 November 2016. The meeting brought together stakeholders in industry, government, non-governmental officials, and academia to review and discuss developments in global navigation systems.[78]

The meeting consisted of three plenary sessions and a series of working group meetings and included presentations to update ICG members, associate members, and observers on various matters of interest. An application and expert seminar entitled "High precision GNSS applications in various fields of the world economy" provided expert's views on GNSS today and their visions for the future. Topics included "United States Position, Navigation and Timing Advisory Board

[76] "United Nations/Austria Symposium on 'Integrated Space Technology Applications for Climate Change'." UNOOSA 20 Mar. 2017 http://www.unoosa.org/oosa/en/ourwork/psa/schedule/2016/symposium_austria_climatechange.html

[77] "United Nations/Kenya Conference on Space Technology and Applications for Wildlife Management and Protecting Biodiversity." UNOOSA 20 Mar. 2017 http://www.unoosa.org/oosa/en/ourwork/psa/schedule/2016/conference_kenya_biodiversity.html

[78] "International Committee on Global Navigation Satellite Systems (ICG): Annual Meeting." United Nations Office for Outer Space Affairs 21 Mar. 2017 http://www.unoosa.org/oosa/en/ourwork/icg/meetings/ICG-2016.html

(PNTAB): current activities and focus" and "Real-time GNSS for earthquake and tsunami early warning" by US representatives; "Qianxun BeiDou Navigation Satellite Systems (BDS) high precision positioning services and applications", by the representative of China; and "High precision GNSS land applications in the economy" and "Prospects of application of high-precision navigation in integrated agriculture" by representatives of the Russian Federation.[79]

The four working groups also met during the Plenary, focussing on systems, signals, and services (co-led by the Russian Federation and the United States); enhancement of GNSS performance, new services, and capabilities (co-led by China, India, and ESA); information dissemination and capacity building (led by the UNOOSA); and reference frames, timing, and applications (co-led by the International Association of Geodesy, the International Federation of Surveyors, and the International GNSS Service (IGS)). In addition to discussing cross-cutting issues relating to the international GNSS monitoring assessment, the groups reviewed progress made in implementing the recommendations made at previous meetings and ways and means of carrying them into the future. The 12th ICG meeting will take place in Kyoto, Japan, from 2 to 7 December 2017.[80]

United Nations Spatial Data Infrastructure (UNSDI)

The United Nations Geographic Information Working Group (UNGIWG) is a comprehensive, decentralised geospatial information network of UN cartography and geospatial information management science professionals that lay the foundations for the United Nations Spatial Data Initiative (UNSDI) and facilitate decision-making mechanisms between member states, regional organisations, and partners that advance social, economic development, environmental, and humanitarian interests. The UNGIWG held its 15th plenary meeting in New York from 3 to 5 August 2015. This latest meeting was co-chaired by representatives from the United Nations Department of Safety and Security (UNDSS) and the United Nations Office for Outer Space Affairs (UNOOSA).[81] Its plenary conclusions confirmed the need to move the voluntary best-effort approach of the group to the next level and to formalise this geospatial coordination mechanism in the UN system, building on models such as UN-Water, UN-Space, or the EMG. It also called for a permanent, dedicated secretariat for the Working Group and highlighted that coordination was required both on the policy level and the technical level. It also found that

[79] United Nations General Assembly. Eleventh meeting of the International Committee on Global Navigation Satellite Systems – Note by the Secretariat. UN Doc. A/AC.105/1134 of 1 Dec. 2016. New York: United Nations http://www.unoosa.org/oosa/oosadoc/data/documents/2016/aac.105/aac.1051134_0.html

[80] "12th Meeting of the International Committee on Global Navigation Satellite Systems." 7 Feb. 2017. ICG 21 Mar. 2017 http://icg12.jp/

[81] "15th UNGIWG Plenary Meeting." UNGIWG 20 Apr. 2016 http://www.ungiwg.org/meetings/15th-ungiwg-plenary-meeting

challenges remained in the availability of core datasets and of commercial very high-resolution satellite imagery.[82]

Conference on Disarmament (CD)

The Conference on Disarmament (CD) is the key multilateral disarmament and arms control negotiating forum within the international community. The 2016 session of the Conference occurred in three parts, the first part from 25 January to 1 April, the second on 16 May to 1 July, and the third on 1 August to 16 September 2016, during which the Conference held 30 formal plenary meetings and 6 informal plenary meetings, in which member states as well as non-member states were invited to participate in the discussions and outlined their views and recommendations. In the space-related substantive work of the Conference during its 2016 session, development continued on the preambular and operative paragraphs of resolutions regarding preventing an arms race in outer space, on no first placement of weapons in outer space, and on transparency and confidence-building measures in outer space activities. Under the agenda item on preventing an arms race in outer space, delegations reaffirmed and further elaborated their respective positions during the general debate of the Conference. On 4 April 2016, representatives from Venezuela and the Russian Federation submitted the text of a joint statement declaring that they will not be the first to place weapons of any kind in outer space, signed in New York on 26 September 2015. On 3 June 2016, the document entitled "Malaysia on behalf of Member States of G-21. Working paper. Prevention of an Arms Race in Outer Space (PAROS)" was submitted to the Conference for review. And on 16 September 2016, the United States submitted a note verbale clarifying its activities on "Implementing the Recommendations of the Report (A/68/189*) of the Group of Governmental Experts on Transparency and Confidence-Building Measures in Outer Space Activities to Enhance Stability in Outer Space".[83] Other topics addressed by the CD included the cessation of the nuclear arms race and nuclear disarmament, prevention of nuclear war including all related matters, effective international arrangements to assure non-nuclear-weapon states against the use or threat of use of nuclear weapons, new types of weapons of mass destruction and new systems of such weapons (radiological weapons), a comprehensive programme of disarmament, and transparency in armaments.

On 19 October 2016, discussions on the nature of the outer space security regime moved from the CD into the UN First Committee (Disarmament and International Security) in its thematic debate on outer space. Delegates from the Committee juxtaposed the merits of legally binding instruments, including China and the Russian Federation's updated draft Treaty on the Prevention of the Placement of Weapons in

[82] "United Nations Geographic Information Working Group | Report to the UN Committee of Experts on Global Geospatial Information Management (GGIM)." 25 Aug. 2015. UNGIWG 20 Apr. 2016 http://ggim.un.org/docs/meetings/GGIM5/UNGIWG%20Introduction.pdf

[83] United Nations General Assembly. Report of the Conference on Disarmament. 2016 session. UN Doc. A/71/27 of 22 September 2016. New York: United Nations https://disarmament-library.un.org/UNODA/Library.nsf/a61ff5819c4381ee85256bc70068fa14/d151f0c3926f060c85257b1a006ea83c/$FILE/A 71 27 Conference on Disarmament.pdf

Outer Space and of the Threat or Use of Force Against Outer Space Objects (PPWT) submitted to the CD in June 2014, and politically binding norms, including the draft International Code of Conduct for Outer Space Activities (ICoC) developed by the European Union in discussion during the same period. While the PPWT was seen to be the most developed proposal provided to the CD so far, it lacks a ban on the development and testing of ground-based antisatellite systems and needs to be comprehensive, effective, and verifiable in order to strengthen trust and confidence between states. Proponents of the ICoC, convinced of the importance of transparency and confidence-building measures, said that international cooperation and agreed standards of responsible behaviour were needed and that a non-legally binding agreement negotiated within the United Nations could be the way forward. Nevertheless, an agreement on the nature of the space security regime remained elusive.[84]

United Nations Institute for Disarmament Research (UNIDIR)
The United Nations Institute for Disarmament Research (UNIDIR) conducts some projects that cover, both directly and indirectly, issues of space security. It seeks to review former proposals and to propose new options for breaking the deadlock in space weaponisation and other emerging security matters at the CD. On 18 July 2016, the UNGA received the report of the director of the UNIDIR on the activities of the Institute for the period from January to December 2015 and the programme of work and financial plan for 2016 and 2017. Activities on space-related emerging security issues included a study entitled "The Realities of Middle Power Space Reliance", which provided a strategic, security-focussed overview of the considerations and options available to Space Middle Powers in order to best position themselves for securing long-term sustainable access to space-based services. During the third Association of Southeast Asian Nations regional forum seminar on international law and space security, UNIDIR hosted a luncheon and capacity-building seminar on outer space to help stakeholders develop their understanding of the dual themes of transparency and confidence-building measures in outer space activities and space security and international law. Lastly, UNIDIR's 14th edition of its space security conference series was held in Geneva, Switzerland, on 28–29 April 2016, on the theme "Sustaining the momentum: the current status of space security".[85] The six panels of the conference sought to facilitate discussion on foundational concepts, provide overviews and updates from a series of space actors and regulatory bodies, examine the key political and legal initiatives currently under consideration by the international community, and discuss the best way forward in terms of a multilateral space security regime.[86]

---

[84] "Debating Proposals on Common Principles to Ensure Outer Space Security, First Committee Delegates Call for Adoption of Legally Binding Treaty." 19 Oct. 2016. United Nations 23 Mar. 2017 https://www.un.org/press/en/2016/gadis3557.doc.htm

[85] United Nations General Assembly. United Nations Institute for Disarmament Research – Note by the Secretary General, on its 71st session. UN Doc. A/71/162 of 18 July 2016. http://undocs.org/A/71/162

[86] "UNIDIR Space Security Conference 2016. Sustaining the Momentum: the Current Status of Space Security." 21 July 2016. UNIDIR 24 Mar. 2017 http://www.unidir.org/files/publications/pdfs/space-security-2016-en-654.pdf

1.2.2 The Group on Earth Observation

The Group on Earth Observation (GEO), with its system-of-systems cooperation platform (GEOSS), is a voluntary partnership of governments and international organisations that facilitates the coordination of current and future Earth observing systems, while also providing structured and sustained data worldwide as a decision support tool for its users. GEO remained active in 2016, conducting around 65 workshops and symposia (including 12 online webinars), which covered an assortment of topics ranging from climate change, water management, disaster risk reduction, and other Earth observation and environmental monitoring topics. Notable conferences included the Global Climate Observation: the Road to the Future conference; the Fourth International Conference on Remote Sensing and Geoinformation of the Environment; the 2nd International Conference on Geographical Information Systems Theory, Applications and Management; the fourth edition of European Space Solutions; Citizen Observatories for Water Management conference; SciDataCon 2016 "Advancing the Frontiers of Data in Research"; the 3rd Biennial Conference of the Southern Africa Society for Disaster Risk Reduction; and the 7th Asia-Oceania/2nd AMS-Asia/2nd KMA Meteorological Satellite Users' Conference. GEO also held its 13th annual plenary meeting in St. Petersburg, Russian Federation, from 7 to 10 November 2016, with sessions addressing Advancing the Vision of GEO; Commercial Sector Engagement; Strengthening Stakeholder Engagement; the GEO Work Programme; and Other Business, Session Outcomes and Closing Remarks.[87]

1.2.3 Europe

1.2.3.1 European Space Agency

ESA's budget increased by 18.5% to €5.253 billion in 2016 from €4.433 billion in the previous year. Direct funding from ESA member states increased by 15.4% to €3.740 billion, while funding from the EU and EUMETSAT grew by 26.7% to €1.510 billion for the year.[88] The biggest budget allocation increase went to launcher development, which grew by 73.0%, reaching €1.051 billion or one-fifth of ESA's spending. Funding for Earth observation increased by 27.8%, reaching €1.604 billion (a 30.5% share of ESA's budget), while spending on Navigation decreased by 8.3% to €609.5 million (11.6%). Funding for space science remained unchanged at €507.9 million (9.7%), while human spaceflight decreased by 1.7% to €365.1 million (7.0%). Next, Telecom and Integrated Applications increased by 16.2% to €359.3 million (6.8%), while Robotic Exploration and Prodex also had an increase

[87] "Past Meetings." GEO 23 Mar. 2017 http://www.earthobservations.org/meetings.php

[88] "ESA Budget 2016." 20 Jan. 2016. ESA 20 July 2017 http://www.esa.int/spaceinimages/Images/2016/01/ESA_budget_2016

of 23.8% reaching €192.8 million (3.7%). Funding for Space Situational Awareness decreased by 7.2% to €12.9 million (0.2%), while the remaining 8.6% of funding was allocated to ESA's basic activities, activities associated with the general budget, and the European Cooperating States Agreement.[89]

ESA Director General Jan Woerner and European Commissioner Elżbieta Bieńkowska signed a "Joint Statement on Shared Vision and Goals for the Future of European Space" on 26 October 2016, coinciding with the release of the European Commission's "Space Strategy for Europe". The document identified the following common interests as guiding principles for future cooperation: maximising the integration of space into European society and economy, fostering a globally competitive European space sector, and ensuring European autonomy in accessing and using space in a safe and secure environment. It also emphasised ESA's and EU's intention to reinforce their cooperation in the future as foreseen in the ESA/EU framework agreement of 2004. These shared vision and goals were also used as a high-level policy guiding element in the preparation of ESA DG Woerner's proposal to ESA member states for the 2016 Ministerial Council in December 2016.[90]

ESA's Ministerial Council meeting took place on 1–2 December 2016, in Lucerne, Switzerland. The meeting was intended to define ESA's objectives based on the vision of a "United Space in Europe in the Era of Space 4.0". Here, Space 4.0 represents the evolution of the space sector into a new era where the increased number of new actors around the world (including the emergence of private companies; participation with academia, industry, and citizens; digitalization; and global interaction) has changed the playing field among governments, the private sector, society, and politics. Moreover, the Space 4.0 motif is analogous and intertwined with Industry 4.0 that represents the Fourth Industrial Revolution of manufacturing and services currently taking place.[91] At the Ministerial, ESA tabled four future goals, in line with the joint statement between ESA and the EU in October 2016, aiming for a total sum of €11 billion in investment; by the conclusion of the meeting, €10.3 billion in investments beyond 2021 had been committed by ministers to reach those future goals. Its first goal, to maximise the integration of space into European society and economy, received the targeted €2.5 billion; the second goal, to foster a globally competitive European space sector, received €1.4 billion, below its €1.5 billion target; the third goal, to ensure European autonomy in accessing and using space in a safe and secure environment, received €1.8 billion, much less than the €2.5 billion target; however the development of a foundation of excellence in space science and technology received €4.6 billion, higher than its €4.5 target.[92]

[89] "ESA Budget 2016 by Domain. 14 Jan. 2016. ESA 20 July 2017 http://www.esa.int/spaceinimages/Images/2016/01/ESA_budget_2016_by_domain

[90] "Shared vision and goals for the future of Europe in space." 26 Oct. 2016. ESA 31 July 2017 http://m.esa.int/About_Us/Welcome_to_ESA/Shared_vision_and_goals_for_the_future_of_Europe_in_space

[91] "Media Backgrounder: ESA's Ministerial 2016 in Lucerne." 14 Nov. 2016. ESA 1 Aug. 2017 http://www.esa.int/About_Us/Ministerial_Council_2016/Media_backgrounder_ESA_s_Ministerial_2016_in_Lucerne

[92] "European Ministers Ready ESA for a United Space in Europe in the Era of Space 4.0." 2 Dec. 2016. ESA 1 Aug. 2017 http://www.esa.int/About_Us/Ministerial_Council_2016/European_ministers_ready_ESA_for_a_United_Space_in_Europe_in_the_era_of_Space_4.0

Within that €10.3 billion package, ESA's science programme will receive €508 million per year between 2017 and 2022, increasing by 1% annually; moreover, its ExoMars mission will get €440 million, with €340 coming from additional contributions by governments, while the remaining balance will be sourced from the ESA's budget for mandatory activities. In commitments to the International Space Station to 2024, ESA agreed to spend a total of €960 million, including €153 million that will go towards experiments on Europe's Columbus laboratory on the station. ESA is also building the service module for NASA's Orion spacecraft to offset its 8.3% annual pro rata share of the ISS's operating costs. The Space Rider mission was allowed to move to the critical design review, which will use the structure of ESA's Intermediate Experimental Vehicle (IEV) to build an orbital reusable vehicle that is capable of being launched vertically on a Vega launcher and landing horizontally on an airstrip. Moreover, ESA has also agreed to spend up to €100 million on the Prometheus engine, intended to form the basis of a reusable first-stage engine for a future rocket. However, ESA's Asteroid Impact Mission (AIM) with NASA did not gain the necessary support to begin.[93] The next ESA Ministerial Council meeting will take place in Spain in late 2019.

1.2.3.2 European Union

With the EU's three flagship space programmes (Copernicus, Galileo, and EGNOS) being well advanced, Europe has shifted its focus from building space infrastructure towards ensuring strong market uptake of their space data and services by the public and private sectors. The European Commission released its space strategy on 26 October 2016 with the overarching aim of building a sustainable space economy. The strategy focusses on the need for investing in start-ups through its Investment Plan for Europe and boosting private investment with the development of a Pan-European Venture Capital Fund-of-Funds programme. It also aims to build on Europe's space situational awareness capability and restated its support for a GovSatCom programme.[94]

The strategy proposes four strategic goals to be set by the European Commission. The first is to "maximise the benefits of space for society and the EU economy" by encouraging the uptake of space services and data and advancing the EU space programmes and meeting new user needs. Here, the Commission will promote the uptake of Copernicus, EGNOS, and Galileo solutions in EU policies; facilitate the use of Copernicus data and information; stimulate the development of space applications; and promote the efficient and demand-driven use of satellite communications to foster ubiquitous connectivity in all member states. The Commission will also maintain the stability of the EU space programmes and prepare the new genera-

[93] De Selding, Peter B. "Europe commits to the space station and ExoMars as part of $11 billion in commitments to ESA." 2 Dec. 2016. SpaceNews 4 Jan. 2017 http://spacenews.com/europe-commits-to-the-space-station-and-exomars-as-part-of-11-billion-in-commitments-to-esa/

[94] "Space." European Commission 31 July 2017 http://ec.europa.eu/growth/sectors/space_en

tions on a user-driven basis; and it will address emerging needs related to climate change, sustainable development, and security and defence.[95]

Its second goal is to "foster a globally competitive and innovative European space sector" by supporting research and innovation and development of skills and fostering entrepreneurship and new business opportunities. Here, the Commission will step up its efforts to support space R&D activities and to boost the competitiveness of the European space sector; it will strengthen the use of innovative procurement schemes to stimulate the demand side of innovation and explore new ways to leverage private sector investments and partnerships with industry; it will promote the use of common technology roadmaps to ensure greater complementarity of R&D projects; and it will include space and Earth observation in its blueprint for sectoral cooperation on skills, addressing new skills requirements in the sector. It will also step up support to space entrepreneurs through EU funding programmes; engage in a dialogue with the European Investment Bank (EIB) and European Investment Fund (EIF) on the support of investment in the space sector; and support space start-ups, including by exploring synergies with the upcoming pan-European fund of funds.[96]

Its third goal is to "reinforce Europe's autonomy in accessing space in a safe and secure environment" by maintaining Europe's autonomous access to space, ensuring access to radio-frequency spectrum, ensuring the protection and resilience of critical European space infrastructure, and reinforcing synergies between civil and security space activities. Here, the Commission will aggregate its demand for launch services to provide visibility to industry and reduce implementation costs; support research and innovation efforts to react to and anticipate disruptive changes, such as in launcher reusability and small launchers; support European launch infrastructure facilities in line with its policy goals; and encourage the development of commercial markets for new space activities. Additionally, the Commission will enhance the current EU space surveillance and tracking (SST) services framework and will raise awareness of risks to critical European space infrastructure, while also considering a more comprehensive space situational awareness service (e.g. space debris, space weather, and cyber alerts) in the long term. Lastly, it will propose a GovSatCom initiative to ensure reliable, secured, and cost-effective satellite communication services for EU and national public authorities and infrastructure and strengthen security requirements when developing EU space systems.[97]

In its fourth goal, the Commission aims to "strengthen Europe's role as a global actor and promoting international cooperation" by pursuing space dialogues with strategic international partners, by taking space policy into account in EU export control dialogues with third countries, and it will use economic diplomacy and trade policy

[95] Commission of the European Communities. Communication from the Commission to the European Parliament, the Council, the European Economic and Social Committee and the Committee of the Regions. COM (2016) 705 final of 26 October 2016. Brussels: European Union https://ec.europa.eu/transparency/regdoc/rep/1/2016/EN/COM-2016-705-F1-EN-MAIN.PDF

[96] Ibid.

[97] Ibid.

instruments to assist European companies active in global markets and to address societal challenges. The Commission will also foster the EU's contribution to GEO and CEOS and will engage with international partners to promote responsible behaviour in outer space and the preservation and protection of the space environment.[98]

1.2.3.3 EUMETSAT

The European Organisation for the Exploitation of Meteorological Satellites (EUMETSAT) is an intergovernmental organisation that supplies weather- and climate-related satellite data to the National Meteorological Services of its Member and Cooperating States in Europe and other users worldwide. EUMETSAT's 85th Council meeting took place in Darmstadt, Germany, on 28 and 29 June 2016. At that meeting, the Council endorsed its new strategy "Challenge 2025" which establishes the framework for its activities in the next decade; the strategy focusses on the smooth transition of the current generation systems to the new MTG, EPS-SG, and Jason-CS satellite systems and the full implementation of the Copernicus Sentinel-3, Sentinel-4, Sentinel-5, and Sentinel-6 missions on behalf of the EU and further cooperation with international partners.[99] As a first step, the Council approved moving the Meteosat-8 spacecraft in GEO orbit over the Indian Ocean to replace its ageing Meteosat-7 spacecraft as it approaches the end of its service in the first quarter of 2017. The approval is EUMETSAT's best-effort contribution to the multi-partner Indian Ocean Data Coverage (IODC) mission, which also involves geostationary satellites from India, Russia, and China; and it relieves a source of strain regarding a gap in coverage with the US Air Force in 2015, as it relied on Meteosat-7 data for cloud characterisation and weather imagery over the war-wracked region, both essential for maintaining battle space awareness.[100] Moreover, implementation of its Polar System Second Generation (EPS-SG) programme progressed following the Council's approval of a cooperation agreement with the DLR for the development of three METimage instruments and of an important ground segment development contract covering all systems required to command and control the operations of the Metop-SG satellites from Darmstadt, Germany. The Council also approved the Third Continuous Development and Operations Phase (CDOP 3) of EUMETSAT's eight Satellite Application Facilities (SAF) covering the period 2017–2022.[101]

---

[98] Ibid.

[99] "EUMETSAT Strategy Challenge 2025 Summary." 15 June 2016. EUMETSAT 31 July 2017 https://www.eumetsat.int/website/wcm/idc/idcplg?IdcService=GET_FILE&dDocName=PDF_BR_COR02SUM_EN&RevisionSelectionMethod=LatestReleased&Rendition=Web

[100] Gruss, Mike. "U.S. Military Faces Weather Coverage Gap over Hot Zones." 13 Mar. 2015. SpaceNews 22 Dec. 2015 http://spacenews.com/u-s-central-command-faces-weather-coverage-gap-over-hot-zones/

[101] "At its 85th session, the EUMETSAT Council endorsed a new strategy establishing the framework for EUMETSAT activities in the next decade." 29 June 2016. EUMETSAT 31 July 2017 https://www.eumetsat.int/website/home/News/DAT_3112252.html

During EUMETSAT's 86th Council meeting, held on 6 and 7 December 2016 in Darmstadt, Germany, Council members agreed to shift the Metop-A polar-orbiting satellite into a drifting orbit in June 2017 and extend its operating life by 2 to 3 years while also conserving enough fuel to deorbit the spacecraft at the end of its service. Metop-A is part of the space segment of EUMETSAT's current EUMETSAT Polar System (EPS) programme. The EPS programme consists of three identical polar-orbiting Metop satellites together with relevant ground systems. The Metop satellites are launched at 6-year intervals between 2006 and 2018. Metop-A was launched on 19 October 2006, joined by Metop-B on 17 September 2012; Metop-C is expected to launch in October 2018.[102] The Council meeting also approved a contract with Thales Alenia Space for the payload data acquisition and processing function of the ground segment for the Metop-SG satellites.[103]

Six Metop Second Generation (Metop-SG) satellites will form the space segment of the Second Generation EUMETSAT Polar System (EPS-SG). The entire programme is budgeted at about €4.1 billion, with EUMETSAT contributing about 80% of the budget, while ESA will cover the rest; the satellites are intended to operate between 2021 and around 2042.[104] The Metop-SG satellites will operate in three pairs, each carrying a different but complementary suite of instruments, and will be manufactured by Airbus D&S under a €1.32 billion contract signed on 16 October 2014.[105] While the first two Metop-SG satellites, Metop-SG A and Metop-SG B, are scheduled to launch in 2021 and 2022, respectively, both satellites have undergone a late redesign to include larger propellant tanks to ensure a controlled deorbit at the end of their service.[106]

EUMETSAT's current Meteosat programme consists of both Meteosat First Generation (MFG) and the Meteosat Second Generation (MSG) satellites operating in geostationary orbit over Europe and Africa. EUMETSAT had added the MSG-4 satellite to its MSG programme on 15 July 2015, and following its commissioning phase, the spacecraft was renamed Meteosat-11 in December 2015. Following its successful launch and commissioning, Meteosat-11 will be placed into in-orbit storage for 2.5 years, meant for use prior to the deployment of EUMETSAT's MTG system. The MSG programme has three other satellites in operation, Meteosat-8 to Meteostat-10, which are expected to end service in 2019, 2021, and 2022, respec-

[102] "Metop." Eumetsat 31 July 2017 http://www.eumetsat.int/website/home/Satellites/CurrentSatellites/Metop/index.html

[103] Henry, Caleb. "Eumetsat will shift Metop-A's orbit to prolong its life, ensure deorbiting." 16 Dec. 2016. SpaceNews 31 July 2017 http://spacenews.com/eumetsat-will-shift-metop-as-orbit-to-prolong-its-life-ensure-deorbiting/

[104] De Selding, Peter B. "ESA, Eumetsat Finalize Contract for Next-gen Weather Satellites." 6 Oct. 2015. SpaceNews 13 Jan. 2016 http://spacenews.com/esa-eumetsat-finalize-contract-for-next-gen-weather-satellites/

[105] De Selding, Peter B. "Airbus Signs $1.7 Billion Contract for Six Metop Weather Satellites." 16 Oct.2015.SpaceNews3June2016http://spacenews.com/42212airbus-signs-17-billion-contract-for-six-metop-weather-satellites/

[106] Henry, Caleb. "Eumetsat will shift Metop-A's orbit to prolong its life, ensure deorbiting." 16 Dec. 2016. SpaceNews 31 July 2017 http://spacenews.com/eumetsat-will-shift-metop-as-orbit-to-prolong-its-life-ensure-deorbiting/

tively. With the health of all Meteosat satellites confirmed as at February 2016, EUMETSAT's Council was then willing to use the residual capacity of Meteosat-8 to support the IODC above the Indian Ocean.[107]

The Meteosat Third Generation (MTG) system is a series of sounding and imaging satellites in geostationary orbit, being developed by France's Thales Alenia Space and Germany's OHB AG, and is aimed at providing services for the 2020 to 2040 timeframe. On 24 July 2015, the EUMETSAT Council approved the contract with Arianespace for the launches of the first three MTG satellites (the MTG-I1, MTG-S1, and one option for MTG-I2).[108] The satellites are scheduled to be launched to GEO orbit within the 2019–2023 timeframe and will operate from 2020 to 2040.[109]

1.2.3.4 National Governments

France

On 1 June 2016, the merged launch directorate of the French space agency (CNES) and the European Space Agency (ESA) presented two reusable projects to France's space policy minister, Thierry Mandon. The first project, Prometheus, is a joint initiative of CNES and Airbus Safran Launchers (ASL) for the development of a cheaper reusable first-stage engine. Currently, the Vulcain cryogenic engine on the first stage of the Ariane 5 launcher, and intended for the Ariane 6 launcher, costs upwards of €10 million for each launch, not including its strap-on boosters. Depending on its configuration, the future Prometheus engine could reduce that cost by half, as each engine will cost €1 million. The second project, Callisto, is a joint CNES-DLR-JAXA project that is intended as a reusable launcher technology demonstrator of Prometheus to launch before 2020.[110] While CNES is the driver in developing the Prometheus engine, ESA agreed to contribute €80 million to the project following the ESA Ministerial Council meeting on 2 December 2016.[111]

[107] De Selding, Peter B. "Eumetsat to continue gap-filler role over Indian Ocean region." 30 June 2016. SpaceNews 10 Nov. 2016 http://spacenews.com/eumetsat-to-continue-gapfiller-role-over-indian-ocean-region/

[108] "EUMETSAT and Arianespace announce signature of launch contract for three MTG satellites." 16 July 2015. Eumetsat 3 June 2016 http://www.eumetsat.int/website/home/News/DAT_2703454.html?lang=EN&pState=1

[109] Henry, Caleb. "Arianespace Wins Fresh Eumetsat Contract After MSG 4 Launch." 16 July 2015. Via Satellite 3 June 2016 http://www.satellitetoday.com/launch/2015/07/16/arianespace-wins-fresh-eumetsat-contract-after-msg-4-launch/

[110] "Thierry Mandon visits CNES's Launch Vehicles Directorate (DLA): presentation of Ariane 6 and Vega-C programmes and their future evolutions." 1 June 2016. CNES 18 July 2017 https://presse.cnes.fr/en/thierry-mandon-visits-cness-launch-vehicles-directorate-dla-presentation-ariane-6-and-vega-c

[111] Henry, Caleb. "ESA kick-starts Prometheus reusable engine with first funding tranche." 22 June 2017. SpaceNews 18 July 2017 http://spacenews.com/esa-kickstarts-prometheus-reusable-engine-with-first-funding-tranche/

ASL's formal bid on 7 May 2015 to ESA for the production of the Ariane 6 launcher came with a desire for ASL to have oversight over the design, production, commercialisation, and operations of the launcher. ASL also sought to increase its 39% stake in Arianespace to include the near 34.7% stake held by the French government through CNES, which would also remove CNES's minority blocking power in launcher development.[112]

The French government agreed to sell its stake in Arianespace to ASL but reserved informal oversight over the company, which will remain intact at its Evry headquarters and will utilise the Ariane 6 launch platform CNES is building in French Guiana.[113] Following negotiations between ESA and ASL, it was agreed that ASL would contribute €400 million to the Ariane 6 development contract, while €200 million will be cut out partly by shaving the cost of certain buildings and facilities that will integrate the launcher horizontally, and an additional €200 million will be removed as unnecessary expenditures. On 30 November 2016, CNES announced that it would sell its stake in Arianespace to ASL; ASL's share capital in Arianespace will increase to 74%, and the three directors representing CNES will be replaced by three directors appointed by Airbus Safran Launchers, while CNES will join ESA as statutory censors of Arianespace's board of directors.[114] ESA signed the Ariane 6 development contract with ASL on 12 August 2015. The contract, now valued at around €2.4 billion, will cover development of the launcher from 2015 to its inaugural flight in 2020.

CNES enhanced its collaborative relationships in 2016. On 25 January 2016, CNES President Jean-Yves Le Gall and ISRO's President A.S. Kiran Kumar signed three cooperation agreements during French President François Hollande's state visit to India. CNES and ISRO will develop a thermal climate-monitoring satellite to map heat exchanges on Earth's surface and to offer support to new applications in agriculture, forest monitoring, soil and groundwater pollution monitoring, and volcanology. ISRO's Oceansat-3 satellite will host France's Argos 4 environmental data collection and location instrument when it is launched in 2018. And the agencies signed a letter of intent on CNES's participation in ISRO's next Mars mission.[115] Also, on 6 September 2016, CNES signed an agreement with the Vietnamese

---

[112] De Selding, Peter B. "New Airbus-Safran Venture Eyes Full Control of Arianespace." 8 Jan. 2015. SpaceNews 12 Jan. 2016 http://spacenews.com/new-airbus-safran-venture-eyes-full-control-of-arianespace/

[113] De Selding, Peter B. "France Giving up Arianespace Ownership, but not Oversight." 19 June 2015. SpaceNews 22 Dec. 2015 http://spacenews.com/france-giving-up-arianespace-ownership-but-not-oversight/

[114] Messier, Doug. "Airbus Safran Launchers Buys CNES Shares of Arianespace." 3 Dec. 2016. ParabolicArc 21 Aug. 2017 http://www.parabolicarc.com/2016/12/03/airbus-safran-launchers-buys-cnes-shares-arianespace/

[115] "List of Agreements/MOUs signed during the State Visit of President Francois Hollande of the French Republic to India (January 25. 2016)." 25 Jan. 2016. Ministry of External Affairs – Government of India 20 Aug. 2017 http://www.mea.gov.in/bilateral-documents.htm?dtl/26294/List+of+AgreementsMOUs+signed+during+the+State+Visit+of+President+Francois+Hollande+of+the+French+Republic+to+India+January+25+2016

Academy of Science and Technology (VAST) to reinforce cooperation between France and Vietnam. France and Vietnam have long-standing relations in this field, with the launches of three Vietnamese satellites having been carried out from the Guiana Space Centre in Kourou, French Guiana.[116]

Germany

At the beginning of the year, Germany's future participation on the ISS was pending an ongoing study to assess the sustainability of the investment and its associated benefits.[117] While the assessment was not published, German delegates to the December 2016 ESA Ministerial Council meeting acknowledged the excellent opportunities for research under space conditions offered by the ISS and benefits derived by German industry in the field of materials research; Germany ended up politically supporting the extension of ISS operations until 2024 and allocated around €346 million to continuing operation of the ISS until 2019.[118]

The DLR formed and reinforced several collaborative relationships in 2016. The first took place on 19 January 2016 in the form of a 3-year agreement between the DLR's Project Management Department and China's National Center for Science and Technology Evaluations (NCSTE) for cooperation in the development of research and innovation, i.e. specifically in terms of quality assurance and the evaluation of research projects, the continued development of national promotion systems, and joint publications on selected topics.[119] On 23 February 2016, the DLR signed an implementing agreement with the Korea Aerospace Research Institute (KARI) to strengthen their partnership in the operation of and reception of data from Korean EO satellites and for increased scientific exchange. And on 25 February 2016, the DLR and the Japan Aerospace Exploration Agency (JAXA) signed an "Inter Agency Arrangement for Strategic Partnership" to achieve common strategic goals, including the development and use of aeronautic and space technologies that contribute to solving global societal challenges, expanding their foundations for scientific excellence in research, and guaranteeing the enhancement of the competitiveness of both countries.[120] The DLR and France's CNES also renewed their 2002

[116] "France/Vietnam: la visite de François Hollande relance la coopération économique." 7 Sept. 2016. Le Moci 19 Aug. 2017 http://www.lemoci.com/actualites/pays-marches/france-vietnam-la-visite-de-francois-hollande-relance-la-cooperation-economique/

[117] De Selding, Peter B. "ESA members give space agency an 18-percent budget boost." 15 Jan. 2016. SpaceNews 10 Nov. 2016 http://spacenews.com/esa-members-give-space-agency-an-18-percent-budget-boost/

[118] "ESA Council meeting at ministerial level in Lucerne – Germany awards approximately two billion euro to space projects." 2 Dec. 2016. DLR 4 Aug. 2017 http://www.dlr.de/dlr/presse/en/desktopdefault.aspx/tabid-10308/471_read-20304/year-all/#/gallery/25207

[119] "Germany and China broaden their collaboration in science management." 22 Jan. 2017. DLR 3 Aug. 2017 http://www.dlr.de/dlr/presse/en/desktopdefault.aspx/tabid-10172/213_read-16485/#/gallery/21731

[120] "DLR is expanding cooperation with Japan and South Korea." 29 Feb. 2016. DLR 3 Aug. 2017 http://www.dlr.de/dlr/en/desktopdefault.aspx/tabid-10857/1527_read-16946/#/gallery/22220

framework agreement for bilateral cooperation on 2 June 2016,[121] and the two agencies signed another agreement for the design, construction, and operation phases of the Methane Remote Sensing LIDAR Mission (MERLIN) environmental satellite on 14 September 2016.[122] And a memorandum of understanding was signed between the DLR and the private Russian university, Skolkovo Institute of Science and Technology, for scientific cooperation covering materials research, life sciences, climate research, space law, and planetary research.[123]

Italy

The Italian Space Agency (ASI) released its Strategic Vision Document for the period 2016–2025 (Documento di Visione Strategica 2016–2025) on 14 December 2016. The document lays out four strategic priorities for ASI over the next decade. Those goals are (1) to promote the development of services and applications for the space economy, (2) to promote the development and use of infrastructures for the space economy, (3) to accelerate and support scientific and cultural progress (science diplomacy), and (4) to raise the country's international prestige (space diplomacy).[124] Nestled within these goals are specific intervention areas and programme areas that indicate the direction in which ASI will operate in order to make progress in satisfying each goal. The first goal focusses on the upstream and downstream components of the space value chain and is centred on "user uptake" of services to facilitate the growth of the space economy over the next decade. The second goal looks at developing infrastructures that facilitate the emergence of an industrial and economic base in which new initiatives can grow. The third goal envisages the preferred approach of defining and coordinating scientific programmes, identifying seven areas of focus: high-energy and space astrophysics; planetology, solar system science, and exoplanetology; cosmology; fundamental physics; Earth sciences; scientific and technological research on the ISS; and disseminating space culture. The fourth goal seeks to raise Italy's global prestige in space diplomacy in three areas: cooperation on European level, cooperation with NASA, and cooperation with other space agencies and institutions in the world.[125]

ASI formed and reinforced several collaborative relationships in 2016. The first took place on 25 January 2016, in the form of a memorandum of understanding

[121] "DLR and CNES renew their framework agreement for bilateral cooperation." 2 June 2016. DLR 3 Aug. 2017 http://www.dlr.de/dlr/en/desktopdefault.aspx/tabid-10081/151_read-18064/year-2016/151_page-4/#/gallery/23231

[122] "Climate protection – DLR and CNES sign an agreement for the construction and operation of the MERLIN environmental satellite." 14 Sept. 2016. DLR 3 Aug. 2017 http://www.dlr.de/dlr/en/desktopdefault.aspx/tabid-10081/151_read-19317/year-2016/151_page-3/#/gallery/24323

[123] "Abkommen zur Wissenschaftskooperation mit der russischen SkolTech-Universität unterzeichnet." 13 June 2016. DLR 3 Aug. 2017 http://www.dlr.de/dlr/desktopdefault.aspx/tabid-10857/1527_read-18243#/gallery/23392

[124] Battiston, Roberto. "Documento di Visione Strategica 2016-2025." 14 Dec. 2016. ASI 2 Aug. 2017 http://www.asi.it/it/news/documento-di-visione-strategica-2016-2015

[125] "STRATEGIC VISION DOCUMENT 2016-2025." 14 Dec. 2016. ASI 2 Aug. 2017 http://www.asi.it/sites/default/files/attach/dettaglio/dvs-ing_web.pdf

(MoU) with the United Arab Emirates (UAE) Space Agency, which provided a broad framework agreement to carry out cooperative space activities between the two entities.[126] On 18 May 2016, the heads of ASI and the Argentinian Space Agency (CONAE) signed a letter of intent to extend and strengthen collaboration in the field of space activities[127]; and on 17 June 2016, another MoU was agreed between ASI and the Russian Space Agency (Roscosmos) to cooperate in the field of remote sensing and Earth observation during the Saint Petersburg Economic Forum. And on 27 October 2016, ASI and the government of Australia signed a collaborative partnership agreement to pursue space activities, promoting future joint research and development, academic exchange, and industry cooperation.[128]

United Kingdom

The UK government's National Space Policy document was released near the end of 2015. It sets out its vision to capture a greater share of the world's thriving space market as the United Kingdom aims to become a future European hub for commercial spaceflight and related space sector technologies.[129] The National Space Policy commits to four key principles in the government's use of space, i.e. it recognises the strategic importance of space, commits to preserving and promoting the safety and security space, supports the growth of the space sector, and commits to international cooperation to deliver maximum benefit from UK investment in space.[130] In that pursuit, the UK government spent much of 2016 working with industry and the science community to develop a new strategy for its implementation, issuing a call for ideas and evidence to help develop a new space strategy on 15 November 2016.[131]

On 7 April 2016, the UK Space Agency published its corporate plan for the period 2016/2017, laying out an overview of the Agency's mandate, strategy, and key targets and milestones. Between 2016 and 2017, it expects to launch its NovaSAR imaging spacecraft. It also hopes to expand its Space for Smarter Government Programme and continue the International Partnership Programme (IPP). Lastly, it hopes to deliver the UK flight instruments for integration in the

[126] Bongiorno, Massimo. "Italy and the United Arab Emirates space agreement." 25 Jan. 2016. ASI 1 Aug. 2017 http://www.asi.it/en/news/italy-and-united-arab-emirates-space-agreement

[127] "Italy – Argentina means space collaboration." 18 May 2016. ASI 1 Aug. 2017 http://www.asi.it/en/news/italy-argentina-means-space-collaboration

[128] "Australia, Italy sign contract for a collaborative partnership in space." 27 Oct. 2016. Geospatial World 1 Aug. 2017 https://www.geospatialworld.net/news/australia-italy-sign-contract-collaborative-partnership-space/

[129] "National Space Policy." 13 Dec. 2015. GOV.UK 27 July https://www.gov.uk/government/uploads/system/uploads/attachment_data/file/484865/NSP_-_Final.pdf

[130] Ibid.

[131] "Call for ideas and evidence: help develop a new space strategy." 15 Nov. 2016. GOV.UK 27 July2017https://www.gov.uk/government/publications/call-for-ideas-and-evidence-help-us-develop-our-new-strategy

ESA's ExoMars rover and Solar Orbiter and to receive the first science survey results from Gaia and the first data from LISA Pathfinder.[132]

The United Kingdom also took a step closer to realising its goal to be a spaceport for orbital and suborbital commercial launch activities. On 12 July 2016, the British government awarded feasibility study contracts for horizontal or vertical launch proposals by five industrial teams: Airbus Safran Launchers, Orbital Access (in association with BAE Systems and Reaction Engines Ltd), Virgin Galactic, Deimos Space UK (in association with Firefly Space Systems of the United States), and Lockheed Martin. As proposals of the last three teams are centred on US-based technology, they would need to clear export restrictions under the multilateral Missile Technology Control Regime and the US ITAR.[133] The total value of the contracts is a modest £1.5 million, but aside from that injection, there were no other government commitments to funding spaceport development in 2016.[134]

1.2.4 United States

The White House released its 2017 NASA budget request to the US Congress on 9 February 2016, seeking \$19.025 billion for 2017, a decrease of 1.3% from the \$19.285 billion it received from the 2016 omnibus spending bill signed by President Obama on 18 December 2015.[135] Exploration programmes saw the sharpest cuts in spending, decreasing from \$4.030 billion in 2015 to \$3.337 billion; the decrease came from a \$689.7 million cut to the SLS programme and \$150.2 million cut in the Orion programme, amounting to \$1.310 billion and \$1.120 billion, respectively. Despite raising concerns in Congress about whether the request will keep NASA on track with its exploration goals, NASA officials believe the budget will support the Exploration Mission (EM) 1 of the first launch of the SLS and Orion spacecraft in 2018 and a second launch of EM-2 as early as 2021.[136] NASA's science programme received nearly the same funding as in 2016; however, Earth science will increase by \$111.2 million to \$2.032 billion, while planetary science will decrease by almost

---

[132] "UK Space Agency Corporate Plan 2016–2017." 7 Apr. 2016. GOV.UK 27 July 2017 https://www.gov.uk/government/publications/uk-space-agency-corporate-plan-2016-2017

[133] De Selding, Peter B. "Britain selects U.S., French, British teams to study spaceport feasibility." 13July2016.SpaceNews10Nov.2016http://spacenews.com/britain-selects-u-s-french-british-teams-to-study-spaceport-feasibility/

[134] De Selding, Peter B. "Britain endorses ESA, promises increased export-credit support for industry." 25 Nov. 2016. SpaceNews 4 Jan. 2017 http://spacenews.com/britain-endorses-esa-promises-increased-export-credit-support-for-industry/

[135] Smith, Marcia S. "NASA Gets Big Boost in Final FY2016 Appropriations Bill." 16 Dec. 2015. Space Policy Online 30 June 2016 http://www.spacepolicyonline.com/news/nasa-gets-big-boost-in-final-fy2016-appropriations-bill

[136] Foust, Jeff. "Bolden defends NASA budget in a House committee farewell." 18 Mar. 2016. SpaceNews 10 Nov. 2016 http://spacenews.com/bolden-defends-nasa-budget-in-a-house-committee-farewell/

the same amount ($112.3 million) to $1.519 billion. The funding requested for NASA's Commercial Crew Programme also decreased by $59 million, reaching $1.185 billion for 2017.[137] A final 2017 appropriation bill was not approved by Congress by the end of 2016, as US House of Representatives passed its final bills of the 114th Congress on 8 December 2016, while the US Senate passed a NASA authorization bill on the following day – leaving the Senate's amended version the "NASA Transition Authorization Act of 2016" open for discussion when the 115th Congress convenes in January 2017.[138]

On 29 December 2016, the incoming administration of President-Elect Trump announced the idea of relaunching the National Space Council (NSC) to oversee US space policy. The NSC had originated during the inception of NASA in 1958 as a mechanism to help guide the US space agenda. Upon taking office, it is likely that the NSC will be established within the White House with Vice President Mike Pence as its chair. Some cross-cutting issues that it might choose to address include export control and acquisition reform, the health of the US space industrial base, space debris mitigation and space traffic management, the facilitation of emerging commercial space industries, and the determination of goals and priorities for space activities beyond LEO.[139]

1.2.5 Canada

On 22 November 2016, Innovation, Science and Economic Development Canada (ISED) Minister Navdeep Bains announced that a new Canadian space strategy would be released in June of 2017. While it was discussed during the AIAC 2016 Canadian Aerospace Space Summit, the announcement hinted that MDA and its partners would continue to be supported with a C$54 million contribution from its Technology Demonstration Programme (TDP) and that Canada's government is committed to revitalising the Space Advisory Board (SAB). Prior to the strategy's release in June, the SAB will consult stakeholders to define the key elements of a space strategy, and similar to Europe's space strategy, it will focus on using space to drive broader economic growth through supporting talent, research, and entrepreneurship in the industry.[140]

---

[137] Foust, Jeff. "White House proposes $19 billion NASA budget." 9 Feb. 2016. SpaceNews 10 Nov. 2016 http://spacenews.com/white-house-proposes-19-billion-nasa-budget/

[138] Foust, Jeff. "In symbolic move, Senate passes NASA authorization bill." 10 Dec. 2016. SpaceNews 4 Jan. 2017 http://spacenews.com/in-symbolic-move-senate-passes-nasa-authorization-bill/

[139] David, Leonard. "Playing the Space Trump Card: Relaunching a National Space Council." 29 Dec. 2016. Space.com 19 Aug. 2017 https://www.space.com/35163-trump-administration-national-space-council.html

[140] Boucher, Marc. "New Canadian Space Strategy Coming in June 2017." 22 Nov. 2016. SpaceRef Canada 21 Aug. 2017 http://spaceref.ca/news/new-canadian-space-strategy-coming-in-june-2017.html

The CSA budget for the fiscal year 2016–2017 covering 1 April 2016 to 31 March 2017 decreased by 10.6% to C$432.39 million from C$483.43 million in the previous year. About 49.7% of the funding was allocated to its space data, information, and services, while 23.0% went to space exploration, 15.3% went to Future Canadian Space Capacity, and the remaining 12.0% share went to internal services. The decrease came mainly from a C$43.3 million reduction related to the RADARSAT Constellation Mission (RCM), due to a change in cash flow requirements. Also its contribution to ESA increased to C$27.031 million in 2016–2017, from C$26.215 million in the previous year.[141] Under its Economic Action Plan, Canada plans to increase its spending by an additional C$30 million for ESA's ARTES programme, distributed over the period of 2016 to 2019. Moreover, Canada has also committed to extending its participation in the ISS to 2024 and has historically provided 2.3% of the ISS's common operating costs.[142]

1.2.6 Russia

On 28 December 2015, Russian President Vladimir Putin signed a decree dissolving Russian Federal Space Agency, known as Roscosmos, transferring the agency's responsibilities to the newly formed "Roscosmos State Corporation".[143] The decree took effect on 1 January 2016, and while the powers and functions of Roscosmos will remain the same, the new organisation will be run as a corporation with control over Russia's entire space industry.[144] The creation of the Roscosmos State Corporation is another step in Russia's reorganisation of its space sector that began in December 2013.[145] In addition to increasing Russia's competitiveness, both in gaining market share and in securing parity and advantage over geopolitical opponents, the reorganisation is intended to strengthen Russia's struggling space industry, which has seen a number of high-profile failures in recent years. In that context, on 19 April 2016, Roscosmos provided a 20 billion rouble cash infusion to the space hardware builder Khrunichev Space Centre to repay its suppliers. As the total debt of the Khrunichev Space Centre

[141] "Main Estimates – 2016–17 Estimates – Canadian Space Agency." 23 Feb. 2017. Government of Canada 20 Aug. 2017 https://www.canada.ca/en/treasury-board-secretariat/services/planned-government-spending/government-expenditure-plan-main-estimates/2016-17-estimates/main-estimates.html#toc7-37

[142] SpaceNews Editor. "Editorial|Canadian Commitment Builds Momentum for Space Station Extension." 5 May 2015. SpaceNews 11 Jan. 2016 http://spacenews.com/editorial-canadian-commitment-builds-momentum-for-space-station-extension/

[143] "Russian space agency gets replaced by state corporation – Kremlin." 28 Dec. 2015 TASS 18 Aug. 2017 http://tass.com/science/847295

[144] Nowakowski, Tomasz. "Russia dissolves its federal space agency, what now?" 30 Dec. 2015. Spaceflight Insider 18 Aug. 2017 http://www.spaceflightinsider.com/organizations/roscosmos/russia-dissolves-federal-space-agency-now/

[145] "Putin Signs Decree on Creation of United Rocket and Space Corporation." 2 Dec. 2013. TASS 18 Aug. 2017 http://en.itar-tass.com/russia/709849

stood at 114 billion as at 2014, Roscosmos will facilitate subsidies from the Russian government and loans from the Russian development bank, Vnesheconombank, to stabilise its accounts as part of the first phase of Khrunichev's 10-year recovery schedule. The second phase, taking place between 2017 and 2020, will include a broad reorganisation of Khrunichev which should yield sustained profitability between 2021 and 2025, in addition to continued increases in both labour productivity and salaries.[146]

On 18 August 2016, Boeing and its Russian and Ukrainian partners in the Sea Launch joint venture reached a preliminary framework agreement to settle a lawsuit initiated by Boeing in February 2013. Boeing had sued Russia's RSC Energia and the Ukrainian launch vehicle manufacturer Yuzhnoye for refusing to pay more than $350 million following Sea Launch's 2009 bankruptcy, leaving Boeing to cover $449 million in loan guarantees to third-party creditors on its own. In its lawsuit, Boeing said that RSC Energia owed at least $222.3 million and that Yuzhnoye owed at least $133.4 million.[147] On 12 May 2016, the US district court in California issued a judgement in favour of Boeing, concluding that RSC Energia owed Boeing more than $320 million in reimbursement obligations as defined in prior agreements. Boeing and RSC Energia expect to reach a final agreement to settle the lawsuit by the end of 2016, wherein the parties have agreed that Boeing will write-off part of the judgement in compensation for a cooperative agreement to develop a docking adaptor that could be used by both Boeing's CST-100 Starliner crew spacecraft and Russia's next-generation crewed spacecraft called "Federation", being developed by Energia for Roscosmos.[148] The settlement still leaves open the status of Sea Launch which is now primarily owned by RSC Energia and has remained docked in port in California since its last mission to launch Eutelsat's Eutelsat 3B to GEO on 26 May 2014.

1.2.7 Japan

Japan's Space Policy Commission published a revised version of the third iteration of its Basic Plan for Space Policy on 11 November 2015.[149] Other than recommending that the IGS system be expanded to eight satellites (plus two relay satellites to support the constellation) from the original two optical and two radar satellites (plus

---

[146] De Selding, Peter B. "Russia bails out debt-ridden Khrunichev." 20 Apr. 2016. SpaceNews 10 Nov. 2016 http://spacenews.com/russian-government-provides-debt-ridden-khrunichev-with-cash-and-loans/

[147] Raymond, Nate. "Boeing sues Sea Launch partners for $350 million." 4 Feb. 2013. Reuters 18 Aug. 2017 http://www.reuters.com/article/2013/02/04/boeingsealaunch-idUSL1N0B31GP20130204

[148] Foust, Jeff. "Boeing and Energia negotiating Sea Launch settlement." 19 Aug. 2016. SpaceNews 18 Aug. 2017 http://spacenews.com/boeing-and-energia-negotiating-sea-launch-settlement/

[149] Kallender-Umezu, Paul. "What's Behind Japan's Sudden Thirst for More Spy Satellites." 13 Nov. 2015. SpaceNews 7 Jan. 2016 http://spacenews.com/whats-behind-japan-sudden-thirst-for-more-spy-satellites/

one on-orbit spare), the revised version was in line with the original third iteration of its Basic Plan for Space Policy published on 9 January 2015. This new 10-year roadmap marks a shift in Japan's priorities from its previous basic plans published in 2009 and updated early in 2013, now focussing on security and commerce rather than its earlier emphasis on the peaceful use of outer space. Another departure in this policy is in the naming of China as a destabilising factor in global security, particularly in its growing counter-space capability and development of antisatellite weapons.[150] The new policy puts greater focus on developing Japan's Information Gathering Satellites (IGS) to further improve the country's surveillance and reconnaissance competencies. Moreover, it looks to increasing cooperation with the United States on an equal basis, while also maintaining and strengthening its own industrial and science and technology sector.[151]

According to the Space Report 2016, Japan's combined space budget was increased by 2.4% to ¥332.4 billion for the fiscal year 2016 (beginning on 1 April 2016 and ending 31 March 2017).[152] The 2015 budget in the same fiscal period was ¥324.5 billion. The 2016 budget, encompassing the space activity of 11 government ministries, saw a 1.7% funding decrease (i.e. ¥179.3 billion) for the Ministry of Education, Culture, Sports, Science and Technology (MEXT) which governs the Japan Aerospace Exploration Agency (JAXA); meanwhile, Japan's Ministry of Defence had a 15% increase in funding (i.e. ¥34.2 billion), likely excluding Japan's ballistic missile defence budget as in the previous year.[153]

In line with its enhanced cooperation with the United States, Japan has agreed to extend its participation in the International Space Station (ISS) through 2024. By December 2015, the Strategic Headquarters for Space Policy of the Cabinet Office in Japan's government officially approved a plan to develop the HTV-X to follow on from JAXA's next three HTV missions to the ISS, not including HTV-6. The HTV-7 will launch in February 2018, followed by HTV-8 in February 2019 and then by HTV-9 in February 2020; the HTV-X will be launched in 2021.[154] Japan spends about ¥40 billion on the ISS annually.[155]

---

[150] Kallender-Umezu, Paul. "Japan Boosts Space Spending In Support of Security Focus." 2 Feb. 2015. SpaceNews 10 Dec. 2015 http://spacenews.com/japan-boosts-space-spending-in-support-of-security-focus/

[151] Rajagopalan, Rajeswari Pillai. "Op-ed.|Japan's Space Policy Shift Reflects New Asian Realities." 23Feb.2015.SpaceNews10Dec.2015http://spacenews.com/op-ed-japans-space-policy-shift-reflects-new-asian-realities/

[152] The Space Report 2017. Colorado Springs: The Space Foundation, 2017: 16.

[153] Ibid.

[154] Gebhardt, Chris. "Japan's HTV-6 resupply vehicle arrives at the ISS." 13 Dec. 2016. NASASpaceflight.com 19 Aug. 2017 https://www.nasaspaceflight.com/2016/12/jaxa-iss-htv-6-resupply-launch-station/.

[155] Foust, Jeff. "Japan Seeks To Become Full Partner with U.S. in Space." 11 Dec. 2015. SpaceNews 7 Jan. 2016 http://spacenews.com/japan-seeks-to-become-full-partner-with-u-s-in-space/

1.2.8 China

China's 13th Five-Year Plan (2016–2020) was preliminarily approved by China's Communist Party on 29 October 2015, while its details were finalised in early 2016.[156] The 13th Five-Year Plan has been formulated around five philosophical tenets, i.e. innovation, coordination, green growth, opening up, and inclusive development.[157] Its major objectives for economic and social development are as follows: (1) maintain a medium-high rate of growth; (2) achieve significant results in innovation-driven development; (3) further coordinate development; (4) improve standards of living and quality of life; (5) improve the overall calibre of the population and the level of civility in society; (6) achieve an overall improvement in the quality of the environment and ecosystems; and (7) ensure all institutions become more mature and better established.[158]

China's Middle and Long Term Development Plan for State Civil Space Infrastructure (2015–2025) was released on 26 October 2015.[159] Under this new roadmap, satellite remote sensing, communications, broadcasting, and navigation systems will be built up during the 13th Five-Year Plan to establish the state civil space infrastructure system. Using those systems, the 2015–2025 space development plan aims to produce comprehensive application demonstrations in 12 fields (including territory, mapping, energy, communications, and environmental protection) to provide core business with timely, accurate, and stable space information services. Additionally, the 2015–2025 plan stresses the importance of investment in its domestic industry and calls for more investment of private capital.[160] For instance, it foresees 100 launches of its Long March launcher family during the 2015–2025 period to meet domestic demand; it also aims to court commercial launch contracts by providing commercial launch services outside of its territory.[161]

---

[156] Magnier, Mark. "China's Communist Party Approves Five-Year Plan." 29 Oct. 2015. The Wall Street Journal 1 Feb. 2016 http://www.wsj.com/articles/chinas-communist-party-approves-five-year-plan-1446124597

[157] "An Overview Of The 13th Five-Year Plan." 7 Dec. 2016. NDRC People's Republic of China 19 Aug. 2017 http://en.ndrc.gov.cn/newsrelease/201612/t20161207_829923.html

[158] "The 13th Five-Year Plan for Economic and Social Development of The People's Republic of China (2016–2020)." 7 Dec. 2016. NDRC People's Republic of China 19 Aug. 2017 http://en.ndrc.gov.cn/newsrelease/201612/P020161207645765233498.pdf

[159] Google Translated. "关于印发国家民用空间基础设施中长期发展规划(2015-2025年)的通知." 26 Oct. 2015. National Development and Reform Commission 2 Feb. 2016 http://www.sdpc.gov.cn/zcfb/zcfbghwb/201510/t20151029_756376.html

[160] "Middle and Long Term Development Plan for Civil Space Infrastructure Issued." 13 Jan. 2016. LexisNexis 2 Feb. 2016 https://hk.lexiscn.com/latest_message.php?access=show_detail&id=184408

[161] De Selding, Peter B. "With Naga-L Rocket, China Would Turn Tables on U.S. Export Ban." 15 Oct. 2015. SpaceNews 24 Dec. 2015 http://spacenews.com/with-naga-l-rocket-china-would-turn-tables-on-u-s-export-ban/

1.2.9 India

In 2016, India's Parliament allocated 75.09 billion rupees to the Indian Space Research Organisation (ISRO) for the fiscal year 2016–2017, beginning on 1 April 2016. The budget was later revised to 80.45 billion rupees, an increase of 15.6% from the 69.6 billion rupees allocated in ISRO's revised 2015–2016 budget. A total of 72.87 billion rupees of the revised 2016–2017 fiscal budget went towards ISRO's Central Sector Schemes/Projects covering space technology, space applications, space sciences, and ISRO's INSAT constellation of communications and meteorological satellites. Space technology accounted for 62.9% of that spending, while ISRO's space applications accounted for 15.2%, and space sciences received just 1.8% of the allocation. Moreover, around 20.1% was allocated towards ISRO's INSAT system.[162]

India's satellite telecommunications market is known for its high barriers to entry for foreign satellite operators wishing to access its large DTH growth market for satellite TV, satellite broadband, and cellular backhaul. Non-Indian operators (including SES, Eutelsat, Intelsat, MEASAT, AsiaSat, Singtel, ABS, and APT Satellite) are only permitted to have an operating licence if India's own INSAT telecommunications satellites, owned and operated by ISRO, do not have sufficient capacity to meet programmers' demand. While the INSAT system is normally unable to keep up with market demand, another hurdle for non-Indian operators is that rather than being allowed to deal with their customers directly, they must sell their own bandwidth to their competitor ISRO, which then resells it to broadcasters at prices set by ISRO. This is not the case for television broadcasters which seek operating licences using India's own INSAT system; moreover, they are given preferential treatment under Indian law.[163]

On 23 May 2016, the Telecom Regulatory Authority of India (TRAI) published a "preconsultation paper" highlighting the savings that satellite television broadcasters could realise if they transmitted popular satellite television programmes using common transponder space shared by multiple DTH providers, rather than transmitting the same programmes individually on different satellites. TRAI went on to emphasise that sharing transponder capacity would help the flow of revenue to non-Indian operators that provide most of the satellite capacity. Nevertheless, the proposal will likely be met with resistance from India's domestic broadcasters and from DTH operators, as it threatens their business models, which focus on differentiation.[164]

[162] Notes on Demands for Grants, 2017-2018. "Department of Space | No.91 | Department of Space." 1 Feb. 2017. Indiabudget.nic.in 5 Aug. 2017 http://indiabudget.nic.in/ub2016-17/eb/sbe84.pdf

[163] De Selding, Peter B. "Satellite operators give negative reviews of Indian regulator's satellite-TV proposal." 8 June 2016. SpaceNews 10 Nov. 2016 http://spacenews.com/satellite-operators-give-negative-reviews-of-indian-regulators-satellite-tv-proposal/

[164] De Selding, Peter B. "Satellite operators give negative reviews of Indian regulator's satellite-TV proposal." 8 June 2016. SpaceNews 10 Nov. 2016 http://spacenews.com/satellite-operators-give-negative-reviews-of-indian-regulators-satellite-tv-proposal/

1.3 Worldwide Space Budgets and Revenues

From the Space Report 2017, total government space expenditure was $76.42 billion in 2016, slightly below the $76.52 billion spent in 2015 – the reduction can be partly explained by a slight decrease in total US spending in civil and military space programmes and by weak currency exchange rates which mask the fact that most other government space budgets increased from 2015.[165] Total government expenditure for civil space programmes grew by 2.5% to $43.42 billion from $42.37 billion in 2015.[166] On the other hand, Euroconsult estimates that government spending in space programmes reached $62.2 billion in 2016, i.e. slightly less than the spending recorded in 2015. Around 65% of that government spending went towards civil programmes, about $39.34 billion, while 35% went to defence programmes; the share of spending on civil programmes has continued to grow as a proportion of global expenditures in recent years.[167]

The Space Report 2017 noted that the total revenue of commercial satellite services, including telecommunications, Earth observation, and positioning services, marginally increased by 0.23% to $126.62 billion in 2016 from $126.33 billion in 2015. However, revenue from space-related commercial infrastructure, including manufacturing of spacecraft and in-space platforms, launch services as well as ground equipment increased by 5.14% to $126.26 billion in 2016 from $120.09 billion in 2015; the growth was mainly driven by the sale of GNSS receivers. Here, total commercial space revenue decreased by 2.6% to $252.88 billion in 2016 from 246.42 billion in 2015.[168] The following section provides a more detailed analysis of institutional budgets.

1.3.1 Overview of Institutional Space Budgets

From the Space Report 2017, total institutional spending on space programmes in 2016, including that of intergovernmental organisations, stayed level with $76.42 billion from $76.52 billion in 2015. While the US institutional space budget decreased slightly to $44.44 billion from $44.57 billion in 2016, non-US institutional spending slightly increased by 0.1% to $31.98 billion from $31.95 billion in 2015; this increase was masked by negative dollar exchange rates for a second year.[169] Around 56.8% of the total institutional space expenditure went towards civil expenditure ($43.43 billion), while 43.2% of the spending went to defence

[165] The Space Report 2017. Colorado Springs: The Space Foundation, 2017: 4.

[166] Ibid.

[167] Smith, Andrew. "Euroconsult: Government Spending in Space Programs Reaches $62 Billion in 2016." 30 May 2017. PRWeb 21 Aug. 2017 http://www.prweb.com/releases/2017/05/prweb14377281.htm

[168] Ibid.

[169] Ibid. at 15.

expenditure ($33.00 billion), a slight increase in the amount of civil space spending compared to defence in 2015.[170]

The Space Report 2017 also estimates that worldwide defence-related expenditure dropped to $33.00 billion in 2016. Here, the United States accounts for 66.7% (or an estimated $22.00 billion) of space security programmes under its Department of Defence (DoD). The US DoD's space budget funds its military space programmes, in addition to organisations such as the National Reconnaissance Office (NRO) and the National Geospatial-Intelligence Agency (NGA). Defence spending by non-US government space actors accounted for the remaining 33.3% (or $11.00 billion), a slight increase from 2015; this increase was also masked by negative dollar exchange rates.[171] Moreover, there is a degree of uncertainty regarding expenditures on defence space activities as not all relevant funding is made public.

While spending among space actors saw significant changes in 2016, they should not be ranked against each other given the fluctuations in currency exchange rates as these budgets are converted to US dollars (Fig. 1.1).[172] The United States continued to have the largest space budget, increasing its civilian space spending to $22.440 billion, while decreasing its defence spending to an estimated $22.000 billion. China's space budget is the second largest and is likely larger than the modest estimate of 28.7 billion yuan ($4.22 billion), if China's spending matched the average expenditure of 0.039% of GDP on space activities, not including the United States or Russia, according to the Space Report 2017.[173] Japan's space budget was ¥332.4 billion ($3.236 billion), followed by France and Germany. Next, as Russia's economic crisis continued in 2016, its space budget contracted 104.500 billion roubles ($1.630 billion); India followed with a revised budget of $1.192 billion (80.45 billion rupees).[174]

ESA's budget was not included within these figures, as the contributions of individual ESA members were included in their own budgets. Nevertheless, ESA's 2016 budget increased by 18.4% to €5.250 billion ($5.820 billion) from €4.433 billion ($4.944 billion) in 2015, following an additional 28.6% increase in EU spending. ESA member state spending increased by 15.4% in 2016, reaching €3.740 billion ($4.15 billion) from €3.241 billion ($3.61 billion) in 2015.[175] Among ESA member states, the five biggest contributors to the total ESA budget remained in the same positions as in 2015, with Germany 16.6%, France 16.1%, Italy 9.8%, the United

[170] Ibid.

[171] Ibid.

[172] N.B.: Figures in this section are based on the Space Report 2017 data (USA, China, Russia, Japan, India and South Korea), while all other values in Fig. 1.1 come from the Euroconsult Government Space Programs. Moreover, the different currencies in other sections of this chapter were not converted to a baseline currency (except for comparison purposes) as recent currency fluctuations skewed the changes in the spending by other countries.

[173] The Space Report 2017. Colorado Springs: The Space Foundation, 2017: 10.

[174] Ibid.

[175] "ESA Budget 2016." 20 Jan. 2016. ESA 9 Aug. 2017 http://www.esa.int/spaceinimages/Images/2016/01/ESA_budget_2016

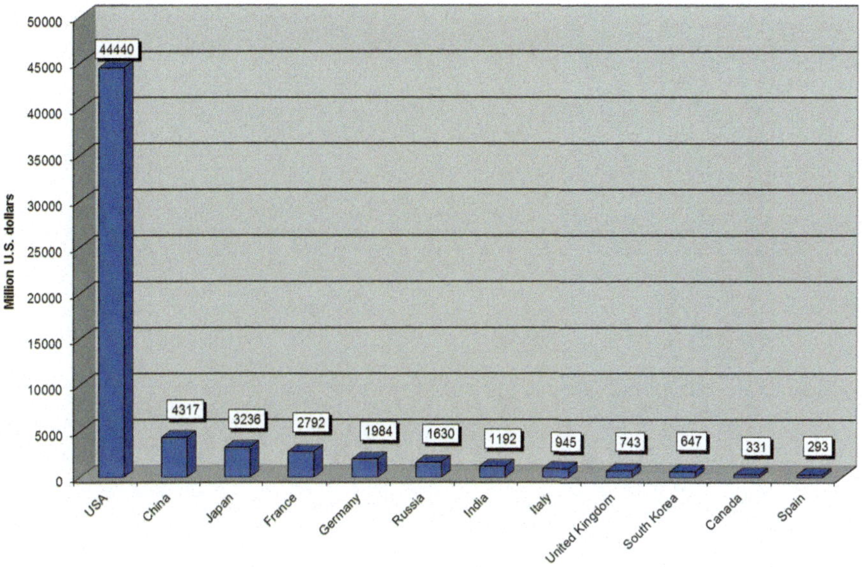

Fig. 1.1 Public space budgets of major space powers in 2016 (Based on Space Report 2017 and Euroconsult data)

Kingdom 6.2%, and Belgium 3.6%. Spain was the next highest contributor at 2.9%, followed closely by Switzerland at 2.8% for seventh position in the 2016 budget.[176]

Additional perspective can be gained by measuring the investment of countries in the space sector with regard to GDP generated in 2016 (Fig. 1.2).

The United States remained in the front position in space spending as a share of GDP in 2016, although its spending decreased to 0.2393% from 0.2483% in 2015. Luxembourg entered into second position in terms of spending as a share of its GDP, reaching 0.1312% from 0.0488% in 2015; this follows Luxembourg's strategy to invest several hundred million euros in space mining ventures. Russia followed close behind in third position, dropping to 0.1273% from 0.2259% in 2015. France moved to the fourth position, while increasing its spending as a share of its GDP to 0.1133% in 2016. Japan came next with 0.0655%, followed by Germany with 0.0572%, India with 0.0528%, and Italy with 0.0511% in space spending as a share of GDP in 2016. Other leading space countries in Europe and the rest of the globe invested less than 0.0500% of their GDP in space activity, while the European Union overall spent an estimated 0.0118% of its GDP on space.

Looking at space spending in terms of per capita investment provides a different picture. Here, the United States is again in first position; however, its spending decreased by 1.1% to $137.22 in 2016 down from $138.75 in 2015. Luxembourg, with its increased space start-up investment in 2016 entered into second position, more than doubling its per capita space spending to $130.00 in

[176] Ibid.

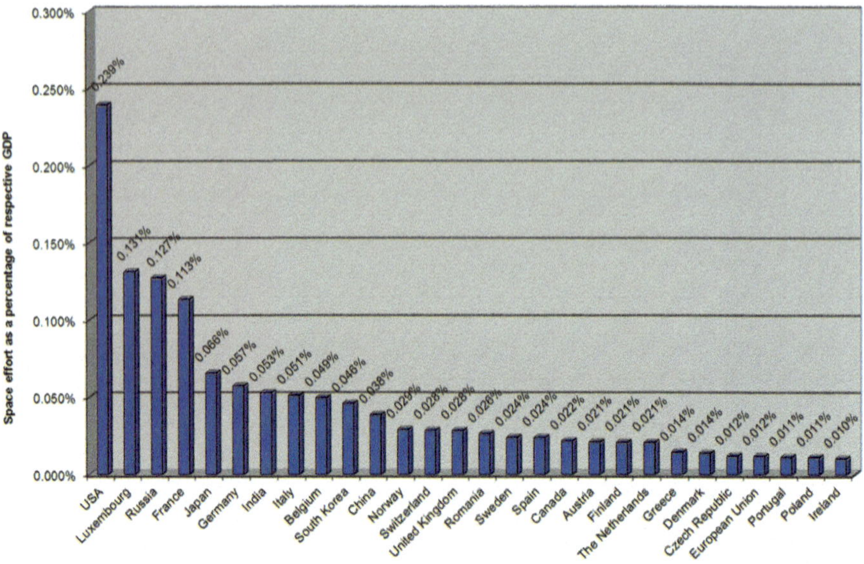

Fig. 1.2 Public space budgets (selection) as a share of nom. GDP in 2016 (Source: The Space Report/Euroconsult/IMF)

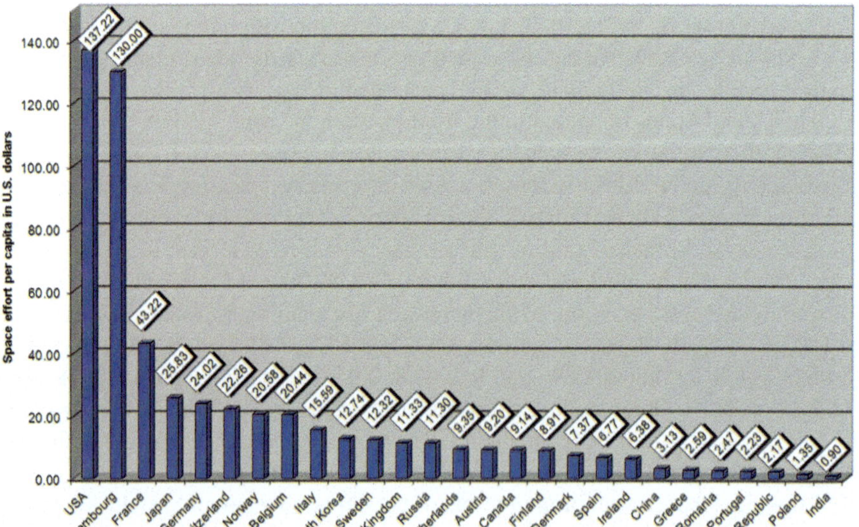

Fig. 1.3 Public space budgets per capita (selection) in 2016 (Source: The Space Report/ Euroconsult/PRB)

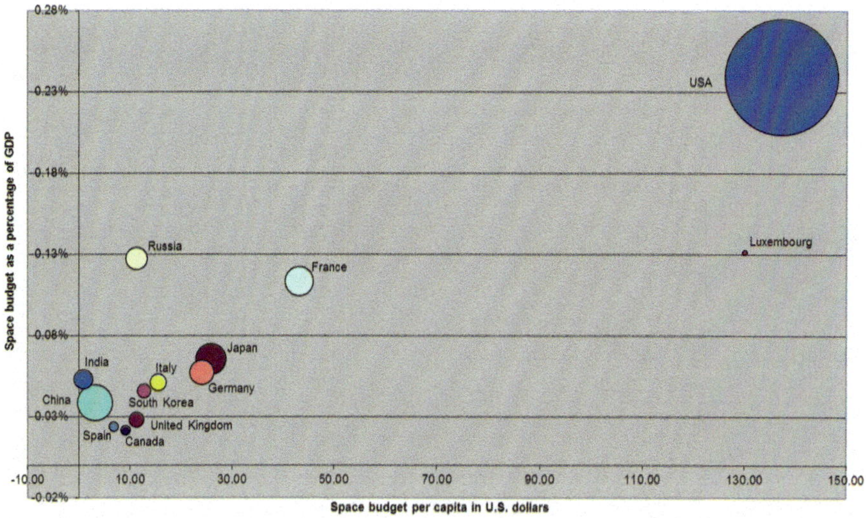

Fig. 1.4 Public space budgets as share of GDP mapped against space budgets per capita in 2016. The bubble size indicates the absolute space budget (Based on the Space Report 2017, Euroconsult, and publicly available data)

2016 from $46.67 in 2015. France came in third place in per capita expenditure, which increased to $43.22 in 2016; it was followed by Japan with $25.83, Germany with $24.02, Switzerland with $22.26, Norway with $20.58, and Belgium with $20.44. While per capita space spending increased among most space actors in 2016, it decreased in Russia, Canada, Finland, the United Kingdom, Sweden, and the United States.

Contrasting the GDP share of public space funds and per capita public space funds provides another picture of institutional investment in space (see Figs. 1.4 and 1.5). While the United States continues to excel in terms of budget size, spending per capita, and a percentage of GDP in 2016, Luxembourg emerged in second place in terms of spending per capita and a percentage of GDP. Meanwhile, Russia's budget size, spending per capita, and percentage of GDP further diminished in 2016. And France holds the fourth position both in space budget per GDP and space budget per capita but trails the United States, China, and Japan in its total space budget. Yet, some caution is needed when considering these figures, due to fluctuating exchange rates and the uncertainty of reported values; and more specifically, these figures likely underestimate both China and Russia's efforts and capability in space.

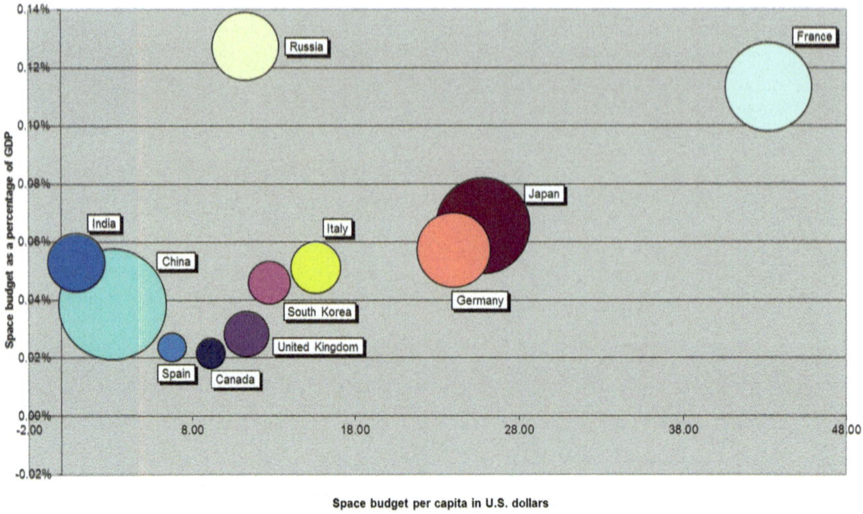

Fig. 1.5 Magnification of public space budgets as a share of GDP mapped against space budgets per capita in 2016, not including the United States and Luxembourg. The bubble size indicates the absolute space budget (Based on the Space Report 2017, Euroconsult, and publicly available data)

1.3.2 Overview of Commercial Space Markets

The Satellite Industry Association (SIA) reported that global industry revenues from satellite services, satellite manufacturing, launch industry, and ground equipment segments grew by 2.2% in 2016 reaching $260.5 billion from $254.8 billion in 2015.[177] But according to the Space Report 2017, the combined revenue from commercial space products and services, and from commercial infrastructure and support industries, increased by 2.6% to $252.88 billion in 2016 from $246.42 billion in 2015.[178] It should be noted that these authorities use different methodologies in reaching their assessments, and there is continued discrepancy in the findings of SIA and the Space Report, resulting in a difference of $7.2 billion in the figures for 2016, from $8.4 billion in 2015. The following section presents key figures and data on commercial space activities divided by field of activity, based primarily on available SIA figures generated by Bryce Space and Technology.

Satellite Services

According to the SIA, the revenue earned from satellite services stayed flat for 2016, increasing by only 0.2% to $127.7 billion in 2016 from $127.4 billion in

[177] "2017 State of the Satellite Industry Report." 15 June 2017. Satellite Industry Association and Bryce Space and Technology 12 July 2017 http://www.sia.org/wp-content/uploads/2017/07/SIA-SSIR-2017_full-1.pdf

[178] C.f. The Space Report 2017. Colorado Springs: The Space Foundation, 2017: 16, and The Space Report 2016. Colorado Springs: The Space Foundation, 2016: 16.

2015; the rate has steadily decelerated in recent years.[179] Satellite services accounted for a 49.0% share of the total revenue earned by the global satellite industry in 2016. Moreover, the bulk of satellite service revenue came from its consumer services subgroup that continues to be a key driver for the overall satellite industry with an 82.0% share of satellite service revenue earned in 2016.[180]

Satellite services can be further distinguished by its subgroups, including consumer services (i.e. satellite TV (DBS/DTH), satellite radio (DARS), and satellite broadband services), fixed satellite services (i.e. transponder agreements and managed services), mobile services (voice and data), and Earth observation. The following is a breakdown of the industry's key developments and trends, according to the nature of the services provided.

Consumer Services

Consumer services include satellite television, satellite radio, and satellite broadband services. The segment's downstream service revenue grew by less than 0.4% to $104.7 billion in 2016 from $104.3 billion in 2015; moreover, $58.3 billion of 2016 revenue was earned outside of the United States, mainly in DBS/DTH services and from some growth in satellite broadband. With around 220 million satellite television subscribers worldwide, DBS/DTH customers, increasingly in emerging markets, are a key driver in consumer service revenue; however, there is the potential for a further slowdown in demand growth for DBS/DTH services as customers opt for Internet-based video services. Demand growth for satellite radio increased by 8.7% to $5 billion in 2016 from $4.6 billion in 2015, coming primarily from US customers, and satellite broadband revenue increased by 5.3% to $2.0 billion in 2016 from $1.9 billion in 2015, with a notable increase outside of the United States.[181] Satellite television accounted for 93.3% of consumer service revenue, followed by satellite radio at 4.8% and satellite broadband services at 1.9%.[182]

Fixed Satellite Services

Fixed satellite service (FSS) refers to the use of spacecraft that utilise land terminals in fixed positions to broadcast. Here, FSS relates to commercial signal agreements, such as transponder agreements and managed network services. The segment contracted by 2.8%, earning $17.4 billion in 2016 from $17.9 billion in 2015, due to a decrease in transponder agreements outside of the United States. Revenue from transponder agreements decreased by 9.7% to $11.2 billion in 2016 from $12.4 billion in 2015; nearly all of the revenue generated is from the non-US market. The decrease in the FSS market was offset by 12.7% revenue growth in managed

[179] "2017 State of the Satellite Industry Report." 15 June 2017. Satellite Industry Association and Bryce Space and Technology 12 July 2017: 11 http://www.sia.org/wp-content/uploads/2017/07/SIA-SSIR-2017_full-1.pdf

[180] Ibid. at 4.

[181] Ibid. at 11.

[182] Ibid. at 11.

network services reaching $6.2 billion in 2016 from $5.5 billion in 2015, driven primarily by the increase in high-throughput capacity and in-flight services.[183]

Mobile Satellite Services
Mobile satellite services (MSS) offer both mobile data service and mobile voice service (including satellite phones). MSS revenue grew by 5.9% to $3.6 million in 2016 from $3.4 million in 2015. Nearly all of the increase came from outside of the United States; MSS revenue from the United States was flat, remaining at $0.5 billion in both 2016 and 2015.[184]

Earth Observation Services
Earth observation services refer to commercial companies that provide optical and radar images to the open market; however, demand for such services is mostly driven by government entities. Nevertheless, new entrants such as Terra Bella and Planet (formerly Planet Labs) have continued to raise capital and have begun to deploy initial constellations. Earth observation service revenue increased by 11.1% to $2.0 billion in 2016 from $1.8 billion in 2015; and about 60% of that revenue was generated outside the United States.[185]

Satellite Manufacturing
The SIA reported that the total revenue of satellite manufacturers that built satellites both for governmental and commercial customers decreased by 13.1% to $13.9 billion in 2016 from $16.0 billion in 2015 (Fig. 1.6). It should be noted that the manufacturing sector, while mostly driven by the telecom sector, remains sensitive downturns which took place despite the emergence of new players which are driving an increased competition. Manufacturers outside the United States were more affected by the drop in revenue, decreasing by 24.2% to $5.0 billion in 2016 from $6.6 billion in 2015. In contrast, US manufacturers experienced a decrease of 5.3% to $8.9 billion in 2016 from $9.4 billion in 2015, shifting its share in global revenues to 64% in 2016 from 59% in 2015. Military surveillance satellites accounted for 44% of the revenue generated for the year, followed by communication satellites which dropped to second position earning 22% of the total revenues (i.e. 16% from commercial communications and 6% for civil/military communications). Next, navigation satellites and Earth observation satellites each accounted for 12%, followed by scientific satellites at 5% and meteorology satellites at 4%, while satellites developed for R&D purposes amounted to 1%. CubeSats continued to represent less than 1% of the total revenue generated for the year.[186]

Commercial Space Launch
There were 21 commercial launch events in 2016; however, two anomalies occurred in the year, including the early shutdown of the first stage of the Atlas V launcher carrying the Cygnus CRS-6 ISS resupply capsule in March 2016, which did not

183 Ibid. at 14.

184 Ibid. at 14.

185 Ibid. at 12.

186 Ibid. at 18.

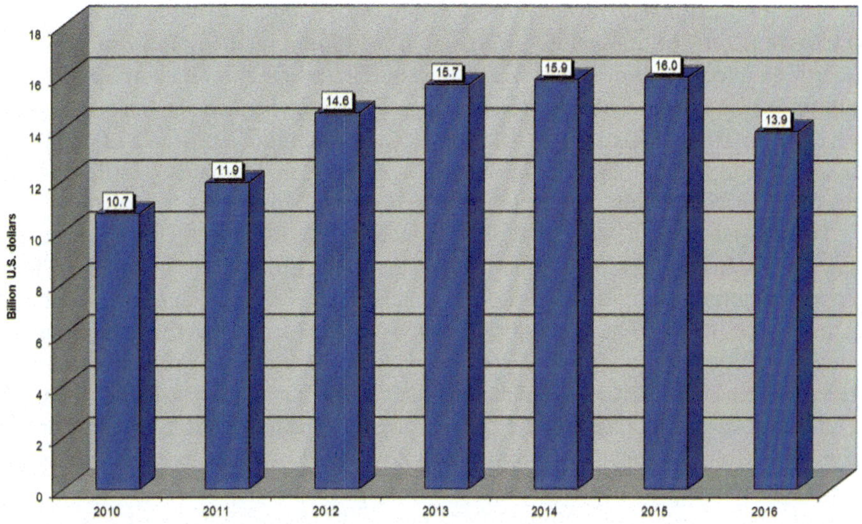

Fig. 1.6 World satellite manufacturing revenue (Source: SIA)

affect the success of the mission, and the explosion of the Falcon 9 FT during a static fire test in September 2016, which occurred days prior to the scheduled launch. The 21 successful commercial launches carried 37 mostly commercial services payloads into orbit.

Commercial launches accounted for 24.7% of the total 85 launches in 2016; and commercial payloads amounted to 35.6% of the 222 payloads launched. Of the 222 payloads that were launched in 2016, 57 were cube satellites launched directly into orbit, while another 49 cube satellites were intended to be released into orbit from the ISS. When not considering cube satellites, the percentage of commercial payloads launched amounted to 25.0% or 29 commercial payloads out of a total of 116 non-cube satellite payloads.

In 2016, US launch providers conducted 11 commercial launches out of a total of 22 launches, including the successful launches of 7 Falcon 9 FT flights, 3 Atlas V flights, and 1 flight by the Antares launcher, amounting to a share of 52.4% of commercial launches for the year. Here SpaceX's Falcon 9 has begun to surpass ULA's Atlas V and Delta 4 launchers in recent years, due partly to its more competitive offer and non-reliance on Russian engines; in order to remain competitive, ULA likely will need to reshape its offer. China conducted 22 non-commercial launches, but did not have any commercial launch activity in 2016. Russia was third in terms of launches in 2016, while only 2 of its 17 launches were for commercial purposes; its share of total commercial launches lowered to 9.5% for 2016. Next, Europe conducted eight commercial launches and three more non-commercial launches in 2016, increasing its share of total commercial launches to 38.1% for the year (not counting the actual number of payloads launched). And finally, India conducted seven non-commercial launches, followed by Japan with four non-commercial launches and Israel and North Korea each with a single non-commercial launch.

The total estimated revenue from commercial launch activities increased by 14.7% to $2.467 billion in 2016, up from $2.15 billion in 2015. The United States nearly doubled its commercial launch revenue, increasing to an estimated $1.185 billion in 2016 from $617 million in 2015. Europe generated the second highest revenue for 2016, increasing by 8.1% to $1.152 billion in 2016 from $1.066 billion in 2015. And Russia held the third position earning an estimated $130 million, a decrease of 55.0% from the $289 million earned in 2015 (Figs. 1.7 and 1.8).[187]

Arianespace conducted a total of 11 launches from its French Guiana spaceport in 2016. Its Ariane 5 ECA launcher had six launches, lifting seven commercial telecommunication satellites (Intelsat 29e, Eutelsat 65 West A, Echostar 18, Intelsat 33e, Intelsat 36, Star One D1, and JCSat-15) and three civil government communication satellite (BRIsat, NBN-Co 1B, and GSAT-18) into GEO orbit. The Ariane 5 ES launcher had one launch, placing four Galileo navigation satellites (Galileo FOC-7, FOC-12, FOC-13, and FOC-14) to medium Earth orbit (MEO) for the European Commission. The Europeanised Soyuz had two launches, one lifting the Sentinel-1B along with several other smaller spacecraft into LEO and the second that lifted the Galileo FOC-10 and FOC-11 into MEO for the European Commission. The Vega launcher conducted another two launches, one which lifted a military remote sensing satellite for the government of Peru and four commercial remote sensing satellites for Terra Bella and one that launched Turkey's Göktürk 1A reconnaissance satellite to LEO.

Ground Equipment

Ground equipment revenue includes infrastructure elements, such as mobile terminals, gateways and control stations, and consumer equipment, such as very small aperture terminals (VSAT), ultrasmall aperture terminals (USAT), DTH broadcast dishes, satellite phones, and digital audio radio satellite (DARS) equipment. Portable navigation devices (PND) form one of the subsegments of end-user electronics that incorporate GPS chipsets.

Ground equipment revenues increased by 7.0% to $113.4 billion in 2016 from $106.0 billion in 2015, driven by growth in consumer equipment for satellite navigation including stand-alone devices and embedded chipsets for smartphones, traffic information systems, and transport vehicles and by increasing demand for network equipment for managed network services. In contrast, consumer equipment for satellite TV, satellite radio, and satellite broadband saw flat growth, offset somewhat by growth in terrestrial broadband and some mobile equipment sales.[188] Consumer equipment for satellite navigation increased by 8.3% to $84.6 billion in 2016 from $78.1 billion in 2015, while network equipment grew by 7.3% to $10.3 billion in 2016 from $9.6 billion in 2015. In contrast, consumer equipment for satellite TV, satellite radio, and satellite broadband grew by just 1.1% to $18.5 billion in

---

[187] Federal Aviation Administration. The Annual Compendium of Commercial Space Transportation: 2017. Washington DC: FAA, Jan. 2017: 40.

[188] "2017 State of the Satellite Industry Report." 15 June 2017. Satellite Industry Association and Bryce Space and Technology 12 July 2017: 29 http://www.sia.org/wp-content/uploads/2017/07/SIA-SSIR-2017_full-1.pdf

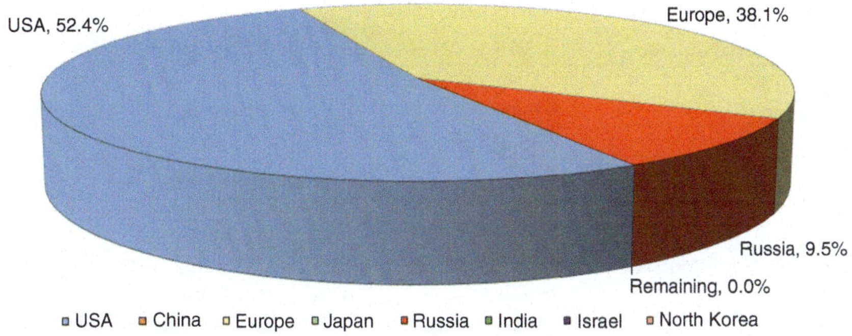

Fig. 1.7 Commercial launch activity by country in 2016 (Source: FAA)

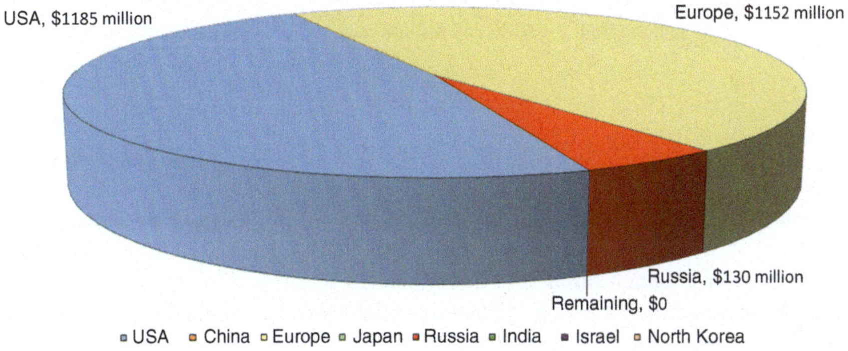

Fig. 1.8 Commercial launch revenues by country in 2016 (Source: FAA)

2016 from $18.3 billion in 2015. Overall, ground equipment revenues accounted for a 43.5% share of $260.5 billion world satellite industry revenue earned in 2016.

The two companies leading the PND market, Garmin and TomTom, switched roles in growth for 2016. Garmin earned $3.019 billion in revenue for the year ending 31 December 2016, an increase of 7.0% from the $2.820 billion earned in 2015. The growth came from increases in revenue in its Outdoor segment by 32.9%, Fitness by 23.7%, Marine by 15.8%, and Aviation by 10.2%; that increase was partly offset by Garmin's largest segment, Automotive, which decreased by 16.9%.[189] In contrast, TomTom earned €987.3 million ($1.040 billion) in revenue for the year ending 31 December 2016, a decrease of 1.9% from the €1.007 billion ($1.100 billion) earned in 2015. While revenue growth came from an 8.5% increase in its Automotive and Licencing segment and a 14.9% increase in its Telematics

[189] "Garmin Reports Solid Fiscal 2016 Revenue and Operating Income Growth." 22 Feb. 2017. Garmin11July2017http://newsroom.garmin.com/press-release/earnings/garmin-reports-solid-fiscal-2016-revenue-and-operating-income-growth

Table 1.1 Understanding TomTom and Garmin variables

	Total revenue	2016	2015
TomTom		€987.329 million ($1.040 billion)	€1.007 billion ($1.100 billion)
Garmin		$3.019 billion	$2.820 billion
	Geographical sales		
TomTom	Europe	€773.235 million ($814.649 million)	€771.491 million ($842.885 million)
	North America	€167.361 million ($176.325 million)	€186.115 million ($203.338 million)
	Rest of the world	€46.733 million ($49.236 million)	€49.001 million ($53.536 million)
Garmin	Europe/Middle East/ Africa	$1.111 billion	$1.013 billion
	Americas	$1.521 billion	$1.469 billion
	Asia-Pacific region	$386.549 million	$337.888 million

segment, it was offset by 9.7% decrease in revenue from its largest segment, Consumer in 2016 (Table 1.1).[190]

Insurance Sector

Insurance premiums for the launch and first year in orbit of satellites reached a historic low of around 5% in 2016, i.e. 60% lower than coverage in 2006, despite the fact that several launch failures in 2015 resulted in either a loss or marginal profit year for insurance underwriters. Moreover, aside from the eventual claim by Intelsat for the shortened in-orbit life of its Intelsat-33e satellite due to a defective propulsion system, 2016 is expected to be a profitable year for underwriters, likely generating between $450 million and $600 million, as no other claims for launch or satellite-related damages are anticipated. The continued success of the Ariane 5 launcher since 2002 has generated consistent profits for underwriters over the years allowing its insurance rate to lower to around 4%, followed by SpaceX whose rates are slightly higher. By comparison, the numerous failures experienced by Russia's Proton-M in the past 5 years have increased its insurance rate to around 12%. However, as several next-generation launchers are in the pipeline for the near future, insurance rates for these new launchers will likely be higher since they might fail more frequently than flight-proven vehicles.[191] The risks associated with launching satellites on refurbished launchers will also need to be addressed by insurance underwriters, in addition to the risks linked to space tourism.

---

[190] "TomTom Annual Report 2016." 31 Dec. 2016. TomTom 12 July 2017: 34 http://files.shareholder.com/downloads/TOMTOM/0x0x928644/59BDDCB7-4F22-45C3-857F-A6153BFD25DB/TomTom_Annual_Report_2016.pdf

[191] De Selding, Peter B. "Space insurers warn that current low rates are not sustainable." 10 Oct. 2016. SpaceNews 9 Nov. 2016 http://spacenews.com/space-insurers-warn-that-current-low-rates-are-not-sustainable/

The prelaunch insurance premiums, normally generating between $10 million and $12 million in total volume annually, will likely increase following the loss of the Amos-6 satellite on 1 September 2016, which was destroyed in the explosion of the Falcon 9 launcher during preparations for a static fire test days prior to its scheduled launch, and the extensive damage sustained by the Superbird-8/DSN-1 communication satellite while being transported to Europe's Guiana Space Centre spaceport in French Guiana in June 2016.[192] Having wiped out more than 20 years of insurance premium for the prelaunch market with the single loss of Amos-6, cargo and marine insurance underwriters may refuse to take on that risk again at the prevailing rates in the prelaunch market.[193] In contrast, in-orbit insurance premiums beyond the first year have decreased substantially, dropping to as low as 0.4% in 2016 from 2.5% in earlier years, despite the $158 million total loss claim of the Amos-5 satellite which stopped communicating from its geostationary orbit on 21 November 2015. With new forms of insurance coverage that require the loss of a few satellites in a constellation before full compensation for the lost spacecraft, these low rates are likely to hold in the near future.[194]

1.3.3 Developments in the Space Industry

1.3.3.1 Industrial Developments in Europe

At the end of 2016, Arianespace reported its revenue to be €1.4 billion annually, and while operating profit was not disclosed for the year, it was expected to break even when counting €100 million in price support from ESA. Arianespace conducted 11 launches in 2016, including 7 Ariane 5 launches (i.e. 6 Ariane 5 ECA and 1 Ariane 5 ES), 2 Europeanized Soyuz launches, and 2 Vega launches. These launches orbited 27 satellites in total, including 10 geostationary telecommunication satellites, 7 EO satellites, 6 navigation satellites, 1 scientific satellite, and 3 CubeSats. Arianespace also won 13 new launch contracts in 2016, valued at slightly more than €1 billion, including 9 Ariane 5 contracts, 2 Vega contracts, and 2 new contracts for the Europeanized Soyuz launcher. Moreover, it has a backlog of 55 launches at the end of 2016, valued at €5.2 billion, and has 12 launches planned for 2017.[195] Moreover,

[192] De Selding, Peter B. "Japan's DSN-1 military communications satellite damaged during transport to launch base." 20 June 2016. SpaceNews 24 July 2017 http://spacenews.com/japans-dsn-1-military-communications-satellite-damaged-during-transport-to-launch-base/

[193] De Selding, Peter B. "Space insurers warn that current low rates are not sustainable." 10 Oct. 2016. SpaceNews 9 Nov. 2016 http://spacenews.com/space-insurers-warn-that-current-low-rates-are-not-sustainable/

[194] Ibid.

[195] "Building on its 2016 successes, Arianespace looks to the future with confidence at the service of its customers." 4 Jan. 2017. Arianespace 22 Aug. 2017 http://www.arianespace.com/press-release/building-on-its-2016-successes-arianespace-looks-to-the-future-with-confidence-at-the-service-of-its-customers/

following the confirmation of the Vega C and Ariane 6 programmes in 2016, whose inaugural flights are expected in 2019 and 2020, Arianespace is now preparing the first offers for these two launchers. And despite the change of its governance structure, with Airbus Safran Launchers as the majority shareholder in Arianespace with a 74% stake, Arianespace will continue to be a full-fledged company acting as the single point of contact for its customers at each stage of their contractual and operational relationship.

Eutelsat earned €1.529 billion in revenue for the year ending 30 June 2016, an increase of 3.6% from the €1.476 billion earned in 2015. Its EBITDA increased by 2.9% to €1.165 billion (76.2% of revenue) by mid-2016 from €1.132 billion (76.7% of revenue) in mid-2015. And while its operating profit had nominal growth at €662.0 million by mid-2016 from €661.5 million in mid-2015, its net profit decreased by 1.9% amounting to €348.5 million from €355.2 million in the previous year.[196] The share of revenue generated from European regions continued to decrease, this time to 56.1% from 59.0% in 2015, while revenue was on the uptick in all other regions, i.e. Americas (21.6%), Middle East (13.3%), Africa (6.3%), Asia (2.6%), and others (0.1%).[197] Moreover, in Eutelsat's first half 2016–2017 results, it earned €755 million, down 2.5% from €774.4 million in the previous year. And Eutelsat's order backlog of guaranteed long-term satellite capacity contracts had decreased from €5.6 billion by mid-2016 to €5.3 billion at the end of the year.[198]During its fiscal year, Eutelsat had successfully launched five satellites: two to address the African broadband market (Eutelsat 8 West B and Eutelsat 36C), along with three more satellites to replace and increase capacities in Europe and the Americas.[199] Eutelsat, in collaboration with Facebook, had also expected to increase its African coverage with the Amos-6, having leased its Ka-band capacity from Israel's Spacecom in October 2015; however, following the loss of the Amos-6 satellite in the prelaunch failure of the Falcon 9 launcher in September 2016, it secured Ka-band capacity from Yahsat to launch its African broadband initiative in the first half of 2017.

SES earned €2.069 billion in revenue for the year ending 31 December 2016, an increase of 2.7% from €2.014 billion earned in 2015. Its EBITDA lowered by 2.9% to €1.452 billion (70.2% of revenue) in 2016 from €1.494 billion (74.2% of revenue) in 2015. And operating profit grew by 47.1% to €1.316 billion in 2016 from €894.6

[196] "EUTELSAT COMMUNICATIONS FULL YEAR 2015-16 RESULTS." 29 July 2016. Eutelsat 3 July 2017 http://news.eutelsat.com/pressreleases/eutelsat-communications-full-year-2015-16-results-1497983

[197] Reference Document 2015–2016. 1 Sept. 2016. Eutelsat 3 July 2017: 157 http://www.eutelsat.com/files/contributed/investors/pdf/AG2016/Eutelsat%20Communications%20Reference%20Document%202015-2016.pdf

[198] Press Release. "Eutelsat Communications First Half 2016-2017 Results." 9 Feb. 2017. Eutelsat 4 July 2017 http://news.eutelsat.com/pressreleases/eutelsat-communications-first-half-2016-2017-results-1794262

[199] "EUTELSAT COMMUNICATIONS FULL YEAR 2015-16 RESULTS." 29 July 2016. Eutelsat 3 July 2017 http://news.eutelsat.com/pressreleases/eutelsat-communications-full-year-2015-16-results-1497983

million in 2015, which came from a €495.2 million gain on disposal of a noncontrolling equity interest prior to the full consolidation of O3b on 1 August 2016. Moreover, its backlog of guaranteed long-term satellite capacity contracts for the year grew to €8.1 billion in 2016 from €7.4 billion in 2015. SES plans to launch six new satellites in 2017; moreover, it plans to add eight new satellites to O3b's unique high-throughput and low-latency MEO constellation beginning in 2018.[200]

Hispasat earned €228.9 million in revenue for the year ending 31 December 2016, an increase of 4.2% from the €219.6 million earned in 2015. Its EBITDA had lowered by 1.8% to €175.7 million (76.8% of revenue) in 2016 from €178.9 million (81.5% of revenue) in 2015, due to extraordinary expenditures for third-party capacity to provide service from the 36° West position until the Hispasat 36W-1 satellite arrives at its orbital position in mid-2017. Moreover, while operating profit grew by 8.1% reaching €99.9 million in 2016 from €92.4 million in 2015, Hispasat's net profit decreased by 59.9% to €25.1 million in 2016 from €62.6 million in 2015; without the extraordinary expenses, its net profit would have otherwise increased by 33.4% to €79.1 million. And its backlog of guaranteed long-term satellite capacity contracts reached €1.485 billion in 2016, i.e. 6.5 times the annual revenue, a leader in the sector. Its revenue from space capacity rentals also grew by 4.2%, reaching €225.5 million in 2016 from €216.4 million in 2015; this year 64.6% of the revenue came from clients in the Americas, while 35.4% came from leasing space capacity to clients in Europe and North Africa.[201]

Telenor Satellite Broadcasting of Norway, a subsidiary of the Telenor Group, changed its name to Telenor Satellite on 25 January 2016.[202] Telenor Satellite earned 955 million kroner (€105.1 million) in revenue for the year ending 31 December 2016, a 5.6% decrease from the 1.012 billion kroner (€105.8 million) earned in 2015, mainly due to declining prices in the data communication market and the termination of a satellite uplink contract in the fourth quarter of 2015. Its EBITDA decreased by 3.0% to 650 million kroner (€71.5 million) (68% of revenue) in 2016 from 670 million kroner (€70.0 million) (66.2% of revenue) in 2015, as the revenue reduction was partly compensated by lower capacity lease and operating costs. Moreover, its operating profit decreased by 19.3% reaching 313 million kroner (€34.4 million) in 2016 from 388 million kroner (€40.6 million) in 2015.[203] It should be noted that total broadcasting revenue (including Nordic DTH subscribers and households in SMATV networks, revenues from satellite services, revenues from terrestrial radio and TV transmission, and sale of encryption and conditional access

200 "SES Annual report 2016 – New Accelerations." 6 Apr. 2017. SES 22 Aug. 2017 https://www.ses.com/sites/default/files/2017-04/SES_AP_A4_EN_web.pdf

201 "HISPASAT increased its total revenue by 4.2% in 2016." 24 Mar. 2017. HISPASAT 27 June 2017 http://www.hispasat.com/en/press-room/press-releases/archivo-2017/257/hispasat-increased-its-total-revenue-by-42-in-2016

202 "Telenor Satellite Broadcasting becomes Telenor Satellite." 25 Jan. 2016. Telenor Satellite 27 June 2017 http://www.telenorsat.com/news/telenor-satellite-broadcasting-becomes-telenor-satellite/

203 Q4-2016. Interim report January–December 2016. 31 Jan. 2017. Telenor Group 27 June 2017: 10 https://www.telenor.com/wp-content/uploads/2017/01/Telenor-Group-Q4-2016-Report-c56b3b5ec5fb3b727eaef25c56a63814.pdf

services for TV distribution) increased by 1.8% to 6.186 billion kroner (€680.8 million) from 6.076 billion kroner (€635.04 million) in 2015.[204]

Inmarsat earned $1.329 billion (€1.261 billion) in revenue for the year ending 31 December 2016, an increase of 4.3% from $1.274 billion (€1.17 billion) in 2015. Its EBITDA increased by 9.5% to $794.8 million (€754.27 million) (59.8% of revenue) from $726.0 million (€664.41 million) (57.0% of revenue) in 2015. Its operating profit increased by 4.8% reaching $447.1 million (€424.30 million) in 2016 from $426.4 million (€390.23 million) in 2015. About 91.0% of that revenue came from its wholesale mobile satellite services (MSS) and other revenue and terminals, amounting to $1.210 billion (€1.15 billion); the remaining 9.0% came from the Ligado agreement (formerly the LightSquared Cooperation Agreement) which earned $119.4 million (€113.31 million) in 2016. While Inmarsat's revenue from its five business units increased by 4.3% to $1.329 billion (€1.26 billion) in 2016, its revenue growth in Government, increasing by 15.3% to $330.5 million (€313.65 million), and Aviation which grew by 12.5% to $142.6 million (€135.33 million) was offset by decreases in Maritime by 3.0% to $575.3 million (€545.97 million), Enterprise 9.3% to $144.6 million (€137.23 million), and Central Services by 14.4% to $16.6 million (€15.75 million) as a result of challenging markets and a decline in legacy product revenue.[205]

The Airbus Group's Airbus Defence and Space (Airbus D&S) division (including Military Aircraft; Space Systems; and Communications, Intelligence and Security (CIS)) earned €11.854 billion in revenue for the year ending 31 December 2016, a decrease of 9.4% from the €13.080 billion earned in 2015, with 31% of that revenue coming from its space systems business line (around €3.674 billion). Its EBITDA dropped to €390 million (3.3% of revenue) in 2016 from €1.39 billion (10.6% of revenue) in 2015,[206] likely coming from the negative impact from a parameter change in its portfolio reshaping as its Airbus Safran Launchers 50:50 joint venture became fully operational on 30 June 2016. Meanwhile, its operating profit (adjusted EBIT) had decreased by 4.7% to €1.002 billion in 2016 from €1.051 billion in 2015, and while its order backlog fell by 3.2% to €41.499 billion in 2016 from €42.861 billion in 2015, its order intake for 2016 grew by 6.6% to €15.393 billion from €14.440 billion in 2015.[207] Its telecom and Earth observation, navigation, and science satellites showed particular success in 2016, with ESA ordering two Sentinel-2 Earth observation satellites and Eutelsat's appointment of Airbus D&S as

---

[204] Annual Report 2016. 29 Mar. 2017. Telenor Group 27 June 2017: 89 https://www.telenor.com/wp-content/uploads/2017/03/Annual-Report-2016-Q-960dfcfa007ceee404c193b48ad20cff.pdf

[205] "Inmarsat plc reports Preliminary Full Year Results 2016." 8 Mar. 2017. Inmarsat 28 June 2017 http://investors.inmarsat.com/wp-content/uploads/2017/02/FY16-results-announcement-final.pdf

[206] Derived from EBIT + depreciation and amortisation; "Financial Statements 2016." 17 June 2017. Airbus 30 June 2017: 27 http://company.airbus.com/dam/assets/airbusgroup/int/en/investor-relations/documents/2017/AGM/Financial-Statements-2016-FINAL0/Airbus%20Financial%20Statements%202016.pdf

[207] "Annual Report 2016." 17 June 2017. Airbus 30 June 2017: 27 http://company.airbus.com/dam/assets/airbusgroup/int/en/investor-relations/documents/2017/AGM/Airbus-Annual-Report-2016/Airbus%20Annual%20Report%202016.pdf

the coprime contractor with ESA for the first Quantum satellite contract, which will be the first satellite capable of adapting its coverage, bandwidth, power, and frequency and its orbital position according to changing customer requirements.

The Thales Group's Aerospace segment, which includes its Avionics and Space Global Business Units e.g. Thales Alenia Space and Telespazio under its Space Alliance strategic partnership with Finmeccanica, earned €5.812 billion in revenue for the year ending 31 December 2016, an increase of 7.9% from €5.387 billion earned in 2015. Its EBITDA increased by 10.7% to €753.8 million (12.8% of revenue) in 2016 from €681.1 million (12.4% of revenue) in 2015.[208] And its operating profit increased by 10.2% to €571 million in 2016 from €518 million in 2015, while the Aerospace segment's order backlog grew to €9.914 billion in 2016 from €9.779 billion in 2015.[209] In 2016, Thales Alenia Space entered into a contract with SES to build its SES-17 telecommunication satellite aimed for the MSS market; this is the second contract for Thales Alenia Space's all-electric Spacebus NEO satellite platform and will provide flexible connectivity services, optimised for commercial aviation, over the Americas. Thales Alenia Space also entered into a phase B contract with LeoSat Enterprise for the development of an 80–120 LEO constellation of satellites offering very high-speed broadband, at a low latency and secure global connectivity. In Earth observation, it signed a contract with ESA and the EU to build the C and D models of the Sentinel-3A environmental monitoring satellite under the Copernicus programme. And in the field of navigation, the European Commission awarded it with a contract to provide system engineering and operational support services for the Galileo programme. Thales Alenia Space also entered into an export contract with the Korean Space Agency (KARI) to provide the Korean Augmentation Satellite System, based on its previously developed EGNOS system. And in science and exploration, Thales Alenia Space has contracted with Orbital ATK to supply nine additional pressurised cargo modules for resupply missions to the ISS, and ESA awarded it the final contract of the ExoMars programme for its 2020 mission.[210]

OHB AG earned €728.39 million in total revenue for the year ending 31 December 2016, a slight decrease of 0.3% from €730.38 million earned in 2015. Its EBITDA increased by 5.6% to €55.08 million (7.6% of revenue) in 2016 from €52.13 million (7.1% of revenue) in 2015. And operating profit grew by 6.2% to €42.70 million in 2016 from €40.21 million in 2015. Moreover, while OHB's Space Systems business unit had generated €559.5 million of the revenue, its order backlog stood at €1.341 billion as at the end of 2016. OHB System delivered Hispasat's

[208] Derived from EBIT + Depreciation and amortisation of property, plant and equipment and intangible assets; "2016 Registration Document Including the Annual Financial Report." 1 Jan. 2017. Thales 29 June 2017: 36 https://www.thalesgroup.com/sites/default/files/asset/document/thales_ddr_2016_va.pdf

[209] "2016 Registration Document Including the Annual Financial Report." 1 Jan. 2017. Thales 29 June 2017: 36 https://www.thalesgroup.com/sites/default/files/asset/document/thales_ddr_2016_va.pdf

[210] "2016 Registration Document Including the Annual Financial Report." 1 Jan. 2017. Thales 29 June 2017: 106 https://www.thalesgroup.com/sites/default/files/asset/document/thales_ddr_2016_va.pdf

Hispasat 36W-1 satellite to Europe's Kourou spaceport in early December 2016; the spacecraft marked its first GEO communication satellite, and it is also OHB's first satellite to be assembled using its modular SmallGEO platform – which is capable of many different configurations to address the needs of institutional and commercial customers in telecommunications and Earth observation. Moreover, 6 additional OHB-built Galileo FOC satellites were launched in 2016, including 4 simultaneously delivered to orbit by an Ariane 5 launcher in November 2016, bringing its total to 14 Galileo FOC in orbit, with 8 more to follow in 2017 and 2018. OHB had also made a significant contribution to ESA's ExoMars' Trace Gas Orbiter, which successfully reached Mars orbit in October 2016. By the end of 2016, OHB's staff numbered 2298, an increase of 242 personnel; the increase was mainly derived from the opening of the OHB "Optics and Science" Space Center in Oberpfaffenhofen, Germany, on 18 April 2016, which brought on around 360 employees. About 65.3% of those employees worked in OHB's Space Systems business unit, while the remainder worked mainly in other aerospace and industrial products. Moreover, 78.5% of OHB's employees were based in companies in Germany, while 15.7% were in other parts of Europe, and the remaining 5.8% worked in the rest of the world.[211]

RUAG Space, a subsidiary of the RUAG Group, earned CHF 345 million (€321.25 million) for the year ending 31 December 2016, an increase of 11.3% from the CHF 310 million (€286.27 million) earned in 2015. Its EBITDA increased by 6.7% to CHF 48 million (€44.70 million) (13.9% of revenue) in 2016 from CHF 45 million (€41.56 million) (14.5% of revenue) in 2015. And operating profit grew by 33.3% to CHF 32 million (€29.80 million) in 2016 from CHF 24 million (€22.16 million) in 2015, with sales and profits exceeding their performance targets. RUAG Space is in the midst of its expansion strategy, already showing success in its launch vehicle business in Europe and the United States and participation in all aspects of European space programmes and a focus on commercial product development and industrial series production. RUAG Space will develop the payload fairing for ASL's Ariane 6 launcher; it will also produce carbon fibre structures for the US United Launch Alliance Atlas V and Vulcan launchers and is also building a production facility in Cape Canaveral, Florida, to manufacture up to 900 structures in series for the OneWeb constellation led by Airbus D&S. Moreover, in May 2016, RUAG Space acquired the German company HTS along with its staff, specialising in engineering services and custom-made components for spaceflight. Subsequently, RUAG Space's presence has expanded from Switzerland, Sweden, Austria, and Finland to also including the United States and Germany. Its staff increased by 53 members to 1257 employees in 2016.[212]

[211] Annual Report 2016. 20 Mar. 2017. OHB 29 June 2017 https://www.ohb-system.de/tl_files/system/pdf/OHB_GB_2016_en_s.pdf

[212] Annual Report 2016. 23 Mar. 2017. RUAG 29 June 2017: 11 https://annualreport.ruag.com/sites/ar16/files/media_document/2017-03/RUAG_GB_2016_EN_Web.pdf

1.3.3.2 Industrial Developments in the United States

Boeing's Network and Space Systems segment, under its Defence, Space and Security division, earned $7.046 billion for the year ending 31 December 2016, a decrease of 9.1% from $7.751 billion earned in 2015. While its EBITDA was not disclosed in its financial results, its operating earnings decreased by 32.1% to $493 million (7.0% of revenue) in 2016, from $726 million (9.4% of revenue) in 2015. Moreover, its order backlog decreased to $5.1 billion at the end of 2016, from $7.4 billion in 2015.[213] Boeing's Network and Space Systems successfully launched the first of six Boeing-built Intelsat Epic satellites, Intelsat 29e, on the Ariane 5 launcher on 27 January 2016. However, its third-quarter performance decreased due to a charge on the Commercial Crew Development programme – largely driven by delays in completion of engineering and supply chain activities. In the second half of 2016, Boeing was awarded a contract for its 702MP satellite with a new digital payload offering twice the capacity of previous designs; and its eighth Wideband Global SATCOM satellite was also launched with an upgraded digital payload. Boeing's Network and Space Systems delivered seven satellites to customers in 2016, i.e. five commercial/civil satellites and two military satellites.[214]

Lockheed Martin's Space Systems segment, whose portfolio includes the development of satellites, strategic and defensive missile systems, and space transportation systems (including its 50% ownership interest in ULA), earned $9.409 billion for the year ending 31 December 2016, an increase of 3.3% from the $9.105 billion earned in 2015. About 13% of the Space Systems revenue earned in 2016 ($1.223 billion) came from satellite products and services, down from 15% of revenue ($1.366 billion) in 2015.[215] It should be noted that in 2016 91% of Lockheed Martin's Space Systems revenue came from US institutional customers, while 5% came from US commercial customers, and the remaining 4% came from international customers. While its EBITDA was not disclosed in its financial results, its total Space Systems operating earnings increased by 10.1%, to $1.289 billion (13.7% of revenue) in 2016 from $1.171 billion (12.9% of revenue) in 2015. Moreover, its order backlog increased to $18.9 billion at the end of 2016, from $17.4 billion in 2015. Lockheed Martin's increase in revenue came mainly from its majority ownership of the AWE Management Limited (AWE) venture (which operates the United Kingdom's nuclear deterrent programme) and an increase in the volume of sales of its fleet ballistic missiles; and while its commercial space transportation programmes earned an additional $150 million due to increased launch-related activities, these

[213] "Boeing Reports Fourth-Quarter Results and Provides 2017 Guidance." 25 Jan. 2017. Boeing 4 July 2017 http://investors.boeing.com/investors/investor-news/press-release-details/2017/Boeing-Reports-Fourth-Quarter-Results-and-Provides-2017-Guidance/default.aspx

[214] "Boeing Reports Fourth-Quarter Results and Provides 2017 Guidance." 25 Jan. 2017. Boeing 4 July 2017 http://investors.boeing.com/investors/investor-news/press-release-details/2017/Boeing-Reports-Fourth-Quarter-Results-and-Provides-2017-Guidance/default.aspx

[215] "Lockheed Martin Reports Fourth Quarter and Full Year 2016 Results." 24 Jan. 2017. Lockheed Martin 4 July 2017 http://news.lockheedmartin.com/2017-01-24-Lockheed-Martin-Reports-Fourth-Quarter-and-Full-Year-2016-Results

increases in revenue were offset by the decrease in the volume of SBIRS and MUOS satellites ordered for US government satellite programmes.[216]

Orbital ATK has three business segments, including its space-related Space Systems Group and Flight Systems Group and its non-space Defense Systems Group. Orbital ATK's Space Systems Group earned $1.237 billion in revenue for the year ending 31 December 2016, an increase of 6.2% from the $1.165 billion earned in 2015. While its EBITDA is unknown, its operating earnings nearly doubled reaching $129.5 million (10.5% of revenue) in 2016, from $46.5 million (4.0% of revenue) in 2015; the 178.5% increase was primarily due to Orbital Sciences Corporation's results prior to its merger with Alliant Techsystems Inc. (ATK) which had been previously excluded when the merger was adjusted as of 9 February 2015 and from the nonrecurrence of profit adjustments that took place in 2015. Moreover, its Flight Systems Group earned $1.497 billion in revenue for 2016, an increase of 1.8% from $1.497 billion earned in 2015. Its operating earnings decreased by 14.1% to $204.7 million (13.7% of revenue) in 2016 from $238.3 million (16.2% of revenue), again due to the nonrecurring adjustments that had taken place in 2015. By the end of 2016, the Flight Systems Group had launched an Antares cargo delivery mission to the ISS and the successful launch of a Pegasus launcher carrying eight small weather satellites to low Earth orbit for NASA. Orbital ATK's total backlog for its three segments (Space Systems, Flight Systems, and Defense Systems) increased to $9.34 billion in 2016, from $8.1 billion at the end of 2015.[217]

Intelsat earned $2.188 billion in revenue for the year ending 31 December 2016, a decrease of 7.0% from the $2.352 billion earned in 2015. Its adjusted EBITDA for 2016 decreased by 10.9% to $1.651 billion (75.4% of revenue) in 2016 from $1.854 billion (78.8% of revenue) in 2015. Moreover, while its operating profit grew to $920.6 million in 2016 from a negative balance of $3.028 billion in 2015, its net profit reached $990.2 million in 2016, from a net loss of $3.933 billion in 2015. And its contracted backlog representing future revenue under existing contracts with customers reached $8.7 billion in 2016 from $9.4 billion in 2015, reflecting a lowering of overall net new contracts. Intelsat's first Epic Next Generation high-throughput satellite, Intelsat 29e, was launched into geostationary orbit on 27 January 2016. Its Intelsat 33e and Intelsat 36 satellites were launched on 24 August 2016; while the Intelsat 36 satellite entered into service in the following month, a malfunction in Intelsat 33e's primary thruster for orbit raising caused its service to be delayed until the beginning of 2017. Intelsat also plans to launch additional Epic NG satellites in 2017, i.e. Intelsat 32e in the first quarter of 2017, Intelsat 35e in the second, and Intelsat 37e in the third.[218]

---

[216] "2016 Annual Report." Lockheed Martin 4 July 2017: 6, 43–44 http://www.lockheedmartin.com/content/dam/lockheed/data/corporate/documents/2016-annual-report.pdf

[217] "Orbital ATK Announces Fourth Quarter and Full Year 2016 Financial Results." 8 Mar. 2017. Orbital ATK 4 July 2017 http://www.orbitalatk.com/news-room/PrinterFriendly.asp?prid=229

[218] "Intelsat Announces Preliminary Fourth Quarter and Full Year 2016 Results." 28 Feb. 2017. Intelsat 5 July 2017 http://www.intelsat.com/wp-content/uploads/2017/03/Intelsat4Q2016EarningsRelease2.29.17.pdf

DigitalGlobe, the commercial high-resolution Earth observation satellite imagery provider, earned $725.4 million in revenue for the year ending 31 December 2016, an increase of 3.3% from the $702.4 million earned in 2015. Its adjusted EBITDA grew by 7.6% to 382.7 million (52.8% of revenue) in 2016 from $355.7 million (50.6% of revenue) in 2015. Moreover, its operating income increased by 67.9% reaching $102.1 million in 2016 from $60.8 million in 2015, while its net income grew by 13.7% to $26.5 million in 2016 from $23.3 million for 2015.[219] DigitalGlobe's main customer is the US government, accounting for 63.7% ($462.2 million) of its revenue for 2016, another 8.2% ($59.7 million) came from commercial customers in the United States, and the remaining 28.0% ($203.5 million) came from international customers.[220]

Following Blue Origin's historic first successful test flight of its reusable New Shepard suborbital launcher on 23 November 2015, which conducted a powered vertical landing, while its unoccupied crew capsule parachuted to a landing after separating at its peak, two more successful flights were conducted in the first half of 2016. Another test flight was already being prepared by mid-2016, which included one intentional parachute failure (one of three parachutes) during the descent of the crew capsule to observe how the capsule will respond in a scenario similar to the Apollo 15 mission in 1971. Blue Origin aims to have its first crewed test flights in 2017 and to conduct commercial services for paying passengers as early as 2018.[221] Additionally, Blue Origin also began constructing its rocket manufacturing facility in Florida, USA, in June 2016. Valued at $205 million, the facility will be complete by the end of 2017; Blue Origin aims to orbital vehicles from launch complex 36 in Cape Canaveral within the next decade.[222] And while financial details of Blue Origin are not available since the company is privately funded, its owner Jeff Bezos stated that he sells $1 billion of shares of Amazon stock each year to put towards Blue Origin.[223]

While SpaceX is a privately held company whose financial performance is normally shrouded from public view, internal documents published by *The Wall Street Journal* at the beginning of 2017 showed that SpaceX experienced a $260 million loss for 2015. The loss resulted from the explosion of its Dragon commercial resupply mission on 28 June 2015, which also grounded the launch company for the next

---

[219] DigitalGlobe 2016 Annual Report. 27 Feb. 2017. DigitalGlobe 11 July 2017: 29, 41 http://investor.digitalglobe.com/phoenix.zhtml?c=70788&p=irol-reportsannual

[220] Ibid. at 74.

[221] Calandrelli, Emily. "Blue Origin continues successful, record-setting year with another NASA contract." 6 June 2016. Tech Crunch 1 Sept. 2017 https://techcrunch.com/2016/06/06/blue-origin-continues-successful-record-setting-year-with another nasa contract/

[222] Richardson, Matthew. "Blue Origin shares construction update for new $205 M rocket factory (Photos)." 6 Dec. 2016. Puget Sound Business Journal 1 Sept 2017 https://www.bizjournals.com/seattle/news/2016/12/06/blue-origin-rocket-factory-florida-space-bezos.html

[223] Fernholz, Tim and Christopher Groskopf. "If Jeff Bezos is spending a billion a year on his space venture, he just started." 12 Apr. 2017. Quartz 1 Sept. 2017 https://qz.com/956607/jeff-bezos-the-worlds-second-wealthiest-human-isnt-spending-billions-on-his-space-venture-blue-origin/

6 months.[224] While financial results for 2016 were not disclosed, the explosion of the Falcon 9 on its launch pad on 1 September 2016 and subsequent grounding of the launch company likely resulted in another losing year, as the aerospace company has a thin bottom line that is vulnerable to accidents; the loss is estimated at up to $740 million for the year.[225] SpaceX's revenue for 2015 was $1.3 billion, which came mainly from a $1 billion investment by Google and Fidelity and advance payments. SpaceX expected to earn $55 million in operating profit following 20 launches in 2016; however, it conducted only eight launches prior to the September 2016 explosion.[226] Optimistically, the documents reveal that SpaceX aims to increase its launches from 27 in 2017 to 52 by 2019.[227]

1.3.3.3 Industrial Developments in Russia

Russia's Vostochny Cosmodrome conducted its first launch using a Soyuz 2.1a on 28 April 2016.[228] Construction of the spaceport is ongoing, with the completion of a second pad for its Angara launcher expected in 2021 and five other pads to come.[229] From the beginning of 2015, signs already indicated that the spaceport would miss its target of being ready before the beginning of 2016. Moreover, there have been many reports of construction being behind schedule, billions of dollars over budget, and cases of embezzlement and the non-payment of workers for months at a time.[230] Yet the importance of the Vostochny Cosmodrome for Russia should be stressed, as it will give Russia an alternative launch port to the Baikonur Cosmodrome that has been leased from Kazakhstan since the Soviet Union's collapse in 1991 at an annual cost of $115 million.[231]

---

[224] Winkler, Rolfe and Andy Pasztor. "Exclusive Peek at SpaceX Data Shows Loss in 2015, Heavy Expectations for Nascent Internet Service." 13 Jan. 2017. The Wall Street Journal 1 Sept. 2017 https://www.wsj.com/articles/exclusive-peek-at-spacex-data-shows-loss-in-2015-heavy-expectations-for-nascent-internet-service-1484316455

[225] Mosher, Dave. "SpaceX lost a quarter of a billion dollars after one of its rockets blew up." 13 Jan. 2017. Business Insider 1 Sept. 2017 http://www.businessinsider.de/spacex-financials-rocket-accident-costs-revenue-2017-1

[226] Wang, Brian. "Spacex financials show it is a big startup and the 4000 satellite internet business is the big payday." 18 Jan. 2017. nextBIGfuture 1 Sept. 2017 https://www.nextbigfuture.com/2017/01/spacex-financials-show-it-is-big.html

[227] Racke, Will. "SpaceX financial records show huge loss in 2015 due to rocket explosion." 13 Jan. 2017. L.A. BIZ 1 Sept. 2017 https://www.bizjournals.com/losangeles/news/2017/01/13/spacex-financial-records-show-huge-loss-in-2015.html

[228] "Russia's brand new cosmodrome launches first-ever rocket." 28 Apr. 2016. RT 15 Aug. 2017 https://www.rt.com/news/341192-vostochny-cosmodrome-first-launch/

[229] Haines, Lester. "First rocket finally departs Russia's Vostochny cosmodrome." 28 Apr. 2016. The Register 15 Aug. 2017 https://www.theregister.co.uk/2016/04/28/vostochny_launch/

[230] Eremenko, Alexey and Alexander Smith. "Vostochny Cosmodrome: Russian Space Project Isn't Going to Plan." 31 July 2016. NBC News 16 Aug. 2017 http://www.nbcnews.com/science/space/vostochny-cosmodrome-russian-space-project-isn-t-going-plan-n618846

[231] Soldatkin, Vladimir. "Putin orders building hastened at new Russian spaceport." 27 Aug. 2014. Reuters 16 July 2015 http://www.reuters.com/article/2014/09/02/us-russia-space-idUSKBN0GX1AV20140902

The Russian Satellite Communications Company (RSCC) and Europe's Eutelsat and Arianespace were caught in the midst of an ongoing dispute between the Russian government and the former shareholders of Russia's Yukos energy company, following a 2014 ruling by an international arbitration panel in The Hague, Netherlands, that awarded the ex-Yukos shareholders $50 billion from the Russian government. The ex-Yukos shareholders' claim was that government's illegal discriminatory treatment led to the company's bankruptcy and dissolution; following the award, the ex-Yukos shareholders sought to collect government assets in legal environments outside of Russia, including France and Belgium, where they were able to freeze payments by Eutelsat to RSCC and by Arianespace to Roscosmos. On 12 April 2016, a court in Evry, France, found that a €300 million payment owed by Arianespace to Roscosmos did not directly belong to the Russian government and should not be held hostage in the Yukos legal action. Another decision issued on 15 April 2016 from the Tribunal de grande instance de Paris also lifted the ex-Yukos shareholders' freeze on €400 million owed by Eutelsat to RSCC, based on a similar ruling.[232] Eutelsat made its first payment of €70 million to RSCC on 23 November 2016 following a further court ruling that omitted restrictions on making payments. However, the ruling involving the €300 million owed by Arianespace and Roscosmos was still in the appeals process. On 21 October 2016, the Russian government sent a formal warning to the French government wanting a resolution of the Roscosmos payment by March 2017, or else Russia would take France to court for violating a 1989 bilateral treaty, and other Euro-Russian space projects would be at risk if the stalemate continued.[233]

1.3.3.4 Industrial Developments in Japan

Mitsubishi Electric Co. (Melco) of Japan develops satellites within its Information and Communication Systems (ICS) business segment; however, as this segment does not separate satellite-related revenue from its telecommunications, information systems, and electronic systems business, it should only be seen as generating a small portion of the total revenue earned by this segment. Melco earned ¥4.394 trillion in total revenue for the year ending 31 March 2016, a 1.6% increase from the ¥4.323 trillion earned in 2015. Melco's ICS segment generated 11.1% of Melco's total revenue, earning ¥561.119 billion in 2016. While revenues from Melco's ICS segment had steadily increased over the past 5 years, its growth was flat in 2016, from the ¥559.521 billion earned in 2015. Its operating income decreased by ¥3.9 billion to ¥14.9 billion (2.7% of revenue) primarily due to a shift in its project

[232] De Selding, Peter B. "French court rulings ease threat to Arianespace, Eutelsat business in Russia." 18 Apr. 2016. SpaceNews 10 Nov. 2016 http://spacenews.com/french-court-rulings-ease-threat-to-arianespace-eutelsat-business-in-russia/

[233] De Selding, Peter B. "Eutelsat, freed by Paris court ruling, makes overdue payment on $424 million RSCC contract." 28 Nov. 2016. SpaceNews 4 Jan. 2017 http://spacenews.com/eutelsat-freed-by-paris-court-ruling-pays-russias-rscc-long-due-424-million/

portfolios of large-scale projects in the defence systems business; however, its electronic systems business saw an increase in orders for large-scale projects in the space business. Additionally, 57.4% of Melco's total revenue for 2016 came from customers in Japan, while other Asian countries accounted for 21.9%, North America 10.2%, Europe 8.4%, and others 2.1%.[234]

With four decades of experience in developing communication subsystems, NEC Corporation of Japan is entering into satellite integration. Its satellite-related revenue is within its Public Business segment under its other Major Products and Services and includes systems integration, maintenance and support, outsourcing/cloud services, and system equipment; hence, its satellite-related business should only be seen as generating a small portion of the total revenue earned by this segment.[235] NEC earned ¥2.821 trillion in total revenue for the year ending 31 March 2016, a 3.9% decrease from the ¥2.936 trillion earned in 2015. NEC's Public Business segment generated 27% of its total revenue, earning ¥766.8 billion in 2016. However, the segment's revenue decreased by 6.7% from the ¥821.9 billion earned in 2015, mainly due to decreased sales from large-scale projects for governmental agencies and public services. Its operating income decreased by 23.1%, reaching ¥57.5 billion in 2016 from ¥74.8 billion in 2015. Additionally, 78.6% of NEC's total revenue for 2016 came from customers in Japan, while China and other Asia-Pacific (APAC) countries accounted for 9.4%, the Americas for 7.1%, and Europe, the Middle East, and Africa (EMEA) for 4.9%.[236]

1.3.3.5 Industrial Developments in China

AsiaSat of Hong Kong earned HK$1.272 billion for the year ending 31 December 2016, a decrease of 3.0% from the HK$1.311 billion earned in 2015. Its gross profit decreased by 12.0% to HK$645.0 million (50.7% of revenue) in 2016 from HK$732.8 million (55.9% of revenue) in 2015. Meanwhile, operating profit decreased by 16.0% reaching HK$511.3 million in 2016 from HK$608.7 million in 2015, and profit attributable to owners decreased by 2.3% to HK$430 million in 2016 from HK$440 million in 2015.[237] In 2016, AsiaSat's fleet of five in-orbit satellites (AsiaSat 4, 5, 6, 7, and 8) performed nominally, while its AsiaSat 3S, launched in 1999, remained operational and provided services to customers for short-term contracts. By January 2016, AsiaSat had successfully regained access to the China video market with its AsiaSat 6 satellite after receiving full regulatory approval in

[234] 2016 Annual Report – For the year ended March 31, 2016. 26 July 2016. Mitsubishi Electric 13 July 2017 http://www.mitsubishielectric.com/company/ir/library/annual_report/pdf/ar2016.pdf

[235] "Annual Report 2016 – Year ended March 31, 2016." 1 July 2016. NEC 13 July 2017 http://www.nec.com/en/global/ir/pdf/annual/2016/ar2016-e.pdf

[236] "Annual Report 2016 (Financials) – Year ended March 31, 2016." 18 Sept. 2016. NEC 13 July 2017 http://www.nec.com/en/global/ir/pdf/annual/2016/ar2016_fin.pdf

[237] 2016 Annual Report. 31 Dec. 2016. AsiaSat 11 July 2017: 62 http://www.asiasat.com/sites/default/files/E01135_AR_0421_0656.pdf

China; AsiaSat had exited China's market in 2007. Moreover, AsiaSat signed a 4-year utilisation agreement with the Israel-based Spacecom in December 2016, where it will relocate its AsiaSat 8 satellite to replace the capacity shortfall created by the loss of the Amos-6 satellite in September 2016, with service expected to commence at the end of February 2017.[238]

Two additional DFH-4 telecommunication satellites were launched into orbit in 2016: Belintersat-1 on 15 January 2016, sold by CASC's China Great Wall Industry Corporation (CGWIC) commercial arm to Belarus in a contract signed in 2011,[239] and Shijian 17 on 3 November 2016 as an experimental geostationary satellite to test new technologies and components, including solar cells and green propellant, and to observe orbital debris at high altitudes, possibly with proximity operations.[240] Of the 17 DFH-4 commercial satellites that have been launched since the platform's development in 2006, CGWIC has sold 9 to international customers; it plans 10 more DFH-4 spacecraft in the coming years.

1.3.3.6 Industrial Developments in India

ISRO's Antrix commercial arm earned ₹1923.63 crore in revenue for the year ending 31 March 2016, an increase of 3.4% from ₹1860.71 crore earned in 2015. Its gross profit decreased slightly by 0.5%, reaching ₹324.07 crore (16.8% of revenue) in 2016 from ₹325.83 crore (17.5% of revenue) in 2015, while net profit increased by 2.0% to ₹209.64 crore in 2016 from ₹205.50 crore in 2015. Antrix earned around ₹230 crore in 2016 through commercial launch services[241]; and by the end of the fiscal reporting year, the PSLV launcher had successfully lifted a total of 57 international customer satellites from 21 countries into orbit.[242]

1.3.3.7 World

Canada's MacDonald, Dettwiler and Associates (MDA Corp.) earned C$2.064 billion for the year ending 31 December 2016, a decrease of 2.5% from C$2.117 billion earned in 2015. Its EBITDA decreased by 1.6% to C$370.7 million (18.0% of revenue) in 2016 from C$376.8 million (17.8% of revenue) in 2015. And operating

---

[238] Ibid. at 6,7.

[239] "Belintersat 1 (ZX 15, ChinaSat 15)." 2 June 2017. Gunter's Space Page 7 Aug. 2017 http://space.skyrocket.de/doc_sdat/belintersat-1.htm

[240] "SJ 17." 26 June 2017. Gunter's Space Page 7 Aug. 2017 http://space.skyrocket.de/doc_sdat/sj-17.htm

[241] "Antrix earned Rs 230 crore during 2015-16 through commercial launch services." 27 July 2016. Indiantelevision.com 13 July 2017 http://www.indiantelevision.com/satellites/satellite-launches/antrix-earned-rs-230-crore-during-2015-16-through-commercial-launch-services-160727

[242] "Annual Report 2015-16." 15 Jan. 2017. Antrix 13 July 2017 http://www.antrix.gov.in/sites/default/files/article-attachments/2015-16-English.pdf

profit decreased by 4.6% to C$211.0 million in 2016 from C$221.1 million in 2015. Moreover, its net profit decreased by 2.2% to C$139.6 million in 2016 from $142.8 million in 2015, while its order backlog also decreased to C$2.386 billion in 2016 from 2.887 billion in 2015.[243] MDA Corp. has two operating segments in its annual report, i.e. Communications and Surveillance and Intelligence. Its Communications segment earnings decreased by 4.4% to C$1.442 billion (69.9% of total revenue) in 2016, with an EBITDA of C$213.3 million (14.8% of Communications revenue); its revenue was impacted by the low number of GEO communication satellite contracts awarded in 2016 and 2015, which is expected to continue until the communication satellite market rebounds in 2017–2018. Its Surveillance and Intelligence segment earnings increased by 2.1% to C$622.2 million (30.1% of total revenue) in 2016, with an EBITDA of C$157.4 million (25.3% of Surveillance and Intelligence revenue); the increase was mainly due to the timing of subcontract activity on larger programmes and from increased contracts in the emerging market sector. In 2016, MDA Corp. booked orders to build four GEO communication satellites, and it signed a contract with OneWeb satellite to build 3600 communication antenna subsystems for integration on the satellite constellation. Moreover, 11 of the geostationary satellites that were launched in 2016 were built by MDA Corp. and its subsidiaries; it was also awarded several contracts from the US government space and defence markets.[244]

COM DEV International (Com Dev) was fully acquired by Honeywell International Inc. in February 2016, in an arrangement valued at $347 million. Prior to its acquisition, the Canadian satellite and space components provider had been coping with fewer tax incentives and a drought of US institutional orders that pushed it to close its facilities in California and bid on US military contracts from Canada; it had also tried to tap the United Kingdom's high-growth space market with acquisitions of MESL Holdings and MESL Microwave of Scotland. Com Dev earned $159 million in revenue for the fiscal year ended 31 October 2015.[245] In the year prior to its acquisition, Com Dev had reported total revenue of C$208.2 million for the fiscal year ended 31 October 2014, down 3.4% from C$215.5 million in 2013.[246]

Canada's exactEarth Ltd., spun off in November 2015 from Com Dev during its acquisition by Honeywell, saw sharply lower earnings in its first year as a standalone company in 2016. It earned C$18.9 million for the year ending 31 October 2016, a decrease of 28.9% from C$26.6 million earned in 2015. Its adjusted

[243] "MDA 2016 Annual Report – A Global Communications and Information Company." 23 Feb. 2017. MDA 5 July 2017: 1 http://mdacorporation.com/docs/default-source/corporate/investor/financial-reports/mda_2016_annualreport.pdf

[244] "MDA 2016 Annual Report – A Global Communications and Information Company." 23 Feb. 2017. MDA 5 July 2017: 14–18 http://mdacorporation.com/docs/default-source/corporate/investor/financial-reports/mda_2016_annualreport.pdf

[245] "Honeywell 2016 Annual Report." 9 Mar. 2017. Honeywell 5 July 2017: 45 http://www.annualreports.com/Click/21194

[246] "COM DEV Announces Fourth Quarter and Year-End Fiscal 2014 Results." 15 Jan. 2015. COM DEV International 10 Feb. 2015 http://www.comdev.ca/images/financial-reports/CDV_Q4_14_Financial_Release2.pdf

EBITDA dropped by 94.2% to C$520 thousand (2.7% of revenue) in 2016 from C$9.033 million (34.0% of revenue) in 2015. And operating profit decreased by 44.5% to C$9.146 million from C$16.486 million. Moreover, exactEarth's net loss dropped to C$35.963 million in 2016, from a loss of C$1.055 million in 2015, due mainly to the impairment and write-down of assets and restructuring of the business in 2016; without these charges, its net loss would have shrunk to $C7.8 million for 2016. The decrease in revenue for 2016 was due to a lower than expected renewal contract with the Canadian government for exactEarth's Automatic Identification System (AIS), which provides maritime traffic information, and the nonrecurring sale of historical AIS data to its alliance partner, Harris Corp. Moreover, exactEarth expects its revenue to increase again after the launch of its second-generation AIS nine-satellite constellation, hosted on Iridium NEXT satellites, beginning in 2017.[247]

Thailand's Thaicom earned 11.517 billion baht in sales and services revenue for the year ended 31 December 2016, a decrease of 7.5% from the 12.453 billion baht earned in 2015. Its EBITDA decreased by 15.6% to 4.860 billion baht (40.9% of revenue) in 2016 from 5.758 billion baht (44.5% of revenue) in 2015. Its operating profit decreased by 35.0% to 1.938 billion baht in 2016 from 2.981 billion baht in 2015, and net profit decreased by 24.0% reaching 1.612 billion baht in 2016 from 2.122 billion baht in 2015.[248] Its satellite and related services amounted to 73.1% of the revenue generated in 2016, down from 74.3% in 2015. Its sales and services revenue by geographic area came from Thailand (62.6%), Australia (9.4%), India (6.1%), Japan (5.5%), Myanmar (4.6%), China (2.6%), and others (9.3%); the Indian market had substantial growth with revenue increasing by almost one-third in 2016, while all other geographical areas saw decreases in revenue from the previous year.[249] On 20 October 2016, Thaicom's subsidiary International Satellite Company Limited (ISC) entered into a satellite procurement contract with China Great Wall Industry Corporation (CGWIC) for the construction and launch of a telecommunication satellite to strengthen Thaicom's broadband and mobility services in the Asia-Pacific region. Expected to be launched by the end of 2019 and valued at 7.280 billion baht ($208 million), all of the satellite's capacity has been leased to an unknown business partner who has also advanced the funds to pay for the satellite's construction.[250]

[247] "exactEarth Reports Fiscal 2016 Financial Results." 19 Jan. 2017. exactEarth 10 July 2017 http://investors.exactearth.com/2017-01-19-exactEarth-Reports-Fiscal-2016-Financial-Results

[248] "Keep Ahead of the Curve – Annual Report 2016." 17 May 2017. Thaicom 10 July 2017: 13 http://thcom.listedcompany.com/ar.html

[249] Ibid. at 74.

[250] Ibid. at 64.

1.3.4 *Industrial Overview*

1.3.4.1 Launch Sector

The launch sector is an enabler rather than a primary economic activity. Yet, with the growth in low-cost launch services, the marginal revenue the launch sector generates is becoming a more important factor to watch.

Launch activity decreased in 2016, with a total of 85 launches conducted by launch providers from the United States, China, Russia, Europe, India, Japan, Israel, and North Korea. Two launch failures and one prelaunch failure occurred in 2016, resulting in the loss of a remote sending satellite, a communication satellite, and an ISS cargo resupply mission. The first failure occurred on 31 August 2016, with the Long March 4C carrying the Gaofen 10 remote sensing spacecraft which did not reach orbit due to a failure of the launcher's third stage. A prelaunch failure of the Falcon 9 occurred during a static fire test on 1 September 2016, 2 days prior to the intended launch date; it was carrying Israeli-based Spacecom's Amos-6 satellite at the time of the explosion. The final failure occurred on 1 December 2016, with the Soyuz U Progress MS-4 ISS cargo resupply mission which lost telemetry just before separating from the third stage prematurely causing a collision between the third stage and spacecraft 382 s into the launch.

When looking into the launches of specific countries (Table 1.2), the United States and China each shared the first position in the number of launches for 2016, conducting 25.9% of total launches. Russia took the next position with a 20.0% share, followed by Europe that had a 12.9% share of the total. India was in fifth position with an 8.2% share, followed by Japan with 4.7%, and Israel and North Korea each accounting for 1.2% of the total launch figure (see Fig. 1.9).[251]

The United States conducted 22 launches using 14 launch system configurations.[252] China also conducted 22 launches but used 10 launch system configurations (including the failed launch of its Long March 4C).[253] Russia used 6 different launch configurations for its 17 launches (including the failed launch of the Soyuz U).[254] Europe relied on its workhorse Ariane 5 ECA launcher, in addition to its Vega launcher and Europeanized Soyuz variants to have 11 launches (6 Ariane 5 ECA, 1 Ariane 5 ECA, 2 Vega, 1 Soyuz STB Fregat-M, and 1 Soyuz STB Fregat-MT). India used three launcher configurations (i.e. 5 PSLV XL, 1 PSLV G, and 1 GSLV Mk.2) for its seven launches, while Japan used three launcher configurations (i.e. 2 H-IIA 202, 1 H-IIB 304, and 1 Epsilon 2) for its four launches. Israel conducted a single

[251] Federal Aviation Administration. The Annual Compendium of Commercial Space Transportation: 2017. Washington DC: FAA, Jan. 2017: 39

[252] That is, Antares 230, Atlas V (401), Atlas V (411), Atlas V (421), Atlas V (431), Atlas V (541), Atlas V (551), Delta 4 Heavy (upgrade), Delta 4 Medium +(5,4) (upgrade), Delta 4 Medium + (4,2), Delta 4 Medium + (5,2) (upgrade), Falcon 9 FT, Falcon 9 v1.1, and Pegasus XL.

[253] That is, Long March 2D, Long March 2F, Long March 3A, Long March 3B, Long March 3C, Long March 4B, Long March 4C, Long March 5, Long March 7, and Long March 11.

[254] That is, Proton-M Briz-M (Ph.3), Soyuz 2.1a, Soyuz 2.1b, Soyuz FG, Soyuz U, and Rokot KM.

Table 1.2 Worldwide launches in 2016 per country, number of launched systems, and commercial status

Launchers	Number of launch systems active in 2016	Total number of launches	Commercial launches	Non-commercial launches
United States	14	22	11	11
China	10	22	0	22
Russia	6	17	2	15
Europe	4	11	8	3
India	3	7	0	7
Japan	3	4	0	4
Israel	1	1	0	1
North Korea	1	1	0	1
Total	42	85	21	64

Source: FAA

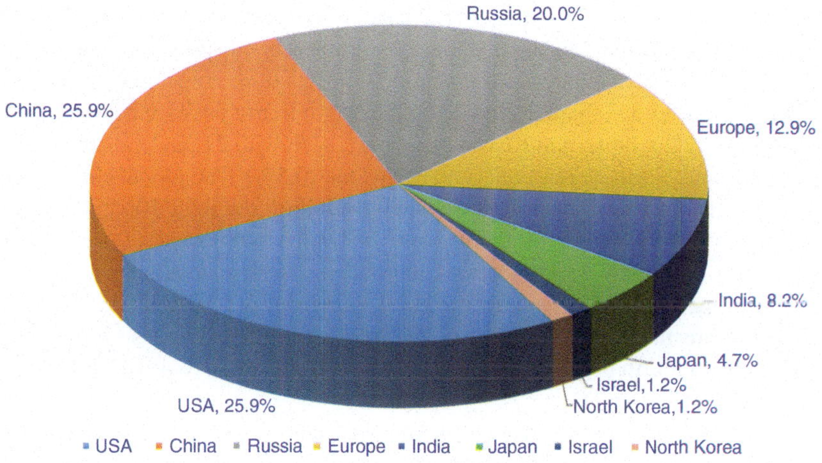

Fig. 1.9 Worldwide launches by country in 2016 (Source: FAA)

launch of its Shavit 2 launcher, while North Korea conducted a failed launch of its Unha-3 launcher.

Launch system utilisation moved to 42 active launch systems in 2016 from 40 used in 2015. The United States, China, and Russia accounted for a combined share of 71.8% of the number of launches for 2016, while launch activity in Europe, India, and Japan either grew or remained the same from 2015. Moreover, this indicator overlooks the fact that certain launchers have dual-launch capabilities such as Europe's Ariane 5, which can lift two standard-sized payloads to geostationary orbit. Hence, the number of launches does not reflect the number of payloads brought to orbit.

Europe was in second position in terms of the number of commercial launches in 2016 and in the amount of commercial launch revenue generated for the year even with its Ariane 5 dual payload capability. Meanwhile, the United States led with the most commercial launches and the highest amount of launch revenue earned in 2016. Russia decreased its number of commercial launches to 2, while it stayed in third position in terms of commercial revenue generated. When considering non-commercial launches, China led the pack with a 34.4% share, followed by Russia with 23.4% of the launches and the United States in the third position with a 17.2% share. The remaining 25% of non-commercial launches was split between India, Japan, Europe, Israel, and North Korea.

While the number of payloads launched decreased to 222 in 2016, from 265 in 2015, the ratio of commercial launches to non-commercial launches remained at about 1:3. Moreover, the ratio of commercial payloads to non-commercial payloads was 1:1.81 in 2016 which can be attributed to large number of commercial cube satellites launched; when considering non-cube payloads, the commercial-to-non-commercial ratio was 1:3 in 2016.

In terms of the global share of payloads launched in 2016 (Fig. 1.10), the United States was in first position with a 33.8% share of the total payloads placed in orbit, i.e. 75 payloads of the 222 total. China was in the second spot, launching 42 payloads, with an 18.9% share. India was in third position, launching 34 payloads, amounting to a 15.3% share, followed by Europe in fourth position having lifted 27 payloads (12.2%). Japan and Russia came next, each with 21 payloads (9.5%), and then Israel and North Korea each with 1 payload (0.5%). The global share of payloads launched changes considerably when excluding the total 106 cube satellite payloads from the assessment. In this case, the United States moves to second position with a decreased 22.4% share; China moves to first position with a 24.1% share. Europe follows closely in the third spot with a 20.7% share. Moreover, Russia's share increases to 16.4% of the total non-cube payloads, while India's share drops to fourth position with 11.2%, followed by Japan (3.4%), Israel (0.9%), and North Korea (0.9%).

There were also some changes in the distribution of payload sizes in 2016 (Fig. 1.11 and Table 1.3). The number of "micro"-sized payloads reduced to 106, accounting for 47.7% of the total payloads launched in 2016. The average mass of the total number of cube satellite payloads was around 9.3 kg, with the sum of their mass reaching 987.6 kg. In 2016, 49 cube satellites were launched to the ISS to be later ejected into orbit; however, one Spire cube satellite failed to deploy properly from the ISS. "Small" and "Large" satellites shared second position, each with 34 payloads, amounting to shares of 15.3%. The "Intermediate" mass class was in third position with 27 payloads at 12.2%. While "Medium" payloads were in the fourth position with 18 payloads, at 8.1%, and 3 "Heavy" payloads accounted for a 1.4% share of the payloads launched in 2016.[255]

[255] Micro payloads have a mass of 91 kg or less and are mainly science satellites, technological demonstrators, or small communication satellites. Small payloads weigh between 92 and 907 kg and are very often Earth observation satellites, similar to the Jason or the RapidEye series. Medium

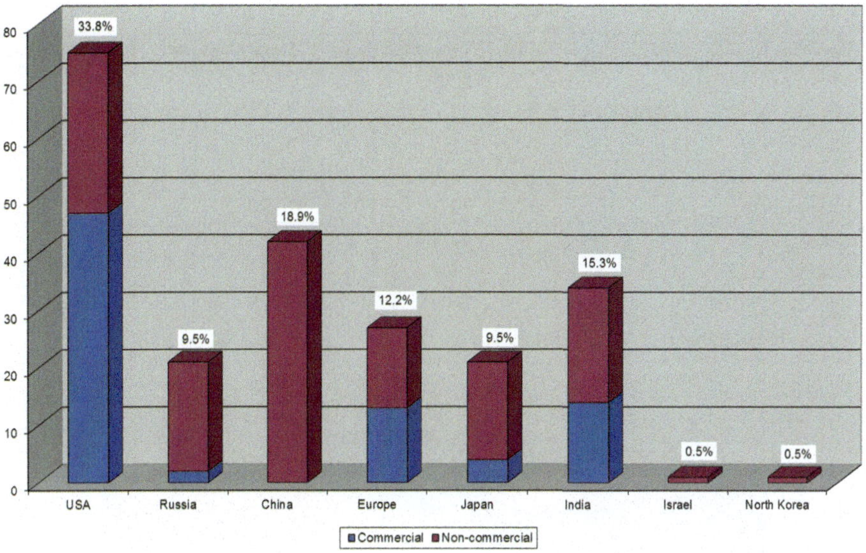

Fig. 1.10 Total payloads launched in 2016 by country, share, and commercial status (Source: FAA)

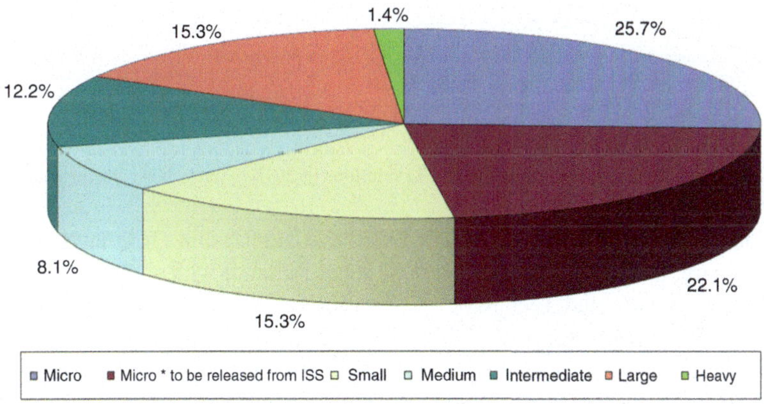

Fig. 1.11 Distribution of the payloads launched in 2016 by mass class (Source: FAA)

---

payloads weigh between 908 and 2268 kg and feature the most diverse set of satellites, including small satcoms in geostationary orbit, Earth observation satellites, and most of the Russian military satellites from the Kosmos series. Intermediate payloads, weighing between 2269 and 4536 kg, comprise medium satcoms and big scientific satellites. Large payloads, between 4537 and 9072 kg, refer to big satcoms, as well as to the Soyuz and Progress spacecraft flying to the ISS. Finally, Heavy payloads, exceeding 9072 kg, are linked to ISS activity, such as the cargo spacecraft, ATV, HTV, etc. *See* Commercial Space Transportation: 2011 Year in Review, 32.

Table 1.3 Distribution of the payloads launched in 2016 by mass class

Payloads by mass class	2016	Percentage (%)	Average mass (kg)	Mass sum (kg)
Micro	57	25.7	13.1	743.9
Micro * to be released from ISS	49	22.1	5.0	243.7
Small	34	15.3	412.9	14,038.0
Medium	18	8.1	1569.8	28,256.0
Intermediate	27	12.2	3226.0	87,102.0
Large	34	15.3	6420.4	218,295.0
Heavy	3	1.4	13,787.3	41,362.0
	222	100	1756.9	390,040.6

Source: FAA and Gunter's Space Page

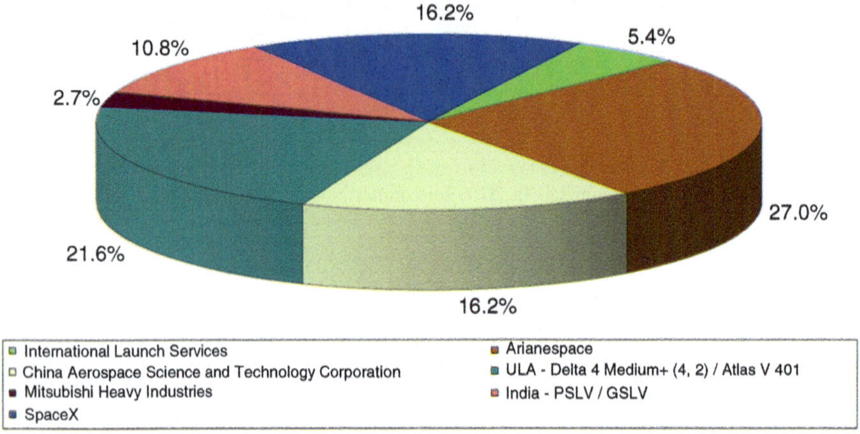

Fig. 1.12 Share of launch contracts for GEO satellites in 2016 by launch service provider

Once again, Arianespace conducted the most launches in GEO in 2016, with a 27.0% share, followed by the US United Launch Alliance with a 21.6% share. China Aerospace Science and Technology Corporation (CASC) and SpaceX each had a 16.2% share, while India had 10.8%, Russia's International Launch Services had 5.4%, and Japan's Mitsubishi had a 2.7% share (Fig. 1.12 and Table 1.4). Arianespace placed nine commercial communication satellites along with India's GSAT-18 communication satellite into GEO orbit using six Ariane 5 ECA launchers. The ULA had seven launches, lifting one signals intelligence satellite (NRO L-37), three military communication satellites (MUOS 5, NRO L-61, and WGS-8), two space surveillance satellites (GSSAP 3 and GSSAP 4), one meteorology satellite for NOAA (GOES-R), and one commercial communication satellite (EchoStar 19) into GEO orbit with various Delta 4 and Atlas V launch configurations. China's CASC had six launches to GEO orbit, i.e. three communication satellites, one BeiDou navigation satellite, one meteorology satellite, and one technology demonstration satellite (Shijian 17) possibly intended for signals intelligence, using mainly Long March 3B and 3C launchers and one Long March 5. And the SpaceX Falcon 9 was used

Table 1.4 Share of launch contracts for GEO satellites in 2016 by launch service provider

Launch service provider	Launches	Payloads	Percentage (%)
Arianespace	6	10	27.03
United launch alliance (ULA)	7	8	21.62
China aerospace science and technology corporation (CASC)	6	6	16.22
SpaceX	5	6	16.22
India	4	4	10.81
International launch services (ILS)	2	2	5.41
Mitsubishi heavy industries	1	1	2.70
	31	37	100.00

Source: FAA

five times, not counting the 1 September 2016 prelaunch explosion, to launch six commercial communication satellites (SES 9, JCSat 14, Thaicom 8, Eutelsat 117 West B, ABS 2A, JCSat 16) into GEO orbit. Three of India's PSLV XL launchers lifted the IRNSS 1E, 1F, and 1G navigation satellites, while its GSLV Mk.2 launched the INSAT-3DR meteorology satellite to GEO orbit. The Russian ILS conducted two Proton-M launches to place two commercial communication satellites (Eutelsat 9B and Intelsat 31) to GEO orbit. And lastly, Japan's H-IIA 202 launcher was used to launch the Himawari 9 meteorology satellite into GEO orbit.

1.3.4.2 Satellite Manufacturing Sector

In 2016, 222 payloads were launched (including an estimated 106 cube satellites; 14 crewed, cargo, or hardware missions to the ISS[256]; and 2 crewed, cargo, or hardware missions to the Tiangong-2 space station). The United States manufactured 47.5% of the launched payloads (including 67 CubeSats), while China accounted for 17.6%, and Russia produced 7.2%. Europe, with its 25 satellites mostly built for navigation and communication purposes, accounted for 11.3% of the payloads launched, while 6.8% of the payloads came from Japan and 5.4% from India. The remaining payloads came from various parts of Asia and the Americas.[257]

Of the 206 satellites launched in 2016,[258] 64.1% were non-commercial. In this field, 52.3% of the 132 non-commercial satellites were built by manufacturers from Asia (including 14 by China's CAST, 8 by India's ISRO, 1 by Japan's Mitsubishi, and 46 by Other Asia/ME). Another 22.0% of the non-commercial satellites were

[256] Including the failed launch of the Progress MS-4 cargo resupply mission on 1 December 2016.

[257] Federal Aviation Administration. The Annual Compendium of Commercial Space Transportation: 2017. Washington DC: FAA, Jan. 2016: 95–99. Payloads are assigned to the nation that commissioned them, not according to the nationality of the manufacturer.

[258] That is, not counting the 14 crewed, cargo, or hardware missions to the ISS and 2 crewed, cargo, or hardware missions to the Tiangong-2 space station.

built by manufacturers in North America (including 4 by Boeing, 3 by Lockheed Martin, 2 by Orbital ATK, 1 by SS/L, and 19 Other North America). Europe's share of non-commercial spacecraft was 18.2% of non-commercial satellites (including 6 by Thales Alenia Space, 6 by OHB Systems, 2 by Airbus D&S, and 10 by other manufacturers in Europe). Russia's share was 6.8% of non-commercial satellites (including two from ISS Reshetnev and seven more by other Russian manufacturers). The remaining non-commercial satellite was developed in South America and accounted for a 0.8% share.

In 2016, 35.9% of the 206 satellites launched were built for commercial purposes. About 97.3% of the 74 commercial satellites came from North America (including 15 by SS/L, 5 by Boeing, 1 by Orbital ATK, and 51 – mostly either Planet or Spire cube satellites – from other parts of North America). Airbus D&S was Europe's only manufacturer to build a commercial satellite, along with another commercial satellite manufactured in Other Asia/ME, each accounting for 1.35% shares (Figs. 1.13 and 1.14).[259]

In 2016, 18.4% of the 206 satellites launched were geostationary satellites (Figs. 1.15 and 1.16). In this field, 64.9% of the 37 GEO satellites were built by manufacturers from North America (including 11 by SS/L, 6 by Boeing, 3 by Orbital ATK, 2 by Lockheed Martin, and 2 by Other North America). Another 32.4% of the GEO satellites was built by manufacturers in Asia (including 5 by China's CAST, 5 by India's ISRO, 1 by Japan's Mitsubishi, and 1 by Other Asia/ME). Airbus D&S was Europe's only manufacturer to have a satellite lifted to GEO orbit, accounting for a 2.7% share.[260]

Among the 169 non-GEO orbiting satellites, North America's share decreased to 45.6% with 77 non-GEO satellites (5 by SS/L, 3 by Boeing, 1 by Lockheed Martin, and another 68 – mostly cube satellites – from other parts of North America). Manufacturers in Asia held a 34.3% share with 58 non-GEO satellites (9 by China's CAST, 3 by India's ISRO, and another 46 developed in Other Asia/ME). Europe's share was 14.2% with 24 non-GEO satellites (6 by Thales Alenia Space, 6 by OHB Systems, 2 by Airbus Defence and Space, and 10 from other European makers). Russia's share was 5.3% with nine non-GEO satellites (two from ISS Reshetnev and the remaining by other Russian manufacturers). The remaining non-GEO satellite was developed in South America and accounted for a 0.6% share.[261]

In 2016, North American prime spacecraft manufacturers had a combined lead in orders for GEO communication satellites, accounting for 52.6% of the 19 contracts awarded, whereas European contracts decreased to 26.3% of the available awards (Fig. 1.17). Boeing Satellite Systems won four commercial satellite orders (AMOS-17, GiSAT-1, ViaSat-3 Americas, and ViaSAT-3 EMEA), as did Space Systems/Loral (Eutelsat 7C, Intelsat 39, SXM 7, and SXM 8). Lockheed Martin won an order to build the JCSat 17, while Orbital ATK won an order for the Eutelsat 5 West

[259] Ibid.

[260] Federal Aviation Administration. The Annual Compendium of Commercial Space Transportation: 2017. Washington DC: FAA, Jan. 2017: 39.

[261] Ibid., 95–99.

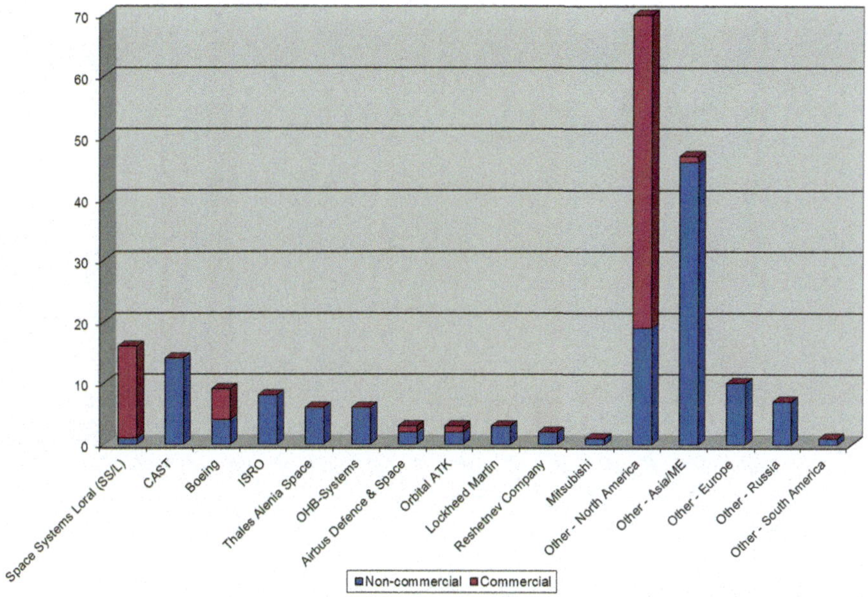

Fig. 1.13 Satellites launched in 2016 by manufacturer and commercial status (Source: FAA)

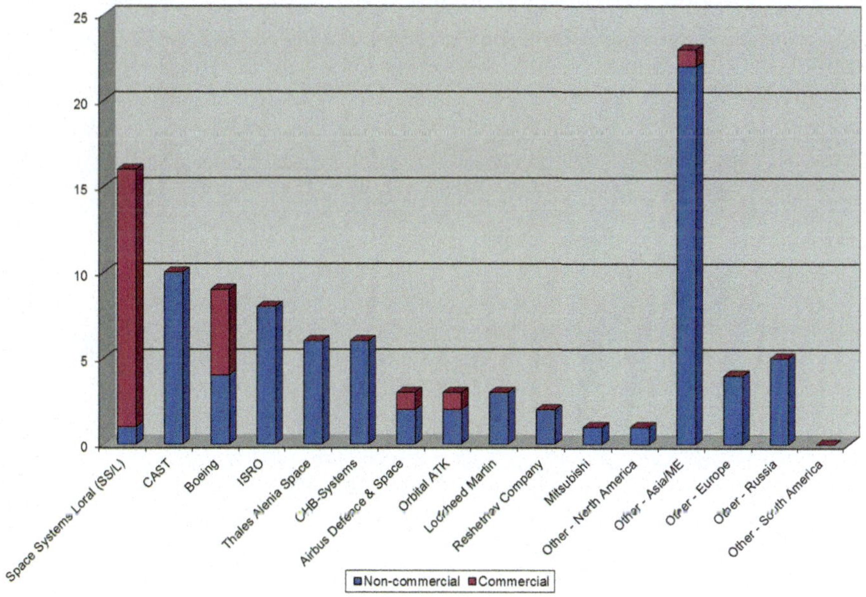

Fig. 1.14 Satellites launched in 2016 by manufacturer and commercial status, not including CubeSats (Source: FAA)

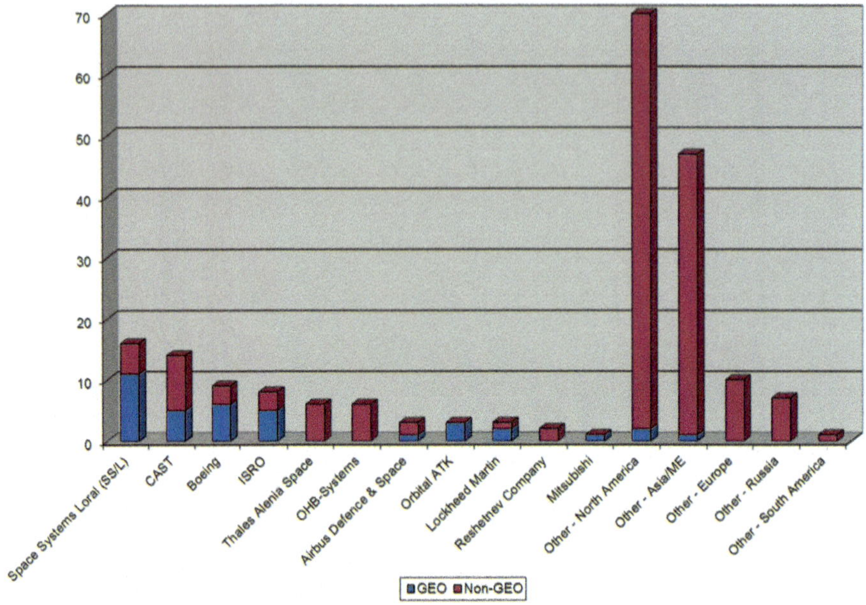

Fig. 1.15 Satellites launched in 2016 by manufacturer and orbit type (Source: FAA)

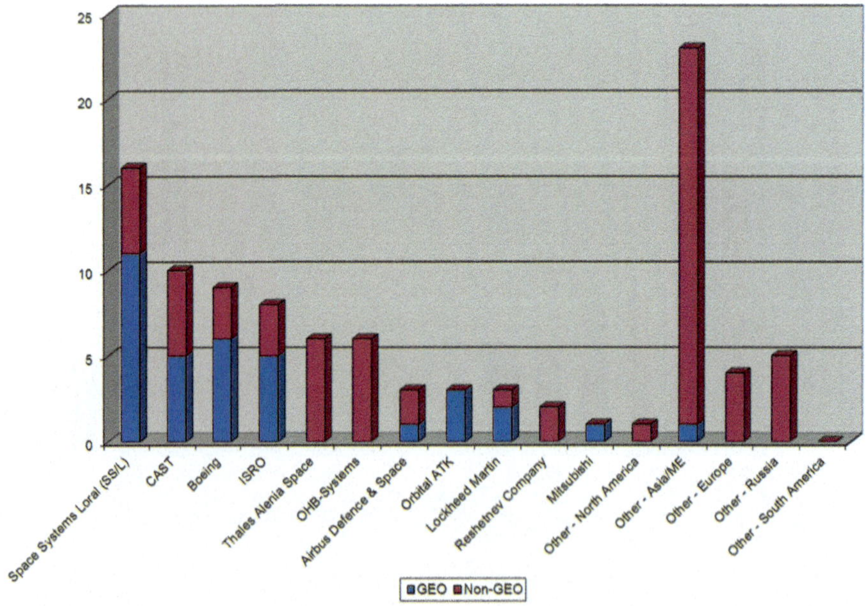

Fig. 1.16 Satellites launched in 2016 by manufacturer and orbit type, not including CubeSats (Source: FAA)

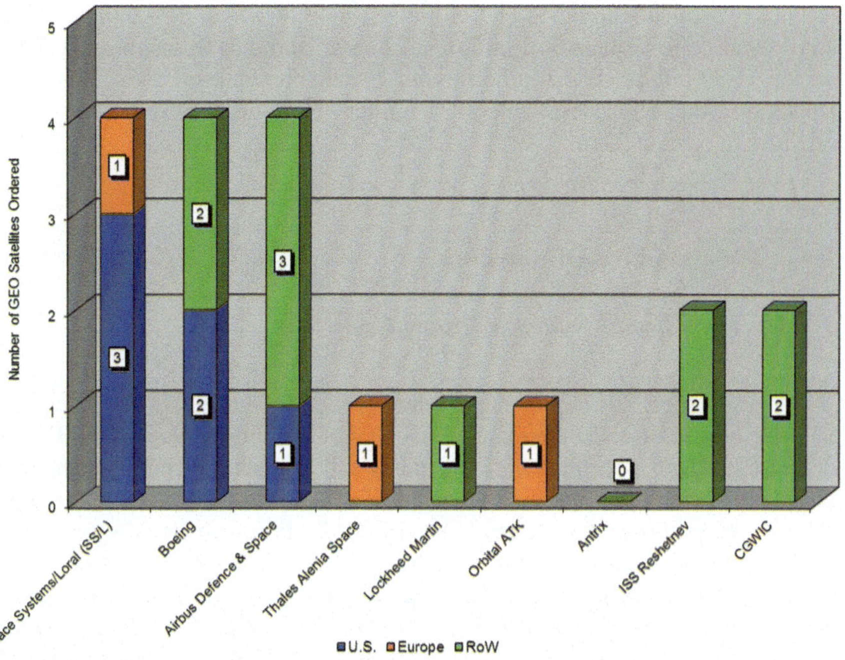

Fig. 1.17 GEO satellite orders in 2016 by manufacturer

B. In Europe, Airbus Defence and Space received four orders (including the DirecTV 16, KMilSatCom 1, SatKomHan 1, and a MilSatCom satellite for Egypt), while Thales Alenia Space received one order for the SES 17. China's CGWIC had a 10.5% share from two contracts (APStar 6D and TCStar 1), as did Russia's ISS Reshetnev (Ekspress 80 and Ekspress 103).[262]

1.3.4.3 Satellite Operators Sector

In terms of revenue earned, the ranking of the five largest satellite fleet operators, Intelsat S.A. (Luxemburg), SES (Luxemburg), Eutelsat (France), EchoStar Satellite Services (United States), and Telesat (Canada), kept remained relatively the same as in 2016.[263] Intelsat was the top FSS operator in 2016, earning $2.19 billion (€1.97 billion) in revenue, with 50 satellites in orbit and 2 satellites to be launched in 2017. SES earned $2.07 billion (€1.87 billion) in revenue, with 43 satellites in orbit and 3 satellites to be launched in 2017, while Eutelsat whose 2016–2017

[262] "Recently awarded GEO-Sat Contracts." 18 Aug. 2017. Gunter's Space Page 20 Aug. 2017 http://space.skyrocket.de/doc_sat/sat-contracts.htm

[263] "WORLD TELEPORT ASSOCIATION PUBLISHES TOP OPERATOR RANKINGS FOR 2016." 6 Jan. 2017. WTA 11 Aug. 2017 http://www.worldteleport.org/news/324926/WORLD-TELEPORT-ASSOCIATION-PUBLISHES-TOP-OPERATOR-RANKINGS-FOR-2016.htm

revenue is estimated to be $1.697 billion (€1.53 billion) has 39 satellites in orbit and 1 satellite to be launched in 2017. EchoStar (through its Hughes subsidiary) earned $1.80 billion (€1.62 billion), with 25 satellites in orbit and 1 expected to launch in 2017, and Telesat with $931 million (€839 million) in revenue had 15 satellites in orbit in 2016.

1.4 The Security Dimension

1.4.1 The Global Space Military Context

The Space Report 2017 estimates military space spending in 2016 to be around $33.000 billion, a 3.4% decrease from the $34.151 billion spent in 2015; this decrease is due to a decrease in US spending and to weak currency exchange rates that mask total military space spending outside of the United States.[264] Meanwhile, Euroconsult estimates that around 35% of the $62.2 billion in global space budgets in 2016 went to military spending, i.e. about $22.86 billion.[265]As is typical with the nature of dual-use technology in space activity, there is a risk that certain military activities have been already included in larger budgets, which can result in double counting. Moreover, while missions, often listed as civil programmes, may also serve dual-purpose military objectives, their expenditure is not included in this section.

According to the Space Report 2017, the United States generated a 66.7% share ($22.000 billion) of global military space spending in 2016, decreasing from 69.0% ($23.572 billion) estimated for 2015. Non-US global military spending increased to a 33.3% share ($11.000 billion) from 31.0% ($10.579 billion) in 2015.[266] However, as in previous years, a direct comparison of the budgets of these countries in fixed dollar values does not present a clear picture of their relative space defence efforts, since fluctuating exchange rates, variations in purchasing power, and different employment costs distort the impact of these investment amounts.

1.4.2 Europe

According to estimates by Eurospace, total funding for European military space programmes was about $753.051 million (€675.2 million) in 2015. France had the highest military budget at $362.696 million (€325.2 million), while the United

[264] The Space Report 2017. Colorado Springs: The Space Foundation, 2017: 15.

[265] Smith, Andrew. "Euroconsult: Government Spending in Space Programs Reaches $62 Billion in 2016." 30 May 2017. PRWeb 21 Aug. 2017 http://www.prweb.com/releases/2017/05/prweb14377281.htm

[266] The Space Report 2017. Colorado Springs: The Space Foundation, 2017: 15.

Kingdom budgeted $317.860 million (€285.0 million), with Germany at $55.765 million (€50.0 million) and Italy at $16.730 million (€15.0 million).

On 30 November 2016, the European Commission released its European Defence Action Plan proposing a European Defence Fund and other actions to support more efficient spending by member states in joint defence capabilities, strengthen European citizens' security, and foster a competitive and innovative industrial base.[267] The three main pillars under this plan include (1) launching a European Defence Fund, (2) fostering investments in defence supply chains, and (3) reinforcing the single market for defence. Moreover, wherever appropriate, the Commission will promote civil and military synergies within EU policies.[268] The business case for defence spending and greater defence cooperation comes from the fact that in 2015, the United States invested more than twice as much as the total spending of EU member states on defence and that China has increased its defence budget by 150% over the past decade. Moreover, in Europe the lack of cooperation between member states in the field of defence and security costs between €25 billion and €100 billion annually, due to inefficiencies, lack of competition, and lack of economies of scale for industry and production. Next, around 80% of defence procurement is run on a purely national basis, resulting in duplication of military capabilities. And finally, there would be a positive spillover effect on the European economy as the European defence industry generates a total turnover of €100 billion per year and 1.4 million highly skilled people are directly or indirectly employed in Europe.[269] The Commission will set up an Implementation Steering Group with member states to monitor and facilitate progress on the actions with the first meeting taking place in the first quarter of 2017.[270]

1.4.3 United States

While the US Department of Defense (DoD) space budget decreased to $22.000 billion in 2016 from $23.572 billion in 2015, the budget request for the US Missile Defense Agency (US MDA) increased to $8.313 billion in 2016 from the $7.868

---

[267] "European Defence Action Plan: Towards a European Defence Fund." 30 Nov. 2016. European Commission Press Release Database 21 Aug. 2017 http://europa.eu/rapid/press-release_IP-16-4088_en.htm

[268] Communication From The Commission To The European Parliament, The European Council, The Council, The European Economic And Social Committee And The Committee Of The Regions. European Defence Action Plan. COM (2016) 950 final of 30 November 2016. Brussels: European Union https://eeas.europa.eu/sites/eeas/files/com_2016_950_f1_communication_from_commission_to_inst_en_v5_p1_869631.pdf

[269] "European Defence Action Plan." 30 Nov. 2016. European Commission 21 Aug. 2017 http://europa.eu/rapid/attachment/IP-16-4088/en/20161130%20Factsheet_EDAP.pdf

[270] "The European Defence Action Plan – FAQs." 30 Nov. 2016. European Commission 21 Aug. 2017 http://europa.eu/rapid/press-release_MEMO-16-4101_en.htm

billion it received in FY 2015[271]; its budget request for Fiscal Year 2017 asked for $7.5 billion.[272] In mid-2015, members of the US Congress' House Armed Services Committee showed renewed interest in a 2009 concept to put several miniaturised kill vehicles on an interceptor missile to overcome the US missile defence system's inability to reliably distinguish between missile warheads and relatively low-tech decoys. The US MDA would make the Multi-Object Kill Vehicle (MOKV) a long-term technology, which would come after the completion of its Redesigned Kill Vehicle (RKV) expected to be ready around 2020.[273] In August 2015, Boeing, Lockheed Martin, and Raytheon each won study contracts worth approximately $9.7 million to develop MOKV concepts that are expected to be completed by May 2016.[274] The US MDA plans to hold its first non-intercept flight test for the new kill vehicle in 2018, followed by an intercept test in 2019, and to field the RKV in[275] 2020. Before that time, they plan to deploy 37 Ground-Based Midcourse Defense interceptors by the end of 2016 in response to provocations from North Korea's growing nuclear capability.

Following the 31 March 2015 launch of Russia's Cosmos 2504 spacecraft, its resulting manoeuvres, including at least one case where the upper stage of its launcher appeared to be nudged to a higher orbit, motivated the United States to early deployment of its once-classified Geosynchronous Space Situational Awareness Program (GSSAP) satellites.[276] Since the GSSAP declassification in February 2014, US defence officials have acknowledged that the satellites will perform their own rendezvous and proximity manoeuvres to allow close-up looks at spacecraft in GEO orbits.[277] The first two GSSAP satellites were launched in July 2014 and had been taken out of test mode twice by 16 September 2015 to make observations of specific objects in geosynchronous orbit. On 18 August 2016, a GSSAP satellite was sent to observe the status of the stalled MUOS-5 satellite that had experienced a failure of its orbit raising propulsion system partway through its

[271] "Missile Defense Agency Fiscal Year (FY) 2017 Budget Estimates | Overview." 16 Jan. 2016 MDA 22 Aug. 2017 http://comptroller.defense.gov/Portals/45/Documents/defbudget/FY2017/budget_justification/pdfs/03_RDT_and_E/MDA_RDTE_MasterJustificationBook_Missile_Defense_Agency_PB_2017_1.pdf

[272] "Missile Defense Agency FY 2017 Budget Request." 10 Feb. 2016. Missile Defense Advocacy Alliance 22 Aug. 2017 http://missiledefenseadvocacy.org/alert/missile-defense-agency-fy-2017-budget-request/

[273] Gruss, Mike. "House Bill Would Revive Dormant Missile Defense Kill Vehicle Project." 11 May 2015. SpaceNews 22 Dec. 2015 http://spacenews.com/house-bill-would-revive-dormant-missile-defense-kill-vehicle-project/

[274] Gruss, Mike. "MDA Pursues Kill Vehicles with Cost-cutting Mandate." 24 Aug. 2015. SpaceNews 24 Dec. 2015 http://spacenews.com/mda-pursues-kill-vehicles-with-cost-cutting-mandate/

[275] Gruss, Mike. "New U.S. kill vehicle will fly in 2018, take on its first target in 2019." 20 Jan. 2016. SpaceNews 22 Aug. 2017 http://spacenews.com/new-u-s-kill-vehicle-will-fly-in-2018-take-on-its-first-target-in-2019/

[276] Gruss, Mike. "Maneuvering Russian Satellite Has Everyone's Attention." 17 July 2015. SpaceNews 20 Mar. 2016 http://spacenews.com/maneuvering-russian-satellite-has-everyones-attention/

[277] Gruss, Mike. "Space Surveillance Sats Pressed into Early Service." 18 Sept. 2015. SpaceNews 11 Jan. 2016 http://spacenews.com/space-surveillance-sats-pressed-into-early-service/

climb to GEO at the end of June 2016; however, it is unclear how the information provided by the GSSAP would allow the US Navy to recover the MUOS-5 from its elliptical orbit.[278] Two more GSSAP satellites were launched on 19 August 2016.[279]

1.4.4 Russia

The Russian government merged its Air Force and its recently formed Aerospace Defence Forces (VKO) under one unified command structure in mid-2015. The move represents an evolution in Russian military thinking from an era where its air and space forces existed as separate branches with little overlap in command authority to one where air and space will be treated more as a seamless theatre of war.[280] By 17 November 2015, Russia had involved ten imagery and electronic warfare reconnaissance satellites, including civilian-use spacecraft, to provide support for its operation against ISIS forces in Syria.[281]

Concerns about the military space capabilities of both Russia and China remained following the annual Worldwide Threat Assessment of the US Intelligence Community to the US Congress on 9 February 2016.[282] In addition to Russia's sophisticated use of imaging and electronic reconnaissance satellites to support military operations in Syria, both countries are developing counter-space capabilities to deny, degrade, or disrupt US space systems. The statement pointed to the 2014 Military Doctrine of the Russian Federation that referenced three space-enabled capabilities (including "global strike", the "intention to station weapons in space", and "strategic non-nuclear precision weapons") as its main external military threats.[283] The statement also referenced Russia's acknowledgement of its deployment of radar imagery jammers and development of laser weapons designed to blind US intelligence and ballistic missile defence satellites and a

[278] Gruss, Mike. "Air Force sent GSSAP satellite to check on stalled MUOS-5." 18 Aug. 2017. SpaceNews 21 Aug. 2017 http://spacenews.com/air-force-sent-gssap-satellite-to-check-on-stalled-muos-5/

[279] Henry, Caleb. "ULA Orbits Second Set of GSSAP Satellites for US Air Force." 22 Aug. 2016. Via Satellite 21 Aug. 2017 http://www.satellitetoday.com/launch/2016/08/22/ula-orbits-second-set-gssap-satellites-us-air-force/

[280] Bodner, Matthew. "Russian Military Merges Air Force and Space Command." 3 Aug. 2015. The Moscow Times 10 July 2016 http://www.themoscowtimes.com/business/article/russian-military-merges-air-force-and-space-command/526672.html

[281] "Russia involves 10 reconnaissance satellites in Syria operation – General Staff." 17 Nov. 2015. TASS 20 Mar. 2016 http://tass.ru/en/science/837273

[282] Clapper, James R. "Worldwide Threat Assessment of the US Intelligence Community." 9 Feb. 2016. Armed-services.senate.gov 11 Aug. 2017 https://www.armed-services.senate.gov/imo/media/doc/Clapper_02-09-16.pdf

[283] "Russian Military Doctrine." 3 May 2016. GlobalSecurity.org 11 Aug. 2017 http://www.globalsecurity.org/military/world/russia/doctrine.htm

2013 recommendation by Russia's Parliament that Russia resume research and development of its antisatellite capabilities.[284]

1.4.5 China

China's military space capability has long attracted heightened speculation, as China's space sector has long been intimately connected to the People's Liberation Army (PLA). Nevertheless, China publically stands against the militarization of space, and its recent Middle and Long Term Development Plan for State Civil Space Infrastructure (2015–2025) appears more focussed on enhancing its space capabilities for domestic purposes – particularly in establishing its state civil space infrastructure system and enhancing its competitiveness.[285]

Nonetheless, concerns about China's military space capabilities remain, especially following the release of a defence white paper on China's Military Strategy by China's Ministry of National Defence (MOD) on 26 May 2015.[286] In addition to reaffirming decades-old commitments not to attack first (but to counterattack strongly) and giving more attention to emerging domains like cyber and space, changes in the white paper were modest and in line with the direction of recent PLA activity. The white paper outlines China's strategic guideline of active defence, along with modernising its military to adapt to new changes in its maritime security environment.[287] And in terms of military space and counter-space activities, aside from a paragraph indicating that China will strengthen its capabilities for strategic deterrence and nuclear counterattack and medium- and long-range precision strikes, it later goes on to say that "China will keep abreast of the dynamics of outer space, deal with security threats and challenges in that domain, and secure its space assets to serve its national economic and social development, and maintain outer space security".[288] While China had conducted a space launch in July 2014 with a similar

---

[284] Clapper, James R. "Worldwide Threat Assessment of the US Intelligence Community." 9 Feb. 2016. Armed-services.senate.gov 11 Aug. 2017 https://www.armed-services.senate.gov/imo/media/doc/Clapper_02-09-16.pdf

[285] Google Translated. "关于印发国家民用空间基础设施中长期发展规划(2015-2025年)的通知." 26 Oct. 2015. National Development and Reform Commission 2 Feb. 2016 http://www.sdpc.gov.cn/zcfb/zcfbghwb/201510/t20151029_756376.html

[286] Minnick, Wendell. "White Paper Outlines China's Ambitions." 27 May 2015. Defense News 22 Jan. 2016 http://www.defensenews.com/story/defense/policy-budget/warfare/2015/05/26/china-us-pentagon-taiwan-report-south-east-sea-islands-reefs-s400-su35-missiles-satellite-space-deterrence/27957131/; *see also* "China's Military Strategy." 26 May 2015. Ministry of National Defense – The People's Republic of China 3 Feb. 2016 http://eng.mod.gov.cn/Database/WhitePapers/index.htm

[287] "III. Strategic Guideline of Active Defense." 26 May 2015 Ministry of National Defense – The People's Republic of China 3 Feb. 2016 http://eng.mod.gov.cn/Database/WhitePapers/2015-05/26/content_4586711.htm

[288] "IV. Building and Development of China's Armed Forces." 26 May 2015 Ministry of National Defense – The People's Republic of China 3 Feb. 2016 http://eng.mod.gov.cn/Database/WhitePapers/2015-05/26/content_4586713.htm

profile to its January 2007 ASAT test, as of 2016 no additional antisatellite programmes had been publicly acknowledged.[289]

1.4.6 Japan

Concern over China's 2007 direct-ascent antisatellite weapon test, and subsequent experiments focussing on jamming and laser-blinding satellites, is said to have motivated Japan's shift towards placing security at the forefront of its national space policy.[290] Concern also exists over North Korea that has continued development of long-range ballistic missiles, one of which provocatively overflew Japan dropping its first stage very near Japanese territory in a test conducted in 1998.[291]

Japan's defence budget has steadily increased from ¥4.65 trillion in 2012 to ¥4.93 trillion in 2016.[292] For the fiscal year 2016 (beginning on 1 April 2016 and ending 30 March 2017), Japan allocated ¥34.2 billion of its ¥332.4 billion in space spending to its Ministry of Defence (MOD). Interestingly, these figures do not include the ¥191.5 billion in space-related ballistic missile defence (BMD) expenditure, which accounts for 87.3% of the total ¥219.3 billion budgeted for Japan's BMD programme; another ¥26.4 billion went to its space activities to strengthen information gathering capabilities and command, control, and communication capabilities by using satellites and implement measures to secure the stable use of outer space.[293] Moreover, Japan plans to expand its QZSS regional navigation system to an enhanced seven-satellite constellation, in addition to its development of a laser-optical data-relay satellite, an advanced Earth observation satellite carrying a ballistic missile early warning sensor as a hosted payload for Japan's Ministry of Defence, and a new line of multipurpose small satellites capable of rapid production and deployment for a range of missions.[294]

[289] "Office of the Secretary of Defense. "Annual Report to Congress – Military and Security Developments Involving the People's Republic of China 2016." 13 May 2016. Defense.gov 10 Aug. 2017 https://www.defense.gov/Portals/1/Documents/pubs/2016%20China%20Military%20Power%20Report.pdf

[290] Kallender-Umezu, Paul. "Japan Boosts Space Spending In Support of Security Focus." 2 Feb. 2015. SpaceNews 10 Dec. 2015 http://spacenews.com/japan-boosts-space-spending-in-support-of-security-focus/

[291] Rose, Frank A. "Commentary | Strategic Stability in U.S.-China Relations." 29 Jan. 2015. SpaceNews 10 Dec. 2015 http://spacenews.com/commentary-strategic-stability-in-u-s-china-relations/

[292] "Defense Programs and Budget of Japan – Overview of FY2016 Budget." Ministry of Defense. 11 Aug. 2017: 11–13 http://www.mod.go.jp/e/d_budget/pdf/280330.pdf

[293] Ibid.

[294] Kallender-Umezu, Paul. "Japan Boosts Space Spending In Support of Security Focus." 2 Feb. 2015. SpaceNews 10 Dec. 2015 http://spacenews.com/japan-boosts-space-spending-in-support-of-security-focus/

1.4.7 India

Separate from the civil space activities of ISRO, India is developing a Ballistic Missile Defence (BMD) Programme through its Defence Research and Development Organization (DRDO) to counter threats raised by Pakistan's Strategic Missile Group, which is developing its own medium-range ballistic missiles. The BMD Programme is a two-tiered system to provide high- and low-altitude cover against incoming ballistic missiles. While the DRDO's Advanced Air Defence (AAD) system is optimised for surface-to-air strikes against aircraft and UAVs at endoatmospheric altitudes between 20 and 40 km, the Prithvi Air Defense (PAD) missile provides exoatmospheric defence at altitudes of 50 to 80 km. On 15 May 2016, the DRDO was reported to have conducted a test of its AAD missile to validate the interceptor in flight mode.[295] However, there are conflicting reports as to whether the test was successful.[296]

It should be noted that the PAD missile can also be seen as a further step for India in developing its own antisatellite capabilities. In this pursuit, the DRDO is looking at the feasibility of developing such an antisatellite vehicle by integrating its Agni-3 missile with its PAD. If it succeeds, the antisatellite missile would have an effective range of about 1400–1500 km and would advance India's missile capabilities to be on a par with the United States and China.[297]

[295] "India's successful AAD makes Pakistan's missile striking capability obsolete." 18 May 2016. Defenceupdate.in 7 Aug. 2017 http://defenceupdate.in/indias-successful-aad-makes-pakistans-missile-striking-capability-obsolete/

[296] Gady, Franz-Stefan. "Did India Hide a Failed Supersonic Missile Test?" 26 May 2016 The Diplomat 9 Aug. 2017 http://thediplomat.com/2016/05/did-india-hide-a-failed-supersonic-missile-test/

[297] "India Contemplates Anti-Satellite Vehicle Integration with Agni-III Ballistic Missile." 15 Oct. 2013. Missile Threat 12 May 2014 http://missilethreat.com/india-contemplates-anti-satellite-vehicle-integration-with-agni-iii-ballistic-missile/

Chapter 2
Developments in Space Policies, Programmes and Technologies Throughout the World and in Europe

Cenan Al-Ekabi

2.1 Space Policies and Programmes

The major space policy developments worldwide were presented in Chap. 1, above, to identify the principal space faring nations' strategies in 2016. In the section below, there will be a brief discussion of developments in technology-related areas, including policies and access to space technologies. The aim of this section is to clarify how these strategies interact with and influence specific space programmes and related research and development projects.

2.2 Space Transportation

2.2.1 Europe

With Europe's launch sector in the midst of a substantial reorganization to increase its competitiveness in the global market, the Arianespace launch consortium aims to increase the launch rate of its Vega launcher to benefit from economies of scale and increased demand. The per-launch cost of the Vega launcher is expected to decrease as its prime contractor Avio SpA ramps up production to enable three launches per year. The Vega launcher is sold by European Launch Vehicle (ELV), the public-private joint venture between Avio (70% share) and ASI (30% share). By the end of 2016, Arianespace had used two of the ten Vega rockets that it purchased from ELV in 2016 to launch five high-resolution optical EO satellites (i.e. the Peruvian government's PerúSat 1 and four SkySat satellites for Google's Terra

C. Al-Ekabi (✉)
European Space Policy Institute (ESPI), Vienna, Austria
e-mail: Cenan.Al-Ekabi@espi.or.at

C. Al-Ekabi, S. Ferretti (eds.), *Yearbook on Space Policy 2016*,
Yearbook on Space Policy, https://doi.org/10.1007/978-3-319-72465-2_2

Bella constellation) on 15 September 2016[1] and Turkey's Göktürk-1 EO satellite on 5 December 2016. The latest launch marked the eighth success of the Vega launcher system since its inauguration in 2012.[2] The next Vega launcher will lift the Copernicus programme's Sentinel-2B satellite into orbit in March 2017. Whereas in 2013, Vega's backlog of four satellite launches was valued at €130 million or €32.5 million per launch; the value of a launch reached €36.4 million per launch as of June 2015.[3]

On 12 August 2015, ESA and CNES began developing the Ariane 6 launch pad and horizontal launcher integration facilities in Europe's Guiana Space Centre in French Guiana Ariane 6 on 12 August 2015. While all previous European launchers have been integrated vertically and then rolled upright to the launch pad by rail, the Ariane 6 will be integrated horizontally enabling time and cost savings derived from fewer crane and hazardous moving operations and a more fluid production flow. Nevertheless, unlike spacecraft launched on other horizontally integrated launchers, including the SpaceX Falcon 9 launcher, which are stored, integrated onto the launcher and then forced to endure additional vibrations while rolled to the launch pad, Ariane 6 payloads will remain vertical, stored in their fairing and then integrated on the now vertical Ariane 6 launcher at the last moment.[4]

By the end of 2016, the Ariane 5 had a total of 76 consecutive successful launches, surpassing the record of its Ariane 4 predecessor, reinforcing its status as the hallmark of reliability – a strong motivator for commercial satellite operators. Yet, the Ariane 5 launcher has usually sold for €150 million, while its launch cost is €170 million, requiring institutional support of Arianespace of around €100 million annually.[5] With orders for the Ariane 5 launcher booked into 2017, Arianespace and ASL aim to decrease the cost of the Ariane 5 by 5–6% in the next batch of launch orders starting in 2019 through improvements in production and operation. They also plan to raise the price for heavier satellites intended for the Ariane 5's upper berth, as demand for that slot has increased following the recent failed launches by competitors ILS and Sea Launch. Moreover, the price for the Ariane 5's lower berth, targeted for small satellites and telecom satellites using electric propulsion to reach geostationary transfer orbit, will be reduced to make it more competitive with SpaceX prices. Another advantage exists in the current EU-USA exchange rate that

[1] De Selding, Peter B. "Europe's Vega launches 1 Peruvian and 4 Terra Bella Earth observation satellites." 16 Sept. 2016. SpaceNews 8 Nov. 2016 http://spacenews.com/europes-vega-launches-1-peruvian-and-4-terra-bella-earth-observation-satellites/

[2] Henry, Caleb. "Vega mission orbits Turkish Earth observation satellite." 5 Dec. 2016. SpaceNews 4 Jan. 2017 http://spacenews.com/vega-mission-orbits-turkish-earth-observation-satellite/

[3] De Selding, Peter B. "Vega Launches Sentinel-2A Observation Satellite." 23 June 2015. SpaceNews 5 Jan. 2016 http://spacenews.com/vega-launches-sentinel-2a-observation-satellite/

[4] De Selding, Peter B. "Airbus Safran Launchers aims for 'the discipline of the flow' in Ariane 6 integration." 7 Apr. 2016. SpaceNews 8 Nov. 2016 http://spacenews.com/airbus-safran-launchers-aims-for-the-discipline-of-the-flow-for-ariane-6-integration/

[5] De Selding, Peter B. "With Revenue Looking Up, Arianespace Seeks To Bring Ariane 5 Costs Down." 21 Oct. 2015. SpaceNews 13 Jan. 2016 http://spacenews.com/with-revenue-looking-up-arianespace-seeks-to-bring-ariane-5-costs-down/

will allow customers to purchase an Ariane 5 launch with a lower value Euro.[6] Once the Ariane 6 becomes operational in 2020, the Ariane 5 will undergo a 3-year phase out.[7] The Ariane 64 will be capable of lifting two telecom satellites with a combined mass of 9500 kg to geostationary transfer orbit and will be sold for €96.34 million per launch (or €48.17 million per customer).[8] The 62 configuration of the Ariane 6 with two boosters for small satellites will be priced around €75 million.[9]

2.2.2 *United States*

NASA's Space Launch System (SLS) heavy-lift launcher is scheduled to launch its first mission, Exploration Mission 1 (EM-1, in late 2018). While uncertainty about the launch date remained due to delays in the delivery of ESA's Service Module, which had completed its critical design review in mid-2016 but had some technical issues that required further study, progress in the development of the various launch vehicle, spacecraft and ground system components needed to support the EM-1 enabled mission planners to confirm that the mission could take place between September and November 2018.[10] Instead of arriving in January 2017, the Service Module is now expected to be delivered in April 2017, giving mission planners a large window to correct any issues faced in integrating all components of the launcher and spacecraft. The Service Module will power the Orion MPCV during its unmanned EM-1 mission. Moreover, the EM-1 mission will carry 13 cubesats as secondary payloads that will be released from a ring linking the SLS's upper stage with the Orion MPCV.[11]

On 23 December 2016, US President Obama signed the National Defense Authorization Act for Fiscal Year 2017 (NDAA-17) which repealed a provision in the previous year's NDAA permitting the use of the Russian-built RD-180 engine

[6] De Selding, Peter B. "Arianespace Assures French Parliament it Can Outcompete SpaceX." 13 May 2015. SpaceNews 5 Jan. 2016 http://spacenews.com/arianespace-assures-french-parliament-it-can-outcompete-spacex/

[7] De Selding, Peter B. "With Revenue Looking Up, Arianespace Seeks To Bring Ariane 5 Costs Down." 21 Oct. 2015. SpaceNews 13 Jan. 2016 http://spacenews.com/with-revenue-looking-up-arianespace-seeks-to-bring-ariane-5-costs-down/

[8] De Selding, Peter B. "With Revenue Looking Up, Arianespace Seeks To Bring Ariane 5 Costs Down." 21 Oct. 2015. SpaceNews 13 Jan. 2016 http://spacenews.com/with-revenue-looking-up-arianespace-seeks-to-bring-ariane-5-costs-down/

[9] Gallois, Dominique. "Ariane 6, un chantier européen pour rester dans la course spatiale." 1 Dec. 2014. Le Monde 9 Mar. 2016 http://www.lemonde.fr/economie/article/2014/12/01/les-europeens-s-appretent-a-mettre-ariane-6-en-chantier_4532259_3234.html

[10] Foust, Jeff. "First SLS mission on schedule for fall 2018 launch." 26 July 2016. SpaceNews 8 Nov. 2016 http://spacenews.com/first-sls-mission-on-schedule-for-fall-2018-launch/

[11] Foust, Jeff. "Spaceflight offers smallsat launch services on government contract schedule." 11 Feb. 2016. SpaceNews 6 Nov. 2016 http://spacenews.com/spaceflight-offers-smallsat-launch-services-on-government-contract-schedule/

for the US Evolved Expendable Launch Vehicle (EELV) programme with the exception of orders or options for RD-180 engines already awarded under a contract signed on 18 December 2013 and for contracts for the use of a total of 18 additional RD-180 engines between the date of the enactment of the NDAA-2017 and ending 31 December 2022.[12] Russia's 2014 incursion in Ukraine has stoked continuing tensions between the USA and Russia that led the US Congress to prohibit US companies from contracting with Russian suppliers of rocket engines or renewing current contracts for space launch activities.[13] Since ULA uses RD-180 engines for the first stage of its Atlas 5 launcher, the restriction limited its use before ULA's follow-on Vulcan launcher – powered by Blue Origin's BE-4 engine – is ready in 2022. The result is that while ULA has an exemption for the 34 RD-180 engines that were ordered while the NDAA-15 was still pending[14] and the 20 more RD-180 engines ULA had ordered by 23 December 2015,[15] the NDAA-17 gives ULA access to 18 more RD-180 engines in order to remain competitive with SpaceX.[16]

2.2.3 *Russia*

On 28 April 2016, Russia's far east Vostochny Cosmodrome, conducted its first launch using a Soyuz 2.1a to lift two small satellites and a cubesat into LEO from its first launch pad.[17] The inaugural launch was delayed by a day due to a fail signal detected on one of the launcher's automatic safety system sensors. While work on the six other pads is ongoing, with the completion of a second pad for its Angara launcher expected in 2021, the Cosmodrome is expected to eventually host ten launches per year.[18] The construction of the Cosmodrome has been accompanied by

---

[12] "Statement by the President on Signing the National Defense Authorization Act for Fiscal Year 2017." 23 Dec. 2016. The White House President Barack Obama 15 Aug. 2017 https://obamawhitehouse.archives.gov/the-press-office/2016/12/23/statement-president-signing-national-defense-authorization-act-fiscal

[13] Section 1608. Carl Leven and Howard P. "Buck" McKeon National Defense Authorization Act for Fiscal Year 2015, Pub. L. no 113–291 (2014) http://www.gpo.gov/fdsys/pkg/CPRT-113HPRT92738/pdf/CPRT-113HPRT92738.pdf

[14] Gruss, Mike. "House-Senate Conference Measure To End Pentagon Use of RD-180." 5 Dec. 2014. SpaceNews 22 July 2015 http://spacenews.com/42701house-senate-conference-measure-to-end-pentagon-use-of-rd-180/

[15] Shalal, Andrea. "ULA Orders 20 More RD-180 Rocket Engines." 23 Dec. 2015. SpaceNews 14 Jan. 2016 http://spacenews.com/ula-orders-20-more-rd-180-rocket-engines/

[16] Gruss, Mike. "HASC doubles Air Force allotment of RD-180 engines, focuses funding on building its replacement." 28 Apr. 2016. SpaceNews 8 Nov. 2016 http://spacenews.com/the-rd-180-amendment-was-one-of-two-launch-related-provisions-that-made-it-into-the-national-defense-authorization-act-of-2017-by-the-time-the-committee-voted-60-to-2-to-send-the-bill-to-the-full/

[17] "Russia's brand new cosmodrome launches first-ever rocket." 28 Apr. 2016. RT 15 Aug. 2017 https://www.rt.com/news/341192-vostochny-cosmodrome-first-launch/

[18] Haines, Lester. "First rocket finally departs Russia's Vostochny cosmodrome." 28 Apr. 2016. The Register 15 Aug. 2017 https://www.theregister.co.uk/2016/04/28/vostochny_launch/

reports of construction being behind schedule, billions of dollars of budget overruns and cases of embezzlement and the non-payment of workers for months at a time. For instance, while the first launch pad was supposed to be completed by mid-2015 and supposed to be just $1.9 billion in today's prices, by mid-2016 work on the first launch pad was still ongoing, and its cost was estimated to have exceeded $2.4 billion. Moreover, around $165 million was embezzled during the construction process, and several contractors have been charged according to Russian prosecutors. Amid the increasing costs, more than 350 workers brought public attention to the fact that there was a 4-month delay in getting paid.[19] The final construction stage of the Vostochny Cosmodrome will develop facilities for a super-heavy launch vehicle that will be capable of delivering 120–150 tons into low Earth orbit by 2020.[20]

While Russia intends to increase the role of the Vostochny Cosmodrome in civilian launches, it also intends to continue its launches from Baikonur, Kazakhstan. On 26 December 2016, Russia and Kazakhstan signed a cooperation agreement extending Russia's use of Kazakhstan's Baikonur Cosmodrome.[21] The agreement regulates continued cooperation at Baikonur and includes an 8-year road map up to the year 2025. It also lists areas where space cooperation can be profitable from an economic standpoint, including remote sensing, communications and satellite navigation technologies and the development of a commercial space launch infrastructure (i.e. spacecraft, launch systems and other space-related equipment and ground-based space infrastructure) to launch spacecraft, scientific equipment and manned space flights.[22] The Baikonur Cosmodrome is jointly managed by Roscosmos and Russia's Aerospace Forces, and it is leased by Russia until 2050 at a cost of $115 million per year.[23] Moreover in 2017, Russia aims to conduct 15 launches from the Baikonur Cosmodrome in Kazakhstan, in addition to 2 launches from its Vostochny Cosmodrome and 6 from its Plesetsk Cosmodrome in Russia.[24]

---

[19] Eremenko, Alexey and Alexander Smith. "Vostochny Cosmodrome: Russian Space Project Isn't Going to Plan." 31 July 2016. NBC News 16 Aug. 2017 http://www.nbcnews.com/science/space/vostochny-cosmodrome-russian-space-project-isn-t-going-plan-n618846

[20] Bodner, Matthew. "Putin Pledges $1 Billion for Completion of New Cosmodrome." 2 Sept. 2014. The Moscow Times 16 July 2015 http://www.themoscowtimes.com/business/article/putin-pledges-1-billion-for-completion-of-new-cosmodrome/506321.html

[21] "Why Russia Won't Be Leaving Kazakhstan's Baikonur Cosmodrome Anytime Soon." 29 Dec. 2016. Sputniknews.com 16 Aug. 2017 https://sputniknews.com/politics/201612291049119821-russia-kazakhstan-baikonur-cooperation/

[22] Google Translated. "Почему Россия не хочет уходить из Байконура – эксперт." 14 Dec. 2016. 365info.kz 16 Aug. 2017 https://365info.kz/2016/12/pochemu-rossiya-ne-hochet-uhodit-iz-bajkonura-ekspert/

[23] Mirovalev, Mansur. "The bumpy road to Vostochny, Russia's new multibillion-dollar spaceport." 4 May 2016. Los Angeles Times 16 Aug. 2017 http://www.latimes.com/world/europe/la-fg-russia-space-20160504-story.html

[24] "Russia to Launch 15 Rockets From Baikonur, 6 From Plesetsk, 2 From Vostochny." 28 Dec. 2016. Sputniknews.com 16 Aug. 2017 https://sputniknews.com/russia/201612281049060014-russia-rockets-launch/

2.2.4 *Japan*

JAXA is developing its H3 next-generation launcher to be more powerful and cost-effective than its current H-2A launcher. On 20 July 2016, JAXA released the basic design of the H3, which aims to be ready for its first launch in 2020. The H3 will have several configurations including two types of fairings, two to three units for the first-stage engines and either zero, two or four solid rocket boosters depending on the payload sizes and their orbits.[25] Its manufacturer, Mitsubishi Heavy Industries, plans to use components from other domestic industries and will apply a flow-line manufacturing system to halve the current cost per orbital launch to around $47 million.[26]

Japan conducted a second launch of its recent three-stage solid-fuel Epsilon launcher on 20 December 2016, with the 'Exploration of energization and Radiation in Geospace' (ERG) scientific satellite on board.[27] The Epsilon launcher is also being developed to cut launcher costs, using the same solid-fuel strap-on booster as on the H-2A.[28] The first launch of the Epsilon launcher was conducted on 14 September 2013, yet Japan has sought to perfect the launch capability of the launcher in the following years. The enhanced Epsilon launcher is designed to be capable of placing a 1.2 ton payload in low Earth orbit (LEO) and 700 kg payload into sun-synchronous orbit (SSO).[29]

2.2.5 *China*

China conducted the maiden launch of its two-stage Long March 7 on 25 June 2016, carrying a prototype version of its future next-generation crew vehicle (NGCV).[30] Launched from the newly constructed Wenchang Satellite Launch Centre on the north-eastern coast of Hainan Island, the Long March 7 has a launch capacity of

[25] "Characteristics of H3 Launch Vehicle." H3 Launch Vehicle 16 Aug. 2017 http://global.jaxa.jp/projects/rockets/h3/

[26] Nowakowski, Tomasz. "Japan's H-3 rocket to be more powerful, cost-effective than predecessor." 26 July 2016. Spaceflight Insider 16 Aug. 2017 http://www.spaceflightinsider.com/organizations/jaxa/japans-h-3-rocket-powerful-cost-effective-predecessor/

[27] "Success of Epsilon-2 Launch with ERG Aboard." 20 Dec. 2016. JAXA 16 Aug. 2017 http://global.jaxa.jp/press/2016/12/20161220_epsilon2.html

[28] Matsuda, Shogo. "Japan's Epsilon rocket shoved aside?" 16 Jan. 2015. Nikkei Asian Review 21 Jan. 2016 http://asia.nikkei.com/Tech-Science/Tech/Japan-s-Epsilon-rocket-shoved-aside

[29] Federal Aviation Administration. The Annual Compendium of Commercial Space Transportation: 2014. Washington DC: FAA, Feb. 2015: 10.

[30] Barbosa, Rui C. "China successfully debuts Long March 7 – Recovers capsule." 25 June 2016. NASASpaceflight.com 16 Aug. 2017 https://www.nasaspaceflight.com/2016/06/china-debuts-long-march-7-rocket/

13.5 metric tons to LEO and will eventually be used to launch taikonauts and cargo resupply missions to China's future Tiangong space station.[31]

2.2.6 India

By the end of 2016, India's PSLV launcher had conducted 6 additional launches, increasing its launch record to 38 consecutive successes out of 39 launches in total.[32] Capable of lifting 3700 kg to LEO and 800 kg to GEO, the PSLV has launched a total of 84 satellites for customers, including 30 satellites launched into LEO and 3 satellites to GEO in 2016 alone. In addition to its successful track record, with just one failure in its inaugural launch in 1993, the launcher's low cost is another drawing feature, dropping to as low as €26 million in December 2015.[33] Incidentally, India's low cost launch capacity, and India's refusal to sign a commercial launch accord with the US government to ensure fair market-based pricing, is the reason US companies have been restricted from using the PSLV in the past decade.[34]

India is also nearing completion of its Geosynchronous Satellite Launch Vehicle (GSLV)-Mark III. ISRO conducted its third and final ground test of its indigenous upper-stage engine CE-20 on 19 February 2016, successfully firing the liquid-hydrogen-fuelled engine for 640 s.[35] While ISRO had hoped to launch its fully developed GSLV-Mark III launcher carrying the GSAT19 weather satellite by December 2016, the launch was postponed and will take place sometime in 2017.[36]

ISRO successfully launched a Reusable Launch Vehicle (RLV) Technology Demonstration mission on 22 May 2016. First conceived by ISRO in 2009, the long delayed first flight test of the scramjet propulsion system took place at the Satish Dhawan Space Centre. Perched atop a HS9 solid booster, the 1750 kg technology demonstrator was launched to an altitude of 70 kms to autonomously test the RLV's hypersonic aerodynamic properties, avionics, thermal protection and control systems and its mission management. This initial mission will be followed by a landing

[31] Clark, Stephen. "China's new Long March 7 rocket successful on first flight." 25 June 2016. Spaceflight Now 16 Aug. 2017 https://spaceflightnow.com/2016/06/25/chinas-new-long-march-7-rocket-successful-on-first-flight/

[32] "PSLV." 13 July 2017. Gunter's Space Page 16 Aug. 2017 http://space.skyrocket.de/doc_lau/pslv.htm

[33] Jayaraman, K.S. "ISRO Launches Six Satellites for Singapore." 16 Dec. 2015. SpaceNews 6 Jan. 2016 http://spacenews.com/isro-launches-six-satellites-for-singapore/

[34] SpaceNews Editor. "Getting the Cubesat Revolution Out of Low Gear." 24 Nov. 2015. SpaceNews 6 Jan. 2016 http://spacenews.com/getting-the-cubesat-revolution-out-of-low-gear/

[35] Jayaraman, K.S. "India's heavy-lift rocket on track for December debut following engine test." 22 Feb. 2016. SpaceNews 16 Aug. 2017 http://spacenews.com/indias-heavy-lift-rocket-on-track-for-december-debut-following-engine-test/

[36] "ISRO to launch weather satellite." 25 July 2016. TheHindu.com 16 Aug. 2017 http://www.thehindu.com/todays-paper/tp-features/tp-educationplus/ISRO-to-launch-weather-satellite/article14506339.ece

experiment, a launch to orbit and return mission and a scramjet propulsion demonstration.[37]

2.3 Space Science and Exploration

In this section, space science is understood to mean using mainly remote observation to make discoveries on the origin, evolution and the future of the universe, its galaxies, our solar system and other celestial bodies, e.g. stars, exoplanets, comets and asteroids. Space exploration, on the other hand, involves human and robotic spaceflight missions. While traditional governmental space agencies dominate in both these fields, progress in the latter category can be seen with the development of exploration involving commercial players and with new space powers demonstrating the technology needed to carry out such missions.

2.3.1 Human Spaceflight Activities

Human spaceflight was focused on low Earth orbit (LEO), with the International Space Station (ISS) at centre stage, following its formal extension to 2024. Russia is currently the sole launch provider capable of transporting crew regularly to the ISS. It also provides ISS cargo resupply services, along with US commercial resupply services (CRS) by SpaceX and Orbital ATK, with Japan's H-II Transfer Vehicle (HTV) providing auxiliary support. Activity with China's space station also occurred in 2016.

ESA astronauts Tim Peake and Thomas Pesquet journeyed to the ISS in 2016. They are among ESA's latest class of astronauts, along with Luca Parmitano, Alexander Gerst, Samantha Cristoforetti and Andreas Mogensen, who had completed their training on 22 November 2010.[38] These latter mentioned astronauts had, respectively, participated on the station in Expeditions 36/37, 40/41 and 42/43 and the tail end of Expedition 44.

ESA astronaut Tim Peake began his 6-month Expedition 46/47 mission on the ISS under the mission banner 'Principia' on 15 December 2015. Days after his arrival, on 21 December 2015, Tim Peake assisted in a spacewalk with NASA astronauts Tim Kopra and Station Commander Scott Kelly to move an equipment carrier on the station.[39] On 15 January 2016, Tim Peake conducted the first spacewalk for a

[37] Gebhardt, Chris. "India successfully launches Reusable Launch Vehicle demonstrator mission." 22 May 2016. NASA Spaceflight.com 16 Aug. 2017 https://www.nasaspaceflight.com/2016/05/india-launch-reusable-launch-vehicle-demonstrator-mission/

[38] "Graduation of Europe's new astronauts." 22 Nov. 2010. ESA 9 May 2017 http://m.esa.int/Our_Activities/Human_Spaceflight/Graduation_of_Europe_s_new_astronauts

[39] "TIM PEAKE SET FOR SPACEWALK." 5 Jan. 2016. ESA 8 May 2017 http://www.esa.int/Our_Activities/Human_Spaceflight/Principia/Tim_Peake_set_for_spacewalk

British astronaut on the ISS, when he and Tim Kopra participated in a near-5-h spacewalk to replace a sequential shunt unit that had been damaged in November 2015. The ISS has a total of eight such units which transfer the electrical power generated by the solar panels; this replacement restored the ISS to 100% of its operational capability. The spacewalk was supposed to last for more than 6 h but was cut short when Tim Kopra reported water in his helmet, while the astronauts conducted their next task of laying cables in advance of new docking ports and the reinstallation of a valve that had been removed for the relocation of the Leonardo module in 2015.[40] In addition to serving as flight engineer and conducting dozens of experiments on the station, Tim Peake assisted in berthing the Cygnus CRS spacecraft to the ISS on 26 March 2016, and he took the lead in berthing the eighth Dragon CRS spacecraft on 10 April 2016.[41] While on the ISS, Tim Peake operated a rover across a simulated Mars terrain in Stevenage, UK; he was also the second ESA astronaut to use the Mares muscle-measurement unit that charts the muscle speed and torque of a bending elbow or knee joint.[42] And while he expected to return to Earth on 5 June 2016, the Principia mission was extended until 18 June 2016 because ground control aimed to keep the ISS operating at full capacity with six crew members closer to when the next Expedition arrived on 7 July 2016.[43]

ESA astronaut Thomas Pesquet began a 6-month Expedition 50/51 mission on the ISS under the mission banner 'Proxima' on 19 November 2016. While Thomas Pesquet is the last of ESA's 2010 astronaut graduates to fly into space, he is the first French astronaut to visit the ISS since the installation of the Columbus module in 2008. Thomas Pesquet will perform about 50 scientific experiments for ESA and CNES and will take part in many research activities for the other station partners.[44] In January 2017, Thomas Pesquet and NASA astronaut Shane Kimbrough will perform two spacewalks on the ISS, under the guidance of fellow ESA astronaut Luca Parmitano at NASA's mission control in Texas, USA, to replace older-technology batteries with newer lithium-ion designs.[45] Moreover, following the completion of

[40] "TIM AND TIM SAFELY BACK IN SPACE STATION AFTER SPACEWALK." 15 Jan. 2016. ESA 8 May 2017 http://www.esa.int/Our_Activities/Human_Spaceflight/Principia/Tim_and_Tim_safely_back_in_Space_Station_after_spacewalk

[41] "BUSY SPACECRAFT AND EXPERIMENT SCHEDULE ON SPACE STATION." 8 Apr. 2016. ESA 8 May 2017 http://www.esa.int/Our_Activities/Human_Spaceflight/Principia/Busy_spacecraft_and_experiment_schedule_on_Space_Station

[42] "New landing date for ESA astronaut Tim Peake." 29 Apr. 2016. ESA 8 May 2017 http://m.esa.int/Our_Activities/Human_Spaceflight/Principia/New_landing_date_for_ESA_astronaut_Tim_Peake

[43] "Tim Peake returns to Earth." 18 June 2016. ESA 8 May 2017 http://m.esa.int/Our_Activities/Human_Spaceflight/Principia/Tim_Peake_returns_to_Earth

[44] "ESA ASTRONAUT THOMAS PESQUET ARRIVES AT THE INTERNATIONAL SPACE STATION." 19 Nov. 2016. ESA 8 May 2017 http://www.esa.int/Our_Activities/Human_Spaceflight/Proxima/ESA_astronaut_Thomas_Pesquet_arrives_at_the_International_Space_Station

[45] "SPACEWALK FOR THOMAS PASQUET." 16 Dec. 2016. ESA 8 May 2017 http://www.esa.int/Our_Activities/Human_Spaceflight/Proxima/Spacewalk_for_Thomas_Pesquet

Thomas Pesquet's mission, ESA astronaut Paolo Nespoli will return to the ISS to participate in Expedition 52/53 in mid-2017.[46]

With ESA's final Automated Transfer Vehicle (ATV) mission to the ISS completed on 15 February 2015, ESA will continue covering its dues for the ISS onward from 2017 by using the knowledge gained from the ATV programme to build the European Service Module for NASA's Orion spacecraft.[47] The Service Module will provide propulsion, electrical power and water and thermal control to the Orion spacecraft and will maintain the oxygen and nitrogen atmosphere for its crew. The first Service Module, built in Turin, Italy, initially underwent a comprehensive series of independent tests at Plum Brook in Ohio, USA, from November 2015 to 14 December 2016; the Service Module will now undergo further testing at vehicle level in conjunction with NASA's Orion Crew Module Structural Test Article.[48] NASA will conduct the first launch of the Orion spacecraft and Service Module on its Space Launch System near the end of 2018 for a month-long un-crewed demonstration mission around the Moon. By 7 December 2016, the agencies had agreed to extend their collaboration in human space exploration, wherein ESA will provide a second Service Module to support NASA's first crewed Orion mission, expected to launch as early as 2021.[49]

Russia launched four expeditions to the ISS in 2016, including the final vehicle in the TMA-M series, and the introduction of the newly upgraded MS variant. The final Soyuz TMA-20 M mission took place on 19 March 2016 with the Expedition 47/48 crew of Aleksey Ovchinin, Oleg Skripochka and Jeffrey Williams.[50] It was followed by the debut launch of the upgraded Soyuz MS-01 spacecraft on 7 July 2016 with the Expedition 47/48 crew of Anatoly Ivanishin, Kate Rubins and Takuya Onishi[51]; MS-02 on 23 September 2016 with the Expedition 49/50 crew of Sergey Ryzhikov, Andrey Borisenko and Shane Kimbrough[52]; and MS-03 on 17 November 2016 with the Expedition 50/51 crew of Oleg Novitskiy, Peggy Whitson and Thomas

[46] "PAOLO NESPOLI'S NEW SOYUZ CREWMATES." 16 Nov. 2016. ESA 8 May 2017 http://www.esa.int/Our_Activities/Human_Spaceflight/Astronauts/Paolo_Nespoli_s_new_Soyuz_crewmates

[47] "Last ATV reentry leaves legacy for future space exploration." 15 Feb. 2015. ESA 30 May 2016 http://m.esa.int/Our_Activities/Human_Spaceflight/ATV/Last_ATV_reentry_leaves_legacy_for_future_space_exploration

[48] "TESTING COMPLETED ON ORION SERVICE MODULE." 14 Dec. 2016. ESA 8 May 2017 http://www.esa.int/Our_Activities/Human_Spaceflight/Orion/Testing_completed_on_Orion_service_module

[49] "ESA TO SUPPLY SERVICE MODULE FOR FIRST CREWED ORION MISSION." 7 Dec. 2016. ESA 8 May 2017 http://www.esa.int/Our_Activities/Human_Spaceflight/Orion/ESA_to_supply_Service_Module_for_first_crewed_Orion_mission

[50] "Mission of Soyuz TMA-20M." 31 Mar. 2016. Russian Space Web 19 May 2017 http://www.russianspaceweb.com/iss-soyuz-tma20m.html

[51] "First Soyuz MS arrives at ISS." 7 July 2016. Russian Space Web 19 May 2017 http://www.russianspaceweb.com/soyuz-ms-01.html

[52] "Mission of Soyuz MS-02 spacecraft." 23 Sept. 2016. Russian Space Web 19 May 2017 http://www.russianspaceweb.com/soyuz-ms-02.html

Pesquet.[53] Russia also conducted three resupply missions (including one failure) to the ISS with its Progress cargo transfer vehicles: Progress MS-02 on 31 March 2016, MS-03 on 17 July 2016 and MS-04 on 1 December 2016 (which failed to reach the ISS due to an anomaly in the launcher's telemetry system following the ignition of its third stage – it crashed in southern Russia 4 h after launch).[54]

The sixth H-II Transfer Vehicle (HTV-6) cargo resupply mission, 'Kounotori 6', was launched to the ISS on 9 December 2016. The launch took place as planned shortly after the failed launch of the Russian Progress MS-04 resupply mission. The HTV-6 carried 5.9 tons of cargo, including 3.9 tons of pressurized cargo (i.e. water and food supplies, system supplies and utilization and experiment-related items) and unpressurized cargo (i.e. ISS battery Orbital Replacement Units and Orbital experiment hardware).[55] The utilization and experiment-related items included seven cubesats to be deployed from the Kibo science module, an upgraded JEM Small Satellite Orbital Deployer (J-SSOD) which doubled the deployment capacity of cubesats from the Kibo module from 6 U to 12 U, the Two-Phase Flow experiment for thermal management systems, the Position-Sensitive Tissue-Equivalent Proportional Chamber (PS-TEPC) for high-precision measurements of ionizing radiation in real-time and the High-Definition TV Camera – Exposed Facility 2 (HDTV-EF2). It also delivered the Orbital Replacement Units (ORU) for the Carbon Dioxide Removal Assembly (CDRA), a crucial system for life support in the ISS, and six lithium battery ORUs to replace the nickel-hydrogen batteries currently used on the ISS, enabling the extension of ISS operations. The HTV-6 also launched the Kounotori Integrated Tether Experiment (KITE) as an on-orbit technological demonstration. In the week following its unberthing from the ISS early in 2017, the KITE will extend from the HTV-6 a 700 m electrodynamic tether (EDT) as a propulsion mechanism for alleviating growing concerns over space debris.[56] Three more HTV missions to the ISS are planned by the beginning of 2020 (i.e. HTV-7 in February 2018, HTV-8 in February 2019 and HTV-9 in February 2020); afterward an upgraded HTV-X vehicle will replace the series in 2021.[57]

Lastly, while ISS operations have been extended to 2024, its utilization beyond 2024 remains open,[58] with some ISS advocates anticipating it will continue at least

[53] "Soyuz MS-03 arrives at ISS." 17 Nov. 2016. Russian Space Web 19 May 2017 http://www.russianspaceweb.com/soyuz-ms-03.html

[54] "Progress MS-04 fails to reach orbit." 31 Dec. 2016. Russian Space Web 19 May 2017 http://www.russianspaceweb.com/progress-ms-04.html

[55] Gebhardt, Chris, and Chris Bergin. "HTV-5 Kounotori sets sail for the ISS." 18 Aug. 2015. NASASpaceFlight.com 28 Jan. 2016 http://www.nasaspaceflight.com/2015/08/htv-5-kounotori-launch-space-station/

[56] "HTV6 Payload." 2 Dec. 2016. JAXA 23 May 2017 http://iss.jaxa.jp/en/htv/mission/htv-6/payload/

[57] Gebhardt, Chris. "Japan's HTV-6 resupply vehicle arrives at the ISS." 13 Dec. 2016. NASASpaceFlight.com 23 May 2017 https://www.nasaspaceflight.com/2016/12/jaxa-iss-htv-6-resupply-launch-station/

[58] Messier, Douglas. "Private Space Stations Could Be a Reality by 2025." 25 Aug. 2015. Space.com 25 Oct. 2016 http://www.space.com/30359-private-space-stations-reality-2025.html

until 2028 with the potential to transition into a commercial station.[59] On 11 April 2016, at the 32nd Annual Space Symposium held in Colorado, USA, United Launch Alliance (ULA) and Bigelow Aerospace announced a partnership agreement to launch two of Bigelow's inflatable B330 space habitats into low Earth orbit on two Atlas V launchers between late 2019 and 2020. The announcement came shortly after Bigelow's BEAM module was launched aboard the SpaceX Dragon CRS-8 mission to the ISS on 8 April 2016; the BEAM module will be attached to the ISS for a 2-year experimental demonstration.[60] NanoRacks has also facilitated the use of the ISS for businesses for over 5 years, brokering the launch to the station and the release of payloads into LEO orbit. And the recently formed Axiom Space LLC, led by former NASA ISS manager, Michael Suffredini, plans to eventually develop its own commercial space station.[61]

China's Tiangong 1 space lab ended its mission in March 2016, after the station stopped sending telemetry data back to Earth.[62] The Tiangong 1 – in sleep mode after its last crew departed in June 2013 – was not planned to be a permanent orbital station; however, in September 2016, the Chinese government confirmed the concerns raised by astronomers that the dormant 8.5 ton space station would have an uncontrolled re-entry into Earth's atmosphere in late 2017.[63] Moving forward, China launched its Tiangong 2 space lab on 15 September 2016 on its Long March 2F rocket from the Jiuquan Satellite Launch Center. The Tiangong 2 uses the same basic module as the Tiangong 1 space lab but will be a test bed for the rendezvous, docking and life-support technologies intended for its future planned space station. The Shenzhou-11 spacecraft, launched on 16 October 2016, delivered the first two taikonauts to the space lab for a 30-day visit. They will be followed by additional crewed and cargo resupply missions in 2017.[64] Construction of China's space station will begin in 2018 with the launch of its experimental core module; China aims to complete construction of the space station by 2022.[65]

[59] Smith, Marcia S. "Panel: Seamless Transition to Commercial LEO Space Station Needed." 22 Sept. 2015. SpacePolicyOnline.com 25 Oct. 2016 http://www.spacepolicyonline.com/news/panel-seamless-transition-to-commercial-leo-space-station-needed

[60] Gebhardt, Chris. "ULA and Bigelow announce partnership for first commercial space stations." 11 Apr. 2016. NASAspaceflight.com 25 Oct. 2016 https://www.nasaspaceflight.com/2016/04/ula-bigelow-partnership-first-commercial-space-stations/

[61] Foust, Jeff. "Former NASA ISS manager planning commercial space station venture." 23 June 2016. SpaceNews 25 Oct. 2016 http://spacenews.com/former-nasa-iss-manger-planning-commercial-space-station-venture/

[62] Jones, Morris. "Has Tiangong 1 gone rogue." 30 Mar. 2016. SpaceDaily 24 May 2017 http://www.spacedaily.com/reports/Has_Tiangong_1_gone_rogue_999.html

[63] Wall, Mike. "Tiangong-1 Space Lab Will Fall to Earth Next Year, China Says." 19 Sept. 2016. Space.com 23 May 2017 http://www.space.com/34089-tiangong-1-fall-earth-2017.html

[64] Reis, Michael. "Tiangong-2: China's Space Lab for Long Missions (Infographic)." 3 Nov. 2016. Space.com 24 May 2017 http://www.space.com/34614-tiangong-2-china-space-lab-explained-infographic.html

[65] "China to Launch Space Station Core Module in 2018." 3 Mar. 2017. Chinese Academy of Sciences 24 May 2017 http://english.cas.cn/newsroom/china_research/201703/t20170303_174538.shtml

2.3.2 Lunar Science

Interest in the Moon is ongoing for both its science and exploration value, in addition to being the finish line for several private space companies competing to win the Google Lunar X Prize. This year, the USA and China continued to progress towards a robotic and human lunar presence; however, budget constraints still have the potential to delay well-intentioned initiatives. Moreover, Google Lunar X Prize competitors have begun to partner up to better their odds of winning the symbolic award.

NASA's Lunar Reconnaissance Orbiter (LRO), launched in June 2009, is scouting the Moon in preparation for future lunar exploration, including finding landing sites; locating resources such as water, ice and hydrogen; and investigating the long-term effects of the lunar environment. The LRO completed its second 2-year extended science mission in September 2016 and was extended for a third 2-year extended science mission, which will run through October 2018. This new 'Cornerstone Mission' will focus on three questions relevant to the US Decadal Survey goals between 2013 and 2022, including (1) Volatiles and the Space Environment, (2) Volcanism and Interior Processes and Impacts and (3) Regolith Evolution.[66] On 23 March 2015, new research published in the journal *Nature* provides evidence that the spin axis of the Moon shifted by about five degrees roughly 3 billion years ago, by examining the distribution of ice at each of the lunar poles. While ice can exist in permanently shadowed areas on the Moon, direct exposure to sunlight would make it evaporate into space; researchers were able to find that the path of the ice that survived this axis shift matched models predicting where the ice could remain stable.[67] On 29 April 2016, data from the LRO data allowed researchers to develop models published in various journals to explain how lunar swirls are formed, i.e. patterns which extend tens of kilometres and are peppered across the Moon's surface in areas where ancient bits of magnetic field are embedded in the lunar crust and appear less weathered than their surroundings. While the swirls were first thought to have formed from plumes of material ejected by comet impacts or by fine dust particles lofted by micrometeorite impacts on the Moon's surface, a third theory which seems to be supported by LRO data is that the less-weathered areas are protected by magnetic field shields embedded in the Moon's surface which create strong electric fields that are able to deflect some of the slower-moving charged solar wind particles.[68] And on 13 October 2016, new observations by the LRO determined that the Moon's surface experiences a heavier bombardment

---

[66] Keller, J. W. and N. E. Petro. "THE LUNAR RECONNAISSANCE ORBITER CORNERSTONE MISSION: A SYNERGISTIC STUDY OF FUNDAMENTAL SOLAR SYSTEM PROCESSES." 11 July 2016. Annual Meeting of the Lunar Exploration Analysis Group (2016) 12 Aug. 2017 https://www.hou.usra.edu/meetings/leag2016/pdf/5036.pdf

[67] "Ancient Polar Ice Reveals Tilting of Earth's Moon." 23 Mar. 2016. NASA 12 Aug. 2017 https://www.nasa.gov/ames/feature/lunar-polar-ice-reveals-tilting-axis-of-earth-s-moon

[68] Steigerwald, Bill. "NASA Research Gives New Insights into How the Moon Got 'Inked'." 29 Apr. 2016. NASA 13 Aug. 2017 https://www.nasa.gov/feature/goddard/2016/lunar-swirls

by small asteroids than previous models had predicted, which implies that a future lunar base may have to be made sturdier than anticipated to withstand secondary debris impacts moving at up to 500 m per second.[69]

China's Chang'e 5-TI test capsule returned to lunar orbit in the week of 12 January 2015.[70] Launched on 23 October 2014, the prototype sample-return capsule reached the Moon within a day, circling it before returning to eject its sample capsule at a higher than average velocity into Earth's atmosphere. Following the successful release of the capsule on 1 November 2015, Chang'e 5-TI began making its way to the Earth-Moon Lagrange (L2) point on the opposite side of the Moon. Reaching L2 by late November 2015, the service module then completed three circles around that point prior to returning to lunar orbit.[71] In addition to testing critical breaking manoeuvres, the Chang'e 5-TI carries a camera system that will help to identify future landing sites for the Chang'e 5 robotic sample-return mission planned for launch in the second half of 2017.[72] This later Chang'e 5 mission will involve a soft landing on the Moon and the collection of 200 g of samples prior to bringing them to Earth.

The Google Lunar X Prize is a competition for a grand prize of $20 million and a second prize of $5 million for the first two privately funded teams to safely land on the Moon, travel at least 500 m across its surface and send high-definition video, images and data back to the Earth. In late 2013, the X Prize Foundation and Google announced a series of interim 'milestone' prizes available to assist the competing teams in accessing finance at a critical point in their mission timeline and to raise public excitement and support for the teams. The Google Lunar X Prize competition was extended to the end of 2017, after two teams had met the X Prize Foundation's 16 December 2014 stipulation requiring at least one team to have made launch arrangements by the end of 2015.[73] Only five teams remained in the Lunar X Prize competition by the end of 2016, out of an initial 33 entrants, as competitors had until the end of the year to submit their own launch documentation to remain in the competition.[74] The Israeli team, SpaceIL was the first to be confirmed by the foundation on 7 October 2015,[75] followed by the American team Moon Express on 8 December

[69] Jones, Nancy and Bill Steigerwald. "Earth's Moon Hit by Surprising Number of Meteoroids." 13 Oct. 2016. NASA 13 Aug. 2017 https://www.nasa.gov/press-release/goddard/2016/lro-lunar-cratering

[70] David, Leonard. "Chinese Spacecraft Enters Orbit around the Moon." 20 Jan. 2015. SpaceNews 8 Jan. 2016 http://spacenews.com/chinese-spacecraft-enters-orbit-around-the-moon/

[71] "Service Module of Chinese Probe Successfully Enters Lunar Orbit." 11 Jan. 2015. Sputnik News 22 Jan. 2016 http://sputniknews.com/science/20150111/1016776437.html

[72] Jones, Andrew. "China's Chang'e-5 Moon sample return mission on course for 2017 launch." 24 Oct. 2016. GBTimes 13 Aug. 2017 http://gbtimes.com/china/chinas-change-5-moon-sample-return-mission-course-2017-launch

[73] Foust, Jeff. "Google Lunar X Prize Extends Competition Deadline." 17 Dec. 2014. SpaceNews 23 July 2015 http://spacenews.com/google-lunar-x-prize-extends-competition-deadline/

[74] Foust, Jeff. "X Prize Verifies Moon Express Launch Contract." 8 Dec. 2015. SpaceNews 6 Jan. 2016 http://spacenews.com/x-prize-verifies-moon-express-launch-contract/

[75] Foust, Jeff. "Israeli X Prize Team Announces Launch Contract for Lunar Mission." 7 Oct. 2015. SpaceNews 5 Jan. 2016 http://spacenews.com/israeli-x-prize-team-announces-launch-contract-for-lunar-mission/

2015.[76] The international team Synergy Moon was the next team to have its launch contract verified by the X Prize Foundation on 30 August 2016, followed by the Indian team TeamIndus on 1 December 2016, and the German team PT Scientists which announced its launch contract on 29 November 2016 to be confirmed later in the year.[77] The American team Astrobotic Technology separated from the X Prize competition by choosing not to secure a launch at the end of 2016; Astrobotic Technology intends to fly its first mission to the Moon in 2019.[78]

2.3.3 Mars Science

For decades the focus for Mars science has been the investigation of the planet's habitability, in a search for the presence of water. The collected data continues to suggest that Mars was once partially covered by large oceans and that life could have been possible in many locations on the planet's surface.

ESA's Mars Express orbiter, launched in June 2003, continued its high-resolution imaging mission of Mars, including the mapping of its mineral composition and atmosphere and determining the structure of the subsurface to a depth of a few kilometres, the effect of the atmosphere on the surface and the interaction of the atmosphere with the solar winds. The images taken by the high-resolution stereo camera on Mars Express in mid-2015 helped to reveal in extraordinary detail complex features such as the Noctis Labyrinthus region on the western edge of Valles Marineris, an entire network of plateaus and fractures spanning around 1200 km that suggests many episodes of tectonic stretching and volcanic activity in Mars' Tharsis region.[79] Moreover, images of the Arda Valles, north of the Holden crater and Ladon Valles, revealed a dendritic drainage system carved by vast volumes of water that once flowed from the southern highlands.[80] Images from Mars' Colles Nili region showed the erosional remnants of a former plateau, whose layered deposits gently sloped away from the sides of the hills and the series of ridges and troughs found around the mounds and inside some of the impact craters on the channel floors. These are thought to be associated with buried ice that has since been covered over by wind-blown dust and local debris.[81] And images of ridges and troughs in the western part

---

[76] Foust, Jeff. "Moon Express Buys Rocket Lab Launches for Lunar Missions." 1 Oct. 2015. SpaceNews 5 Jan. 2016 http://spacenews.com/moon-express-buys-rocket-lab-launches-for-lunar-missions/

[77] Foust, Jeff. "Indian X Prize team secures launch contract with ISRO." 2 Dec. 2016. SpaceNews 4 Jan. 2017 http://spacenews.com/indian-x-prize-team-secures-launch-contract-with-isro/

[78] Thornton, John. "Graduating from the Google Lunar X Prize." 19 Dec. 2016. SpaceNews Magazine 14 Aug. 2017 http://www.spacenewsmag.com/commentary/graduating-from-the-google-lunar-x-prize

[79] "MARTIAN LABYRINTH." 28 Jan. 2016. ESA 24 May 2017 http://www.esa.int/Our_Activities/Space_Science/Mars_Express/Martian_labyrinth

[80] "FOOTPRINTS OF A MARTIAN FLOOD." 18 Feb. 2016. ESA 24 May 2017 http://www.esa.int/Our_Activities/Space_Science/Mars_Express/Footprints_of_a_martian_flood

[81] "BURIED GLACIERS ON MARS." 13 Oct. 2016. ESA 24 May 2017 http://www.esa.int/Our_Activities/Space_Science/Mars_Express/Buried_glaciers_on_Mars

of Acheron Fossae, 1000 km north of Olympus Mons and other volcanic giants in the Tharsis bulge, suggest a complex history, as its pattern of cross-cutting faults imply that the region experienced stresses from different directions over time; as Acheron Fossae has been likened to Earth's continental rift system and is associated with plate tectonics, rifts are important for studies of the general evolution of the crust as well as the thermal evolution of the deeper subsurface.[82] Additionally, plasma and solar wind measurements from Mars Express' Analyzer of Space Plasmas and Energetic Atoms (ASPERA-3) plasma instrument suite and the Mars Advanced Radar for Subsurface and Ionospheric Sounding (MARSIS) instrument revealed that the cloud-like plume that reached an altitude of 250 km in 2012 was likely the result of a coronal mass ejection (CME) from the Sun that was large enough to impact Mars and increase the escape of plasma from the planet's atmosphere – adding an important angle on the potential role of space weather in how Mars may have lost much of its atmosphere in the past.[83]

The ExoMars programme is a joint endeavour between ESA and the Russian Space Agency, Roscosmos, to addresses whether life has ever existed on Mars. The programme is comprised of two missions: the first began in 2016 with the launch of ESA's Trace Gas Orbiter (TGO) and the Entry, Descent and Landing Demonstrator Module (EDM) 'Schiaparelli'; and the second mission will take place in 2020 and comprises a rover and surface science platform. Both missions will be launched to Mars using Roscosmos' Proton launcher.[84] The first ExoMars mission began its 7-month journey to Mars on 14 March 2016; on reaching a distance of 900,000 km from the planet on 16 October 2016, the TGO and Schiaparelli separated, with the TGO moving into a 4-day elliptical orbit around Mars, while Schiaparelli veered into the atmosphere to descend to Mars' surface.[85] After separation, the TGO began a series of complex aerobraking manoeuvres that will continue over the course of 13 months to lower its elliptical orbit of 250 km by 98,000 km to a circular orbit of 00 km before its main scientific mission to analyse rare gasses in the atmosphere begins. It will also act as a data relay for surface rovers, providing two to three over-flights of each Mars rover every day to send signals back to Earth. Despite being placed into safe mode for a brief period after faulty configuration of the TGO's main engine caused a temporary glitch during preliminary testing, the TGO was performing nominally by the end of 2016.[86] Nevertheless, on 19 October 2016, Schiaparelli

[82] "A RECORD OF ANCIENT TECTONIC STRESS ON MARS." 3 Nov. 2016. ESA 24 May 2017 http://www.esa.int/Our_Activities/Space_Science/Mars_Express/A_record_of_ancient_tectonic_stress_on_Mars

[83] "ARE MYSTERY MARS PLUMES CAUSED BY SPACE WEATHER?" 23 May 2016. ESA 24 May 2017 http://www.esa.int/Our_Activities/Space_Science/Mars_Express/Are_mystery_Mars_plumes_caused_by_space_weather

[84] "What is ExoMars?" 29 Mar. 2017. ESA 26 May 2017 http://www.esa.int/Our_Activities/Space_Science/ExoMars/What_is_ExoMars

[85] ExoMars on its way to solve the Red Planet's mysteries." 14 Mar. 2016. ESA 26 May 2017 http://www.esa.int/Our_Activities/Space_Science/ExoMars/ExoMars_on_its_way_to_solve_the_Red_Planet_s_mysteries

[86] "Skimming an alien atmosphere." 16 Dec. 2016. ESA 26 May 2017 http://www.esa.int/Our_Activities/Operations/Skimming_an_alien_atmosphere

experienced an anomaly as it descended into Mars's atmosphere; while its radar Doppler altimeter functioned correctly and the measurements were included in the guidance, navigation and control system, a saturation of the inertial measurement unit (IMU) occurred shortly after the parachute deployed, causing its navigation system to estimate that it was already below ground level and triggering the final landing stages, while the vehicle was at an altitude of around 3.7 km. ESA has called for an external independent inquiry board to conduct a technical investigation into the anomaly and expects to provide a full report early in 2017.[87]

NASA's Mars Odyssey mission, launched on 7 April 2001, is the longest-operating spacecraft to be sent to Mars. The Mars Odyssey mission marked a turning point for NASA in Mars exploration, after the failure of two preceding missions in 1999 prompted an overhaul of NASA's exploration plans. While its prime mission to make the first global map of the amount and distribution of numerous chemical elements and minerals on Mars surface was completed in August 2004, its mission had been extended for 12 additional years by 2016; Mars Odyssey is also a communication relay for rovers and landers on the planet.[88] In March 2016, NASA published its most detailed gravity map of Mars, derived from Doppler and range tracking data collected by NASA's Mars Odyssey, Mars Global Surveyor and the Mars Reconnaissance Orbiter spacecraft. The improved resolution allows researchers to better interpret how the crust of the planet has changed over Mars' history in many regions. Moreover, by analysing tides in the crust and mantle caused by the gravitational pull of the Sun and the two moons of Mars, researchers confirmed that Mars has a liquid outer core of molten rock. Researchers also determined that during the winter season for both hemispheres within Mars' 11-year orbit around the Sun, approximately 3–4 trillion tons of carbon dioxide, i.e. 12–16% of the mass of the entire Martian atmosphere, freezes out of the atmosphere onto the northern and southern polar caps, respectively.[89] On 26 December 2016, the Mars Odyssey spacecraft briefly put itself into safe mode due to uncertainty about its orientation with regard to Earth and the Sun – similar to a fault experienced in December 2013; following a reset of the inertial measurement unit and the circuit card that serves as interface between that sensor, the flight software and the star tracker, for determining spacecraft attitude, the orbiter's knowledge of its orientation was restored. By 30 December 2016, Mars Odyssey had resumed communication relay assistance to Mars rovers, with science observations of Mars following shortly afterward.[90]

---

[87] "Schiaparelli landing investigation makes progress." 23 Nov. 2016. ESA 26 May 2017 http://www.esa.int/Our Activities/Space Science/ExoMars/Schiaparelli_landing_investigation_makes_progress

[88] "2001 Mars Odyssey." NASA 29 May 2017 https://mars.jpl.nasa.gov/programmissions/missions/present/odyssey/

[89] "New Gravity Map Gives Best View Yet Inside Mars." 21 Mar. 2016. NASA 29 May 2017 https://mars.jpl.nasa.gov/news/new-gravity-map-gives-best-view-yet-inside-mars

[90] "NASA Mars Odyssey Orbiter Resumes Full Operations." 28 Dec. 2016. NASA 29 May 2017 https://mars.jpl.nasa.gov/news/2016/mars-odyssey-mission-status-report-orbiter-recovering-from-precautionary-pause-in-activity

NASA's Mars Reconnaissance Orbiter (MRO) continued operations in 2016, providing data for the purpose of determining whether life has ever existed on Mars, characterizing the climate and geology and preparing for future human exploration. In the decade following its insertion into Mars' orbit in March 2006, the MRO completed 45,000 orbits of Mars[91]; while its primary science mission ended in November 2008, the MRO is currently in its third extended mission, which began in October 2012.[92] In March 2016, new evidence from the MRO's Compact Reconnaissance Imaging Spectrometer for Mars (CRISM) instrument suggested that Mars' Sisyphi Montes region, studded with flat-topped mountains, formed when volcanoes erupted beneath an ice sheet billions of years ago; the research helps show that there was extensive ice on ancient Mars and adds information about an environment combining heat and moisture that might have provided favourable conditions for microbial life.[93] Moreover, 'radargrams' produced with data from MRO's Shallow Subsurface Radar (SHARAD) instrument confirmed previous 2003 and 2007 models that indicate that Mars' most recent ice age ended about 400,000 years ago, as the poles began to cool and thicken relative to Mars' equator.[94] Other clues about Mars' climate history were revealed when images of the northern Arabia Terra region from the Context Camera and High Resolution Imaging Science Experiment camera showed the formation of runoff lakes and streams that appeared roughly billion years after an earlier era of wet conditions on ancient Mars.[95]

On 5 August 2012, NASA's Mars Science Laboratory (MSL) rover 'Curiosity' began its mission to address whether Mars ever had the right environmental conditions to support microbial life. It seeks to determine whether life ever arose on Mars and characterize its climate and geology to help prepare for human exploration. Its biological objectives are to determine the nature and inventory of organic carbon compounds, conduct an inventory of the chemical building blocks of life and identify features that may represent the effects of biological processes; its geological and geochemical objectives are to investigate the chemical, isotopic and mineralogical composition of Martian geological materials and to interpret the processes that have formed and modified rocks and soils; its planetary process objectives include assessing 4-billion-year timescale atmospheric evolution processes and determining the present state, distribution and cycling of water and carbon dioxide; and its surface radiation objective is to characterize the broad spectrum of surface radiation, including galactic cosmic radiation, solar proton events and secondary neutrons.[96]

---

[91] "Ten Years of Discovery by Mars Reconnaissance Orbiter." 9 Mar. 2016. NASA 30 May 2017 https://mars.nasa.gov/news/ten-years-of-discovery-by-mars-reconnaissance-orbiter

[92] "Mission Timeline." NASA 30 May 2017 https://mars.nasa.gov/mro/mission/timeline/

[93] "Found: Clues about Volcanoes Under Ice on Ancient Mars." 3 May 2016. NASA 30 May 2017 https://mars.nasa.gov/news/found-clues-about-volcanoes-under-ice-on-ancient-mars

[94] "NASA Radar Finds Ice Age Record in Mars' Polar Cap." 26 May 2016. NASA 30 May 2017 https://mars.nasa.gov/news/nasa-radar-finds-ice-age-record-in-mars-polar-cap

[95] "Some Ancient Mars Lakes Came Long After Others." 15 Sept. 2016. NASA 30 May 2017 https://mars.nasa.gov/news/some-ancient-mars-lakes-came-long-after-others

[96] "Mars Science Laboratory: Mission Objectives." NASA 31 May 2017 https://mars.nasa.gov/msl/mission/science/objectives/

Curiosity began its third Martian year on 11 May 2016, having recorded environmental patterns through two full cycles of Martian seasons while crossing the Gale Crater's rugged terrain; each year on Mars lasts 668.6 sols, with each sol lasting 39.6 min longer than an Earth day.[97] The rover continues to redefine researchers' assumptions of Mars' volcanic history after an analysis of a rock sample using an X-ray diffraction instrument detected significant amounts of tridymite, created when large amounts of silica are heated at extremely high temperatures; the finding suggests that Mars once had explosive volcanoes that led to the presence of the mineral.[98] Another surprise came when a study of active sand dunes at the Bagnold Dunes of Mars' Mount Sharp showed sinuous crest lines of ripples similar to underwater ripples on Earth, with each ripple steeper than the face on the other side, instead of impact ripples formed by wind-carried sand grains colliding with other sand grains along the ground; this led researchers to believe that the mechanism forming ripples on Mars was its thicker atmosphere in the past which worked like a fluid to form smaller wind-drag ripples.[99] On 2 July 2016, Curiosity placed itself into safe mode as a precaution after experiencing an unexpected mismatch between camera software and data-processing software in the main computer; the rover had also gone into safe mode three times in 2013. The rover was taken out of safe mode in the following week and resumed full operation on 11 July 2016.[100] Curiosity began its second 2-year mission extension on 1 October 2016, to explore key sites on lower Mount Sharp including a ridge capped with material rich in the iron-oxide mineral hematite and an exposure of clay-rich bedrock; and additional extensions for exploring farther up Mount Sharp appear on the horizon.[101]

NASA's Mars Atmosphere and Volatile EvolutioN (MAVEN) mission aims to explore the planet's upper atmosphere, ionosphere and interactions with the Sun and solar wind, which will be used to determine the role that the loss to space of volatile compounds from the Mars atmosphere has played in the history of Mars' habitability.[102] In March 2016, new research published in the journal Geophysical Research Letters showed how the Comet Siding Spring plunged Mars' magnetic field into chaos as it passed within a distance of 140,000 km in October 2014. While sensitive equipment aboard MAVEN was turned off during the flyby, its magnetometer remained on to observe how the charged plasma in Mars upper atmosphere interacted with Siding Spring's charged coma as it washed over the planet, flooding it

---

[97] "Second Cycle of Martian Seasons Completing for Curiosity Rover." 11 May 2016. NASA 1 June 2017 https://mars.nasa.gov/news/second-cycle-of-martian-seasons-completing-for-curiosity-rover

[98] "NASA Scientists Discover Unexpected Mineral on Mars." 22 June 2016. NASA 1 June 2017 https://mars.nasa.gov/news/nasa-scientists-discover-unexpected-mineral-on-mars

[99] "NASA Rover's Sand-Dune Studies Yield Surprise." 30 June 2016. NASA 1 June 2017 https://mars.nasa.gov/news/nasa-rovers-sand-dune-studies-yield-surprise

[100] "Curiosity Mars Rover Resumes Full Operations." 11 July 2016. NASA 6 June 2017 https://mars.nasa.gov/news/curiosity-mars-rover-resumes-full-operations

[101] "NASA's Curiosity Rover Begins Next Mars Chapter." 3 Oct. 2016. NASA 6 June 2017 https://mars.nasa.gov/news/nasas-curiosity-rover-begins-next-mars-chapter

[102] "MAVEN." University of Colorado Boulder 14 Jan. 2013. http://lasp.colorado.edu/home/maven/

with additional charged particles and temporarily merging the comet's magnetic field with Mars' weak field generated in Mars' upper atmosphere; the effect of the plasma tide was similar to that of a strong but short-lived solar storm and likely fuelled a temporary surge in the amount of gas escaping from Mars' upper atmosphere.[103] New images from MAVEN's Imaging Ultraviolet Spectrograph (IUVS) published on 19 October 2016 provided an unprecedented ultraviolet view of Mars' nightside atmosphere verifying the presence of a 'nightglow' emitted from nitric oxide; the phenomenon begins with chemical reactions on Mars' dayside as ultraviolet light from the Sun breaks down carbon dioxide and nitrogen in the upper atmosphere that are then blown around the planet and eventually descend to lower altitudes on the nightside where nitrogen and oxygen then collide to form the nitric oxide molecules that generate the ultraviolet glow from the release of additional energy. Prior to these images, the nitric oxide glow on Mars was predicted by researchers, and its presence had been detected in prior missions.[104]

ISRO's Mars Orbiter Mission (MOM) 'Mangaalayan' is the first Asian interplanetary space probe to reach the planet Mars, joining the ranks of ESA, NASA and Roscosmos; ISRO is also the first space agency to have been successful in its maiden attempt and at the frugal total cost of ₹450 crore ($73 million).[105] MOM lifted into space on 5 November 2013 and inserted in Mars' orbit on 24 September 2014. While MOM's primary mission is intended as a technology demonstration, its secondary mission is to study Mars' surface features, morphology, mineralogy and atmosphere with its suite of five indigenous scientific instruments: the Mars Colour Camera, the Methane Sensor for Mars, the Thermal Infrared Imaging Spectrometer, the Mars Exospheric Neutral Composition Analyser and the Lyman Alpha Photometer.[106] While ISRO announced the first science results of the mission on 11 November 2015, on 8 December 2016, it was made known that the highly anticipated measurements of atmospheric methane (a biomarker whose presence indicates the current or historical presence of life) from MOM's Methane Sensor for Mars (MSM) instrument might never arrive due to a flaw in the sensor design of the Fabry-Pérot interferometer, which didn't allow for methane data alone to be extracted from carbon dioxide and other gasses in Mars' atmosphere.[107] Nevertheless,

[103] "Close Comet Flyby Threw Mars' Magnetic Field Into Chaos." 9 Mar. 2016. NASA 7 June 2017 https://www.nasa.gov/feature/goddard/2016/close-comet-flyby-threw-mars-magnetic-field-into-chaos

[104] "NASA's MAVEN Mission Gives Unprecedented Ultraviolet View of Mars." 17 Oct. 2016. NASA 7 June 2017 https://mars.nasa.gov/news/2016/nasas-maven-mission-gives-unprecedented-ultraviolet-view-of-mars

[105] "Mars Mission: India creates history as Mangalyaan successfully enters Mars orbit in first attempt." 24 Sept. 2014. The Economic Times 22 Mar. 2015 http://economictimes.indiatimes.com/news/science/mars-mission-india-creates-history-as-mangalyaan-successfully-enters-mars-orbit-in-first-attempt/articleshow/43299562.cms

[106] Laxman, Srinivas. "Mars Orbiter Mission activates all science instruments as NASA, ISRO form joint Mars working group." 1 Oct. 2014. Planetary.org 22 Mar. 2015 http://www.planetary.org/blogs/guest-blogs/2014/10010914-mars-orbiter-mission.html

[107] Mukunth, Vasudevan. "ISRO Mars Orbiter Mission's Methane Instrument Has a Glitch." 15 Dec. 2016. The Wire 8 June 2017 https://thewire.in/85859/methane-isro-msm-spectroscopy/

the MSM is still effective as an albedo mapper and in measuring reflected sunlight.[108] Subsequent science data collected by all instruments of the spacecraft is still being studied and prepared for publication. Lastly, plans for a second MOM spacecraft are already underway with ISRO's 'announcement of opportunity'; the follow-up mission is expected to be launched in March 2018.[109]

2.3.4 *Mercury Science*

The BepiColombo programme is a joint endeavour between ESA and the Japanese space agency, JAXA, to measure and understand the composition, geophysics, atmosphere, magnetosphere and history of the planet Mercury. BepiColombo will be Europe's first mission to Mercury and is based on two spacecraft: the ESA-led Mercury Planetary Orbiter (MPO) and the JAXA-led Mercury Magnetospheric Orbiter (MMO). The MPO will use its instrument suite of 11 experiments and instruments to study the planet's geology, composition, inner structure and exosphere, while the MMO has five experiments and instruments to study the planet's magnetic field, atmosphere, magnetosphere and inner interplanetary space.[110] The mission is expected to be launched in October 2018, having been delayed by 6 months from its earlier intended April 2018 window, after a major electrical problem in the missions' Mercury Transfer Module (MTM) was encountered prior to a thermal test. While the postponement will have no impact on the science return of the mission, it will extend the flight time to 7.2 years, arriving in December 2025 – 1 year later than previously anticipated.[111] BepiColombo will follow in the footsteps of NASA's MErcury Surface, Space ENvironment, GEochemistry and Ranging (MESSENGER) mission that launched in August 2004 and orbited Mercury from March 2011 until it impacted the surface of Mercury on 30 April 2015.[112]

[108] Klotz, Irene. "India's Mars Orbiter Mission Has a Methane Problem." 8 Dec. 2016 Space.com 8 June 2017 http://www.space.com/34943-india-mars-orbiter-mission-methane-detector-flawred-planet.html

[109] Mukunth, Vasudevan. "ISRO Plans Return to Mars with Mangalyaan 2.0." 12 Aug. 2016. The Wire 8 June 2017 https://thewire.in/58415/mars-orbiter-isro-planetary/

[110] "Factsheet." 1 Dec. 2016. ESA 12 June 2017 http://sci.esa.int/bepicolombo/47346-fact-sheet/

[111] "BEPICOLOMBO LAUNCH RESCHEDULED FOR OCTOBER 2018." 25 Nov. 2016. ESA 12 June 2017 http://sci.esa.int/bepicolombo/58591-bepicolombo-launch-rescheduled-for-october-2018/

[112] Nowakowski, Tomasz. "BepiColombo mission to Mercury on track for April 2018 launch." 21 July 2016. Spaceflight Insider 12 June 2017 http://www.spaceflightinsider.com/missions/solar-system/bepicolombo-mission-mercury-april-2018-launch/

2.3.5 *Jupiter Science*

NASA's Juno mission aims to answer questions about Jupiter's origins by observing its gravity and magnetic fields, atmospheric dynamics and composition and evolution. Launched on 5 August 2011, Juno inserted into an elliptical polar orbit around Jupiter on 4 July 2016 and aimed to conduct 37 science orbits of Jupiter over the course of 20 months, before ending its mission in February 2018.[113] Its first close flyby took place on 27 August 2016, with Juno reaching a distance of 4200 km from Jupiter's clouds, this time with its suite of eight science instruments activated to transmit data over the following days.[114] The next close flyby took place in safe mode on 19 October 2016, after the spacecraft's telemetry indicated that two important helium check valves in its main engine had not operated as expected, opening for several minutes rather than a few seconds as intended, during the execution of a period reduction manoeuvre (PRM); while Juno rebooted successfully in the following week and is in healthy operation, no science data was collected on that flyby.[115] The next flyby took place on 11 December 2016, with data collected from seven of Juno's eight science instruments – not including its Jovian Infrared Auroral Mapper (JIRAM) which was receiving a software patch at the time.[116] NASA had decided to also forego the PRM in that third flyby, finding that it would not have affected the quality of the science collected by Juno during its closest approach; however, as the manoeuvre would have reduced Juno's orbital period from 53.4 days to 14 days, maintaining its current path will result in around 12 science orbits instead of 37, unless the mission is extended beyond February 2018.

The next spacecraft to orbit Jupiter will be ESA's Jupiter Icy moon Explorer (JUICE) in development under ESA's Cosmic Vision 2015–2025 plan. Its foreseen launch date is in 2022 aboard an Ariane 5 launcher, and it should arrive at the Jovian planet in 2030. The nearly 5500 kg spacecraft will make a careful investigation of Jupiter's three biggest moons, using the gravity of Jupiter to initiate a series of close flybys around Callisto, Ganymede and Europa, and then finally settle into an orbit around Ganymede for an 8-month study.[117] As all three moons are suspected of having oceans of water beneath their icy crusts, scientists are trying to understand whether there is any possibility that these moons could host microbial life.[118] On 16

---

[113] "Mission to Jupiter – Juno – ABOUT THE MISSION." NASA 12 June 2017 https://www.jpl.nasa.gov/missions/juno/

[114] "NASA's Juno Successfully Completes Jupiter Flyby." 27 Aug. 2016. NASA 12 June 2017 https://www.missionjuno.swri.edu/news/juno-completes-jupiter-flyby

[115] "Juno Spacecraft in Safe Mode for Latest Jupiter Flyby. Scientists Intrigued by Data from First Flyby."19 Oct. 2016. NASA 12 June 2017 https://www.missionjuno.swri.edu/news/mission-status-report-10-19-2016

[116] "NASA Juno Mission Prepares for December 11 Jupiter Flyby." 9 Dec. 2016. NASA 12 June 2017 https://www.missionjuno.swri.edu/news/dec11-jupiter-flyby

[117] De Selding, Peter B. "Airbus To Build ESA's Jupiter-bound Juice Orbiter." 17 July 2015. SpaceNews 11 Jan. 2016 http://spacenews.com/airbus-to-build-esas-jupiter-bound-juice-orbiter/

[118] "ESA Selects 1bn-Euro JUICE Probe to Jupiter." 2 May 2012. BBC News 27 Mar. 2014 http://www.bbc.com/news/science-environment-17917102

July 2015, ESA selected Airbus D&S for a €350.8 million ($374 million) contract to build the JUICE Orbiter. The contract was formalized on 9 December 2015.[119]

2.3.6 *Saturn Science*

The Cassini-Huygens mission, a joint NASA, ESA and ASI mission, was launched in 1997. Reaching Saturn in 2004, Cassini went on to drop the Huygens probe onto Saturn's moon, Titan. From there, the Cassini Solstice Mission was supposed to end in June 2008; however, funding was provided to allow continued operation to provide new insights on Saturn and its moons.[120] The Cassini mission began its final year on 14 September 2016, after more than 12 years of studying Saturn, its rings and moons.[121] On 23 January 2016, Cassini took its second of five large propulsive manoeuvres using gravity assists from Titan to take the spacecraft from an equatorial orbit around Saturn to an inclination above its poles for Cassini's 'Grande Finale' that will involve passes through Saturn's rings and will end with a plunge into its atmosphere.[122] Cassini's final mission phase began shortly before its first close dive through the centre of Saturn's outermost F ring on 4 December 2016; the spacecraft will make 20 of these passes until 22 April 2017 and then transition between Saturn's innermost ring and atmosphere for an additional 22 passes, ending its mission on 15 September 2017.[123]

A new study published in the Journal of Geophysical Research independently confirmed that Ligeia Mare, the second largest sea on Titan, is rich in methane rather than ethane, which is produced in abundance in Titan's atmosphere. A possible explanation for the sea's nearly pure methane composition could be that the sea is regularly replenished with fresh methane rainfall or that the denser ethane somehow drains into the larger adjacent sea, Kraken Mare. Using radar data collected from a 2013 experiment that bounced radio signals off Ligeia Mare, researchers were also able to find that its seabed is likely covered by a sludge layer of organic-rich compounds.[124] Another recent paper in the journal Geophysical Research Letters described how researchers using data from Cassini's radar imager were able to find the first direct evidence of liquid-filled channels on Titan, along

[119] De Selding, Peter B. "ESA, Airbus Formalize Jupiter Icy Moons Contract." 11 Dec. 2015. SpaceNews 11 Jan. 2016 http://spacenews.com/esa-airbus-formalize-jupiter-icy-moons-contract/

[120] Mason, Betsy. "Cassini Gets Life Extension to Explore Saturn Until 2017." 3 Feb. 2010. WIRED 18 Dec. 2012 http://www.wired.com/wiredscience/2010/02/cassini-life-extension-2017/

[121] "Cassini Begins Epic Final Year at Saturn." 14 Sept. 2016. NASA 8 June 2017 https://saturn.jpl.nasa.gov/news/2939/cassini-begins-epic-final-year-at-saturn/

[122] "Cassini Heads for 'Higher Ground' at Saturn." 24 Jan. 2016. NASA 8 June 2017 https://saturn.jpl.nasa.gov/news/2801/cassini-heads-for-higher-ground-at-saturn/

[123] Dyches, Preston. "Cassini Makes First Ring-Grazing Plunge." 4 Dec. 2016. NASA 9 June 2017 https://saturn.jpl.nasa.gov/news/2974/cassini-makes-first-ring-grazing-plunge/

[124] Dyches, Preston and Markus Bauer. "Cassini Explores a Methane Sea on Titan." 25 Apr. 2016. NASA 9 June 2017 https://saturn.jpl.nasa.gov/news/2897/cassini-explores-a-methane-sea-on-titan/

with the first observation of canyons hundreds of meters deep, indicating that it was formed from a process that was active for a long time or that the channels had eroded down much faster than other areas on Titan's surface.[125]

2.3.7 Solar Observation

Continued observation of the Sun's external activity has the benefit of improving our understanding of its interior, its corona, the monitoring of solar wind and its consequences on Earth and its neighbouring planets. Coronal mass ejections (CMEs) from the Sun emit surges of charged particles in directions that may cross Earth's path and can damage satellites, impede space-based services and affect the terrestrial electrical infrastructure.

ESA's Project for Onboard Autonomy (PROBA)-2 microsatellite mission tracks spikes in CMEs ejecting from the Sun that have previously been seen to just skim Earth, typically bringing with them a burst of radio energy.[126] Building on nearly 8 years of successful PROBA-1 experience, the PROBA-2 continues as part of ESA's In-Orbit Technology Demonstration Programme, whose missions are dedicated to the demonstration of innovative technologies; its payload carries 17 new technological developments and four scientific experiments.[127] Launched in November 2009, PROBA-2's mission was once again extended on 22 November 2016; its third operational mission will continue from 1 January 2017 to 31 December 2018.[128] On 9 March 2016, PROBA-2 was able to observe a partial solar eclipse with its SWAP imager from its 800-km-altitude polar orbit; the SWAP views the solar disc at extreme ultraviolet wavelengths to capture the turbulent surface of the Sun and its swirling corona.[129] As the Sun's 11-year activity cycle drew closer to its minimum, where the number of sun spots, active regions, solar flares and eruptions diminish, PROBA-2 was in the right position to view an annular solar eclipse on 1 September 2016. Orbiting the Earth once every 90 min, PROBA-2's SWAP imager and its LYRA radiometer, which measures solar irradiance, managed to view four partial eclipses.[130] ESA is also developing the PROBA-3, as a pair of satellites

[125] Dyches, Preston. "Cassini Finds Flooded Canyons on Titan." 9 Aug. 2016. NASA 9 June 2017 https://saturn.jpl.nasa.gov/news/2927/cassini-finds-flooded-canyons-on-titan/

[126] "Small Sun-Watcher Proba-2 Offers Detailed View of Massive Solar Eruption." 9 June 2011. ESA 17 Apr. 2013 http://www.esa.int/Our_Activities/Technology/Small_Sun-watcher_Proba-2_offers_detailed_view_of_massive_solar_eruption

[127] "About PROBA-2." 17 Dec. 2012. ESA 13 June 2017 http://www.esa.int/Our_Activities/Space_Engineering_Technology/Proba_Missions/About_Proba-2

[128] "Two-year extensions confirmed for ESA's science missions." 22 Nov. 2016. ESA 12 June 2017 http://sci.esa.int/director-desk/58589-two-year-extensions-confirmed-for-esa-s-science-missions/

[129] "Pacific solar eclipse seen from Proba-2." 9 Mar. 2016. ESA 13 June 2017 http://www.esa.int/spaceinvideos/Videos/2016/03/Pacific_solar_eclipse_seen_from_Proba-2

[130] D'Huys, Elke. "Another eclipse!" 13 June 2016. ESA/ROB 13 June 2017 http://proba2.sidc.be/node/316

maintaining a fixed configuration to form a 150-m-long solar chronograph to study the Sun's faint corona closer to the solar rim than previously achieved.[131] The two satellites, a coronagraph spacecraft and an occulter, will be launched together in late 2018 into a highly elliptical tandem orbit, repetitively demonstrating acquisition, rendezvous, proximity operations, formation flying, coronagraph operations, separation and convoy flying.[132]

NASA's Solar Dynamics Observatory (SDO) mission seeks to determine how the Sun's magnetic field is generated and structured and how this stored magnetic energy is released in the form of the solar wind, energetic particles and variations in the solar irradiance. The spacecraft is comprised of three scientific experiments: the Atmospheric Imaging Assembly (AIA), the EUV Variability Experiment (EVE) and the Helioseismic and Magnetic Imager (HMI).[133] Located in a geosynchronous orbit around the Earth, SDO's global view of the Sun facilitates research that focuses on the previously unrecorded real fine structure of the star.[134] SDO continuously observed the Sun's activity throughout 2016, capturing images of coronal holes and solar flares and occasionally catching the transit of a planet passing in front of the star. On 9 May 2016, SDO (along with the NASA/ESA SOHO mission and the JAXA/NASA/UK Hinode mission) studied the planet Mercury as it transited between Earth and the Sun – an event that happens 13 times or so in a century. The transit helped SDO to both align and calibrate its space instruments, enabling researchers to mitigate the effect of stray light in SDO's instruments that should otherwise have viewed the profile of Mercury transiting the Sun without a glow surrounding its circumference. SOHO used the transit opportunity to measure the Sun's rotation axis.[135] On 2 August 2016, SDO entered inertial mode as the Moon transited the Sun; while its HMI and EVE instruments came back online within 2 days of the transit, a temporary glitch in SDO's AIA instrument delayed full science mode operations until 10 August 2016.[136] And in research published on 11 October 2016, scientists using SDO and IRIS data were able to observe certain frequencies of solar seismic waves channelling upwards through the chromosphere and corona atmospheric layers into the Sun's photosphere. The technique gives scientists a new tool to understand the Sun's lower atmosphere and also might help to address a long-standing question in solar physics regarding excess heat in the Sun's corona,

---

[131] "About PROBA3." 19 Nov. 2012. ESA 1 Apr. 2014 http://www.esa.int/Our_Activities/Technology/Proba_Missions/About_Proba-3

[132] "Fact Sheet." ESA 13 June 2017 http://esamultimedia.esa.int/docs/Proba/Proba-3_fact-sheet_final.pdf

[133] "SDO | Solar Dynamics Observatory." NASA Goddard Space Flight Center 4 Mar. 2013 http://sdo.gsfc.nasa.gov/

[134] SpaceNews Staff. "NASA Boasts Big Results from 5-minute Spaceflight." 28 Jan. 2013. SpaceNews 4 Mar. 2013 http://www.spacenews.com/article/nasa-boasts-big-results-from-5-minute-spaceflight

[135] "Satellites to See Mercury Enter Spotlight on May 9." 3 May 2016. NASA 13 June 2017 https://www.nasa.gov/feature/goddard/2016/satellites-to-see-mercury-enter-spotlight-on-may-9

[136] "SDO Status Update – Aug. 10, 2016." 10 Aug. 2016. NASA 13 June 2017 https://www.nasa.gov/feature/goddard/2016/sdo-status

which is about 100 times hotter than the chromosphere below, with waves possibly from reflecting back and contributing to the heat in some way.[137]

The Solar and Heliospheric Observatory (SOHO) mission is a joint collaboration between ESA and NASA to study the Sun from its deep core to the outer corona and the solar wind. Launched in December 1995 for a 3-year mission that was meant to end in 1998, its success prompted several mission extensions with operations continuing throughout 2016.[138] SOHO orbits around the Sun in step with the Earth, at a distance of 1.5 million kilometres from Earth, enabling an uninterrupted view of the star.[139] SOHO's mission was once again extended on 22 November 2016 and will continue from 1 January 2017 to 31 December 2018.[140] In anticipation of Mercury's 9 May 2016 transit past the Sun, two of SOHO's twelve instruments, the Extreme ultraviolet Imaging Telescope (EIT) and the Michelson Doppler Imager (MDI), were restored to full operation, after a 5-year dormancy, to take measurements of the event.[141] On 4 August 2016, SOHO caught the demise of a Kreutz-type comet that plunged too close to the Sun in its highly elliptical orbit.[142] Lastly, a paper appearing in Astronomy and Astrophysics on 6 June 2016 summarized research on a new model, based on data collected from SOHO and STEREO, to map out where solar energetic particles (SEPs) might be found as they spread out and travelled away from the Sun; the model takes into consideration the fact that turbulence in solar material can cause magnetic field lines to wander, showing SEPs taking a much wider path than previous models predicted and explaining how SEPs can reach the far side of the Sun.[143]

NASA's Solar Terrestrial Relations Observatory (STEREO) mission, launched on 26 October 2006, provides the first-ever stereoscopic measurements of the Sun and its CMEs. Made up of two space-based observatories, i.e. STEREO-A travelling in a smaller and faster orbit (ahead of Earth's orbit) and STEREO-B trailing behind with a larger and slower orbit, the mission aims to understand the causes and mechanisms of CME initiation, characterize the propagation of CMEs through the heliosphere, discover the mechanisms and sites of energetic particle acceleration in

[137] Frazier, Sarah. "Tracking Waves from Sunspots Gives New Solar Insight." 20 Oct. 2016. NASA 13 June 2017 https://www.nasa.gov/feature/goddard/2016/tracking-waves-from-sunspots-gives-new-solar-insight

[138] "SOHO Overview." 5 June 2013. ESA 13 June 2017 http://www.esa.int/Our_Activities/Space_Science/SOHO_overview2

[139] About the SOHO Mission. "SOHO Fact Sheet." SOHO – Solar and Heliospheric Observatory 4 Mar. 2013 http://sohowww.nascom.nasa.gov/about/docs/SOHO_Fact_Sheet.pdf

[140] "Two-year extensions confirmed for ESA's science missions." 22 Nov. 2016. ESA 12 June 2017 http://sci.esa.int/director-desk/58589-two-year-extensions-confirmed-for-esa-s-science-missions/

[141] "Satellites to See Mercury Enter Spotlight on May 9." 3 May 2016. NASA 13 June 2017 https://www.nasa.gov/feature/goddard/2016/satellites-to-see-mercury-enter-spotlight-on-may-9

[142] "ESA, NASA's SOHO Sees Bright Sungrazer Comet." 4 Aug. 2016. NASA 14 June 2017 https://www.nasa.gov/image-feature/goddard/2016/esa-nasa-s-soho-sees-bright-sungrazer-comet

[143] Tran, Lina. "Wayward Field Lines Challenge Solar Radiation Models." 17 Oct. 2016. NASA 14 June 2017 https://www.nasa.gov/feature/goddard/2016/wayward-field-lines-challenge-solar-radiation-models

the low corona and the interplanetary medium and improve the determination of the structure of the ambient solar wind.[144] While NASA mission operations had lost communication with STEREO-B on 1 October 2014, during a test of the spacecraft's command loss timer as it neared solar conjunction, NASA managed to re-established contact with STEREO-B on 21 August 2016, following several months of attempts to contact the spacecraft without the Sun's interference.[145] Despite intermittent contact with STEREO-B since that time, the operations team is still in the process of assessing the spacecraft's health, re-establishing attitude control and evaluating all subsystems and instruments, which could take months or even years.[146] STEREO-A continued collecting data, having resumed normal science operations on 17 November 2015, after undergoing side lobe repointing operations on 20 August and 1 December 2014 which had the spacecraft transmit lower-resolution data for most of 2015.[147]

JAXA's Hinode mission, formerly Solar-B, is a joint collaboration with NASA and the UK to measure solar magnetic fields; study the generation, transport and dissipation of magnetic energy from the photosphere to the corona; and record how energy stored in the Sun's magnetic field is released as the field rises into the Sun's outer atmosphere.[148] Launched in September 2006 for a 3-year mission, the polar, Sun-synchronous orbiting spacecraft has allowed scientists to study solar phenomena, from solar explosions to the delicate motion of solar spicules, in great detail for more than 10 years, continuing beyond 2016.[149] In research published on 19 April 2016, scientists using Hinode, SDO and STEREO-A observations from a December 2013 solar flare were provided speed, temperature, density and size measurements that strengthened scientists' understanding that the electromagnetic phenomenon called a 'current sheet' is the result of magnetic reconnection on the Sun. A current sheet is a very fast, very flat flow of electrically charged material that forms when two oppositely aligned magnetic fields come in close contact, creating very high magnetic pressure that is unstable and can lead to new configurations; with the heat and light from the transformation producing a solar flare. As current sheets and magnetic reconnection are so closely associated, such detailed observations bolster the idea that magnetic reconnection is the force behind solar flares.[150]

---

[144] "About the STEREO Mission." 9 Feb. 2013. NASA 14 June 2017 https://stereo.gsfc.nasa.gov/mission/mission.shtml

[145] "NASA Establishes Contact With STEREO Mission." 22 Aug. 2016. NASA 14 June 2017 https://www.nasa.gov/feature/goddard/2016/nasa-establishes-contact-with-stereo-mission

[146] "STEREO: 10 Years of Revolutionary Solar Views." 25 Oct. 2016. NASA 14 June 2017 https://www.nasa.gov/feature/goddard/2016/stereo-10-years-of-revolutionary-solar-views

[147] "Q & A for Operations of STEREO During Superior Conjunction." 3 July 2014. NASA 14 June 2017 https://www.nasa.gov/content/goddard/q-a-for-stereo-during-superior-conjunction

[148] "Hinode: Mission to the Sun." 31 July 2015. NASA 19 June 2017 https://www.nasa.gov/mission_pages/hinode/mission.html

[149] Frazier, Sarah. "Highlights from a Decade of JAXA and NASA's Hinode Solar Observatory." 22 Sept. 2016. NASA 19 June 2017 https://www.nasa.gov/feature/goddard/2016/highlights-from-a-decade-of-jaxa-and-nasa-s-hinode-solar-observatory

[150] Frazier, Sarah. "Seeing Double: NASA Missions Measure Solar Flare from 2 Spots in Space." 19 Apr. 2016. NASA 19 June 2017 https://www.nasa.gov/feature/goddard/2016/seeing-double-nasa-missions-measure-solar-flare-from-2-spots-in-space

NASA's Interface Region Imaging Spectrograph (IRIS) satellite is a Small Explorer mission to observe how solar material moves, gathers energy and heats up as it travels through the Sun's lower atmosphere. Its mission, launched in June 2013 and operating in a polar, Sun-synchronous orbit, complements the SDO and the Hinode missions to explore the Sun's variable atmosphere and how it impacts Earth; that is, while SDO and Hinode monitor the photosphere (solar surface) and corona (outer atmosphere), IRIS observes the chromosphere and transition region between. In addition to being where most of the Sun's ultraviolet emission is generated, this region powers the Sun's million-degree atmosphere and drives the solar wind. Like with the SDO spacecraft, IRIS used the transit of Mercury in front of the Sun to help recalibrate its telescope to correct any changes that might have occurred during its launch into orbit.[151] IRIS observed a mid-level solar flare on 24 July 2016, capturing how large amounts of magnetic energy are released, heating the Sun's atmosphere and releasing energized particles out into space, which in turn drives post-flare loops of plasma (i.e. coronal rain) to puzzlingly rapidly cool from millions down to a few tens of thousands of kelvins while descending to the photosphere.[152] While IRIS's prime mission was for 2 years, it has been extended through September 2018, with the possibility of additional extensions afterward.[153]

NASA's Reuven Ramaty High-Energy Solar Spectroscopic Imager (RHESSI) focuses on the basic physics of particle acceleration and energy release in solar flares. Launched in February 2002 for a 2-year mission, RHESSI has operated for more than 14 years and underwent its fifth month-long detector anneal from 23 February to 29 April 2016. An annual procedure involves heating up detectors to rejuvenate them from accumulated radiation damage and then cooling them back down to operating temperatures for the spacecraft to resume collecting solar X-ray and gamma-ray data.[154]

2.3.8 Solar System Science

The dwarf planet Ceres and near-Earth objects (NEOs) including comets and asteroids were the central focus for exploration in 2016.

NASA's Dawn mission studies the asteroid Vesta and dwarf planet Ceres, the two largest bodies in the asteroid belt between Mars and Jupiter (with average diameters of 525 km and 950 km, respectively), to characterize the early solar system and the

[151] Fox, Karen C. "IRIS Releases New Imagery of Mercury Transit." 18 May 2016. NASA 14 June 2017 https://www.nasa.gov/feature/goddard/2016/iris-releases-new-imagery-of-mercury-transit

[152] Tran, Lina. "IRIS Spots Plasma Rain on Sun's Surface." 5 Aug. 2016. NASA 14 June 2017 https://www.nasa.gov/feature/goddard/2016/iris-spots-plasma-rain-on-suns-surface

[153] Tran, Lina. "NASA's Sun-Observing IRIS Mission." 1 Dec. 2016. NASA 14 June 2017 https://www.nasa.gov/feature/goddard/2016/nasas-sun-observing-iris-mission

[154] "RHESSI STATUS." 29 Apr. 2016. NASA 19 June 2017 https://hesperia.gsfc.nasa.gov/rhessi3/news-and-resources/status/index.html

processes that dominated its formation. Launched on 27 September 2007, Dawn inserted into orbit around Ceres on 6 March 2015.[155] In the following months, Dawn mapped Ceres with increasing detail as it lowered its orbit from 4400 km on 7 June 2015 to 1470 km on 25 August 2015 to its final orbit of 380 km on 26 October 2015.[156] Capturing images at a resolution of 35 m per pixel, Dawn's visible and infrared mapping spectrometer (VIR) enables scientists to identify specific minerals by their reflected wavelength, while data from its gamma ray and neutron detector (GRaND) provides details on Ceres' composition and on the abundances of elements in its surface.[157] Observations of the domed 3-km-tall Ahuna Mons in Ceres' bright Occator crater revealed an abundance of bright material on some of its slopes and less on others.[158] While ice has been located near the surface and in cold traps in permanent shadow on Ceres, the bright material in the crater is likely to be highly reflective salts.[159] The intricate geometry of the crater's interior and of Ahuna Mons suggests that geologic activity occurred in the recent past; moreover, studies of Ahuna Mons have led researchers to conclude the domed mountain to be the first known example of a cryovolcano that erupts a liquid made of volatiles such as water in the form of a salty mud mix, instead of silicates. Researchers have also found evidence that Ceres might once have had a weak temporary atmosphere.[160] Dawns' prime mission ended on 30 June 2016; its extended mission phase began in the following month with an elliptical orbit of 7200 km to view the dwarf planet from a higher vantage.[161]

ESA's Rosetta mission to perform a detailed study of the comet 67P/Churyumov-Gerasimenko concluded on 30 September 2016. Launched on 2 March 2004 on Europe's Ariane 5 launcher, the Rosetta orbiter and its Philae lander had reached the comet after a decade-long journey and became the first mission to successfully orbit a comet on 6 August 2014.[162] On 12 November 2014, Philae accomplished the first successful landing on a comet, albeit bouncing twice due to a misfire of its harpoons and landing in the shadow of a cliff. Philae spent nearly 57 h performing its science objectives, managing to transmit all of the results from its final sequence of measurements

---

[155] "NASA Spacecraft Becomes First to Orbit a Dwarf Planet." 6 Mar. 2015. NASA 24 May 2016 http://dawn.jpl.nasa.gov/news/news-detail.html?id=4503

[156] "Dawn Heads Toward Final Orbit." 26 Oct. 2015. NASA 24 May 2016 http://dawn.jpl.nasa.gov/news/news-detail.html?id=4751

[157] "New Details on Ceres Seen in Dawn Images." 12 Jan. 2016. NASA 19 June 2017 https://dawn.jpl.nasa.gov/news/news-detail.html?id=4817

[158] "Dawn's First Year at Ceres: A Mountain Emerges." 7 Mar. 2016. NASA 19 June 2017 https://dawn.jpl.nasa.gov/news/news-detail.html?id=5745

[159] "Where is the Ice on Ceres? New NASA Dawn Findings." 15 Dec. 2016. NASA 19 June 2017 https://dawn.jpl.nasa.gov/news/news-detail.html?id=6703

[160] "Ceres' Geological Activity, Ice Revealed in New Research." 1 Sept. 2016. NASA 19 June 2017 https://dawn.jpl.nasa.gov/news/news-detail.html?id=6611

[161] "Where is the Ice on Ceres? New NASA Dawn Findings." 15 Dec. 2016. NASA 19 June 2017 https://dawn.jpl.nasa.gov/news/news-detail.html?id=6703

[162] Algar, Jim. "Rosetta's lander Philae snaps selfie with comet." 4 Oct. 2014. Tech Times 20 June 2017 http://www.techtimes.com/articles/17887/20141014/rosetta-s-lander-philae-snaps-selfie-with-comet.htm

before its battery ended.[163] From there, contact was established sporadically before the comet's closest approach to the Sun in August 2015. In the following months, ESA and the DLR extended Rosetta's mission into September 2016 to monitor the comet's evolution as it journeyed deeper into the solar system. In a study published in the journal Nature, researchers using data from Rosetta's RSI instrument confirmed that the comet's low density was likely due to a porous mixture of dust particles and ice rather than a cavernous interior, consistent with earlier results from Rosetta's CONSERT radar experiment.[164] Additional studies published throughout 2016 addressed the detection of magnetic field-free cavities and its association with outgassing[165] and how the colour of the comet changes following a pass around the Sun, as surface dust is ejected – lifted by the sublimated ice beneath it.[166] Prior to the end of its mission, on 2 September 2016, Rosetta's OSIRIS narrow-angle camera had captured images of the Philae lander at 5 cm per pixel, providing the precise location and a contextual conclusion to the data generated in Philae's journey.[167] In the following weeks, the Rosetta spacecraft was set on a collision course with the Ma'at region of the comet's small lobe which plays an important role in the comet's activity, to study the comet's gas, dust and plasma environment near its surface and for very high-resolution images.[168]

NASA's reboot of the Wide-field Infrared Survey Explorer (WISE) mission, this time to discover and characterize near-Earth objects (NEOs) with infrared light, was reactivated in September 2013 for 3 additional years of service.[169] The original WISE mission was completed early in 2011 and was placed in 31 months of hibernation after surveying the whole sky twice in infrared light.[170] Now dubbed NEOWISE, soon after the mission began its third year of operation at the start of 2016, NASA released its latest data from the spacecraft, having detected and characterized 72 NEOs (eight of which being potentially hazardous asteroids (PHA)), bringing the total number of known NEOs to 439.[171] The NEOWISE mission con-

[163] Beatty, Kelly. "Philae Wins Race to Return Comet Findings." 15 Nov. 2014. Sky & Telescope 20 June 2017 http://www.skyandtelescope.com/astronomy-news/philae-lander-success-11152014/

[164] "Inside Rosetta's comet." 4 Feb. 2016. ESA 20 June 2016 http://m.esa.int/Our_Activities/Space_Science/Rosetta/Inside_Rosetta_s_comet

[165] "Rosetta finds magnetic field-free bubble at comet." 11 Mar. 2016. ESA 20 June 2017 http://m.esa.int/Our_Activities/Space_Science/Rosetta/Rosetta_finds_magnetic_field-free_bubble_at_comet

[166] "The colour-changing comet." 7 Apr. 2016. ESA 20 June 2017 http://m.esa.int/Our_Activities/Space_Science/Rosetta/The_colour-changing_comet

[167] "Philae found!" 5 Sept. 2016. ESA 20 June 2017 http://m.esa.int/Our_Activities/Space_Science/Rosetta/Philae_found

[168] "Mission Complete: Rosetta's Journey Ends in Daring Descent to Comet." 30 Sept. 2016. ESA 20 June 2017 http://www.esa.int/Our_Activities/Space_Science/Rosetta/Mission_complete_Rosetta_s_journey_ends_in_daring_descent_to_comet

[169] "NASA Spacecraft Reactivated to Hunt for Asteroids." 21 Aug. 2013. NASA 2 Apr. 2014 http://www.jpl.nasa.gov/news/news.php?release=2013-257

[170] "NASA's Asteroid Hunter Spacecraft Returns First Images after Reactivation." 19 Dec. 2013. NASA 9 Mar. 2015 http://www.jpl.nasa.gov/wise/newsfeatures.cfm?release=2013-373

[171] Zolfagharifard, Ellie. "All of the asteroids near Earth in one video: Nasa releases latest data from Neowise 'hunter' spacecraft." 7 Apr. 2016. MailOnline 21 June 2017 http://www.dailymail.co.uk/

tinued its search through 2016, spotting a comet C/2016 U1 (NEOWISE) on 21 October 2016 and another '2016 WF9' NEO likely with cometary origins on 27 November 2016 – the trajectories of both objects appear not to threaten Earth in the near future.[172] In January 2016, a new study published in the Astrophysical Journal Letters added to findings on the most luminous galaxy, W2246-0526, which belongs to a new class of extremely luminous infrared galaxies (ELIRG). Discovered using WISE data in 2015, W2246-0526 shines with the light of more than 300 trillion suns, some 12.4 billion light-years from Earth, and is thought to have a behemoth black hole at its centre that is heating surrounding gasses to temperatures of millions of degrees and blasting out high-energy, visible, ultraviolet and X-ray light, as it sucks in matter. The researchers studying W2246-0526 found large amounts of ionized carbon in a very turbulent state throughout the entire galaxy, instead of flowing in specific directions from the black hole's accretion disc, suggesting that the momentum and energy of the particles of light deposited in the gas are so great that they are pushing the gas out in all directions; should all the gas and dust eventually blow out of the hot, dust-obscured galaxy (hot DOG), the quasar surrounding the black hole would likely be visible.[173]

Commercial interest in NEOs has increased in recent years with several private US companies already investing millions of dollars to be the first to mine asteroids. As an initial step, Planetary Resources and Deep Space Industries (DSI) business models focus on developing prospecting spacecraft that will image and characterize promising asteroids. On 16 July 2015, Planetary Resources launched its Arkyd 3 Reflight (A3R) spacecraft from the ISS to validate several core technologies that will be incorporated into a future fleet of Arkyd spacecraft that will be launched into the solar system.[174] Similarly, DSI is developing its FireFly spacecraft; but rather than testing its technologies in Earth orbit, the first-generation spacecraft will be launched directly towards promising asteroids. The main focus in asteroid mining is in the potential availability of rare metals that could be returned to Earth or water that could be converted into propellant and oxygen for deeper space exploration. While initial estimates of the value of some known asteroids are upwards of $100 trillion, to reach these caches, both companies will need to invest substantial resources in developing the necessary mining technology, in addition to successfully rendezvousing and securing a spacecraft onto an asteroid's surface.[175]

sciencetech/article-3528830/All-asteroids-near-Earth-one-video-Nasa-releases-latest-data-Neowise-hunter-spacecraft.html

[172] "NASA's NEOWISE Mission Spies One Comet, Maybe Two." 29 Dec. 2016. NASA 21 June 2017 https://www.jpl.nasa.gov/wise/newsfeatures.cfm?release=2016-328

[173] "Most Luminous Galaxy Is Ripping Itself Apart." 15 Jan. 2016. NASA 21 June 2017 https://www.jpl.nasa.gov/wise/newsfeatures.cfm?release=2016-011

[174] "Planetary Resources' First Spacecraft Successfully Deployed, Testing Asteroid Prospecting Technology in Orbit." 16 July 2015. Planetary Resources 27 Nov. 2015 http://www.planetaryresources.com/2015/07/planetary-resources-first-spacecraft-deployed/

[175] Calandrelli, Emily. "The Potential $100 Trillion Market For Space Mining." 9 July 2015. Techcrunch.com 26 Nov. 2015 http://techcrunch.com/2015/07/09/the-potential-100-trillion-market-for-space-mining/

2.3.9 Outer Solar Science

ESA's Gaia mission is measuring the positions and motions of more than 1 billion stars of the roughly 100 billion stars in our galaxy to create the most accurate map yet of the Milky Way. Launched on 19 December 2013 and operating from the L2 Lagrange point, the spacecraft completed the second year of a 5-year survey on 16 August 2016.[176] On 14 September 2016, ESA published the first catalogue based on data collected during its first 14 months of science operations, which feature the density of stars across the entire sky and the parallax, i.e. the apparent motion of a star against a distant background, of more than two million stars. While the image contained stripes and other artefacts, these will gradually fade in subsequent scans as the trace amounts of ice deposits which remained in the spacecraft following its commissioning will be outgassed using heaters beneath Gaia's mirrors.[177] In the lead up to the publication, ESA released a sonification, converting astronomical data into sound, to portray the status of astrometric catalogues prior to the advent of Gaia and demonstrate the remarkable progress that was being made in the field of astrometry.[178] Researchers scrutinizing Gaia data have detected over a thousand transient bright astronomical sources, due to stars undergoing a major outburst or supernova, but in July and August 2016, they observed two rare instances of gravitational microlensing, where the gravity of a massive object between a star and observer causes the path of its light to distort. The first instance, classified as Gaia16aua, observed a faint star of magnitude 19 suddenly brighten by two magnitudes, while the second instance, classified as Gaia16aye, observed the anomalous peaks and troughs in brightness of a magnitude 14.5 star; by pairing Gaia measurements with ground-based data, researchers will be able to estimate the position and mass of objects (e.g. a star or black hole) causing the gravitational microlensing with high precision.[179]

NASA's Kepler mission to discover hundreds of Earth-size and smaller planets in or near the habitable zone of more than 150,000 stars and determine the fraction of the hundreds of billions of stars in our galaxy that might have such planets began in May 2009. By May 2013, the loss of two of the four reaction wheels on the spacecraft brought an end to Kepler's 4-year science mission as the spacecraft had lost its ability to precisely point at the original field of view; but by May 2014, NASA had approved the Kepler 2 (K2) community-driven mission which repurposed Kepler to accurately point at target sky fields along the ecliptic plane of Earth's orbit using the Sun to maintain its stability. The K2 mission is expected to continue operations into

---

[176] "Gaia's Second Anniversary Marked by Successes and Challenges." 16 Aug. 2016. ESA 21 June 2017 http://sci.esa.int/gaia/58135-gaia-s-second-anniversary-marked-by-successes-and-challenges/

[177] "Gaia's Billion-Star Map Hints at Treasures to Come." 14 Sept. 2016. ESA 22 June 2017 http://www.esa.int/Our_Activities/Space_Science/Gaia/Gaia_s_billion-star_map_hints_at_treasures_to_come

[178] "From Hipparchus to Hipparcos: A Sonification of Stellar Catalogues." 22 Sept. 2016. ESA 22 June 2017 http://sci.esa.int/gaia/58311-from-hipparchus-to-hipparcos-a-sonification-of-stellar-catalogues/

[179] "Gaia Spies Two Temporarily Magnified Stars." 27 Oct. 2016. ESA 22 June 2017 http://sci.esa.int/gaia/58546-gaia-spies-two-temporarily-magnified-stars/

2018.[180] By the end of 2016, continued analyses of Kepler data had revealed more than 5100 planet candidates, with more than 2500 verified planets[181]; K2 added another 447 candidates, with 154 verified planets by mid-year.[182] As at 24 August 2016, 22 planets were known to be Earth-sized and orbiting within the habitable zone of their nearest stars.[183] In May 2016, a new study of the Kepler-223 star system showed its four planets to have the same configuration as that of Jupiter, Saturn, Uranus and Neptune in the early history of our Sun's solar system; but whereas our the orbit of planets in the solar system have evolved since its birth 4.6 billion years ago, the much older Kepler-223 system appears to have maintained a single orbital configuration resonating around its star for far longer. As resonances are extremely fragile, it's possible that interactions with numerous asteroids and planetesimals may have dislodged the Sun's giants from their own resonance.[184] Another recent study published in the Astrophysical Journal measured the orbits of 19 heartbeat star systems that were identified by the Kepler mission; because these binary systems are in elongated elliptical orbits, the diameters of the stars tend to rapidly fluctuate at the point of their closest encounter due to the tidal forces caused by each star's gravitational pull. While the tidal stretching of these heartbeat stars should have quickly caused their systems to evolve into circular orbits, the researchers postulate that third or fourth stars might exist in these systems that have gone undetected, which may be maintaining these highly stretched-out, elliptical orbits.[185]

NASA's Spitzer Space Telescope, launched in August 2003, studied the early universe, young galaxies and forming stars and was used to detect dust discs around stars.[186] After running out of the coolant needed to chill its longer-wavelength instruments in 2009, Spitzer was repurposed to track exoplanets around stars with the use of infrared light. As exoplanets cross in front of their stars, they block out a fraction of the light, allowing the size of the planet to be revealed, in addition to giving clues

[180] "Mission overview." 8 Feb. 2017. NASA 23 June 2017 https://www.nasa.gov/mission_pages/kepler/overview/index.html

[181] "NASA Kepler Visionary Honored By American Association for the Advancement of Science." 21 Nov. 2016. NASA 23 June 2017 https://www.nasa.gov/feature/kepler/nasa-kepler-visionary-honored-by-american-association-for-the-advancement-of-science

[182] "Mission Manager Update: K2 Marches On." 9 June 2016. NASA 23 June 2017 https://www.nasa.gov/feature/ames/kepler/mission-manager-update-k2-marches-on; and "NASA's Kepler Confirms 100+ Exoplanets During Its K2 Mission." 18 July 2016. NASA 23 June 2017 https://www.nasa.gov/feature/ames/kepler/nasa-s-kepler-confirms-100-exoplanets-during-its-k2-mission

[183] "Kepler's Small Habitable Zone Planets." 11 May 2016. NASA 23 June 2017 https://www.nasa.gov/mission_pages/kepler/multimedia/images/keplers-small-habitable-zone-planets; "ESO Discovers Earth-Size Planet in Habitable Zone of Nearest Star." 24 Aug. 2016. NASA 23 June 2017 https://www.nasa.gov/feature/jpl/eso-discovers-earth-size-planet-in-habitable-zone-of-nearest-star

[184] "Kepler-223 System: Clues to Planetary Migration." 17 May 2016. NASA 23 June 2017 https://www.nasa.gov/feature/jpl/kepler-223-system-clues-to-planetary-migration

[185] "'Heartbeat Stars' Unlocked in New Study." 21 Oct. 2016. NASA 23 June 2017 https://www.nasa.gov/feature/jpl/heartbeat-stars-unlocked-in-new-study

[186] "Mission Overview." 30 July 2008. NASA 23 June 2017 https://www.nasa.gov/centers/jpl/missions/spitzer.html

about the planet's atmosphere by the infrared light that they also emit.[187] In March 2016, Spitzer data helped researchers to create the first temperature map of an Earth-like exoplanet that is double in size and tidally locked to its star; rather than having a thick atmosphere with winds moving heat around the planet as previously thought, the exoplanet was seen to inefficiently transport heat, and lava flows likely warmed the nightside of the planet.[188] Another planet-finding technique used by Spitzer is called microlensing, which occurs when the light of a distant star is magnified and brightened by the gravity of another star that passes in its foreground. Should the closer star have a planet in its orbit, the planet might cause a blip in the magnification. By mid-2015, Spitzer had viewed a total of 142 microlensing events[189]; moreover, in late 2016 NASA's Spitzer and Swift space telescopes were used to observe a microlensing event of a newly discovered brown dwarf, OGLE-2015-BLG-1319, marking the first time two space telescopes have collaborated to observe a microlensing event, rather than pairing observations with a ground observatory.[190] Spitzer's mission was extended for 2.5 years beginning on 1 October 2016 and continuing through the commissioning phase of the James Webb Space Telescope in early 2019.[191] This 'Beyond' phase will explore a wide range of topics in astronomy and cosmology, as well as planetary bodies in and out of our solar system.[192]

China's Dark Matter Particle Explorer (DAMPE) mission was successfully launched to a 500 km Sun-synchronous orbit on 17 December 2015. Its 3-year mission aims to shed new light on the nature of dark matter – a hypothetical kind of matter (along with dark energy) that could explain where the remaining estimated 85% of the total mass energy in the universe exists. The DAMPE mission searches for dark matter by measuring the properties of particles that annihilate or decay in space, as with experiments conducted inter alia by the Alpha Magnetic Spectrometer (AMS) on board the ISS or the CALorimetric Electron Telescope (CALET) recently attached to Japan's Kobo module. However, DAMPE extends the search into the multi-TeV region with an energy resolution of 1.5% at 100 GeV (i.e. at least three

[187] "How Engineers Revamped Spitzer to Probe Exoplanets." 24 Sept. 2013. NASA 3 Apr. 2014 http://www.spitzer.caltech.edu/news/1560-feature13-07-How-Engineers-Revamped-Spitzer-to-Probe-Exoplanets

[188] "NASA's Spitzer Maps Climate Patterns on a Super-Earth." 30 Mar. 2016. NASA 23 June 2017 http://www.spitzer.caltech.edu/news/1869-ssc2016-01-NASA-s-Spitzer-Maps-Climate-Patterns-on-a-Super-Earth

[189] "NASA's Spitzer Spots Planet Deep Within Our Galaxy." 14 Apr. 2015. NASA 22 May 2016 http://www.spitzer.caltech.edu/news/1746-feature15-05-NASA-s-Spitzer-Spots-Planet-Deep-Within-Our-Galaxy

[190] "NASA Space Telescopes Pinpoint Elusive Brown Dwarf." 10 Nov. 2016. NASA 26 June 2017 http://www.spitzer.caltech.edu/news/1918-feature16-18-NASA-Space-Telescopes-Pinpoint-Elusive-Brown-Dwarf

[191] "Spitzer Operations to Continue into 2019." 9 June 2016. NASA 23 June 2017 http://www.spitzer.caltech.edu/news/1882-feature16-09-Spitzer-Operations-to-Continue-into-2019

[192] "Spitzer Space Telescope Begins 'Beyond' Phase." 25 Aug. 2016. NASA 26 June 2017 https://www.nasa.gov/feature/jpl/spitzer-space-telescope-begins-beyond-phase

times higher than international peers) and will also take precise measurements of the flux of nuclei with a spectrum up to above 100 TeV (i.e. nine times wider than the AMS), to gain insight into the origin and propagation of high-energy cosmic rays.[193] In the initial 2 years of its mission, DAMPE will scan space in all directions, followed by another year or more that will focus on areas where the potential signatures of dark matter are most likely to be observed.[194] By 19 December 2016, the spacecraft had collected 1.8 billion cosmic rays, with more than 1 million high-energy electrons among the collected particles. The mission's first results are expected to be published in early 2017.[195]

ISRO's Astrosat mission, launched on 28 September 2015, is India's first astronomy satellite to provide optical, ultraviolet and X-ray images of black holes and other related phenomena over the course of 5 years.[196] Its first scientific results and the future scope were presented on 29 September 2016 at the Inter-University Centre for Astronomy and Astrophysics (IUCAA), Pune, India; with the spacecraft having orbited the Earth more than 5400 times, executing 343 individual pointings to 141 different cosmic sources.[197] Included among the findings, Astrosat's Large Area X-ray Proportional Counter (LAXPC) instrument observed for the very first time rapid variability of high-energy (particularly >20 keV) X-ray emissions from the enigmatic 'GRS 1915 + 105' black hole system. These quasi-period oscillations are thought to occur because the inner part of the disc surrounding the black hole wobbles as the spinning black hole drags the space-time fabric around it, as predicted by Einstein's general theory of relativity. While these oscillations have been studied in the past in low-energy X-rays, observing phenomenon also in high-energy X-rays from higher energy photons that are emitted closer to the black hole allows researchers to measure the arrival time difference between the energy bands and provides clues to the geometry and dynamic behaviour of the gas swirling around a black hole.[198] The spacecraft began its Open Phase observations based on proposals by numerous institutions on 1 October 2016.

[193] "DAMPE joins the search for dark matter in space." 12 Feb. 2016. CERN COURIER 26 June 2017 http://cerncourier.com/cws/article/cern/63969

[194] Yue, Huang. "China Launches Country's First Dark Matter Satellite." 17 Dec. 2015. CRI English News 22 Jan. 2016 http://english.cri.cn/12394/2015/12/17/4201s908753.htm

[195] Jones, Andrew. "After a year in orbit and 1.8 billion cosmic rays, China's dark matter probe hopes for 'unexpected' finding." 19 Dec. 2016. gbtimes 26 June 2017 http://gbtimes.com/china/after-year-orbit-and-18-billion-cosmic-rays-chinas-dark-matter-probe-hopes-unexpected-findings

[196] De Selding, Peter B. "PSLV Rocket Launches India's 1st Astronomy Satellite, 4 Spire Cubesats." 28 Sept. 2015. SpaceNews 5 Jan. 2016 http://spacenews.com/pslv-rocket-launches-indias-1st-astronomy-satellite/

[197] "A science meet to commemorate one year of AstroSat in orbit." 29 Sept. 2016. ISRO 27 June 2017 http://www.isro.gov.in/update/29-sep-2016/science-meet-to-commemorate-one-year-of-astrosat-orbit

[198] "AstroSat observes the high energy X-ray variability of a black hole system." 31 Aug. 2016. ISRO27June2017http://www.isro.gov.in/pslv-c30-astrosat-mission/astrosat-observes-high-energy-x-ray-variability-of-black-hole-system

Part II
Views and Insights

Edited by Stefano Ferretti

Chapter 3
Space for Sustainable Development

Stefano Ferretti

3.1 Introduction

3.1.1 Sustainable Development

The Sustainable Development Goals (SDGs) were approved by the United Nations (UN) in 2015 as a response to the current and anticipated challenges facing the global community until 2030. In contrast with their predecessors, the Millennium Development Goals (MDGs) – which were geared primarily towards developing countries – the SDGs present a holistic and global perspective which calls upon all relevant actors, both at national and regional levels, in the public and private sectors, to unite to come up with solutions to the 17 identified challenges, thus improving life across the world.[1] According to Prof. Sachs, these challenges arise from current developments in technology, particularly the digital revolution, which have pushed the boundaries of planetary resource utilisation, such as fossil fuels, land and water, to unsustainable levels.[2] Diminishing poverty rates have been fuelled by practices that can no longer be maintained due to rising populations, coupled with the finite resources that our planet offers. Agenda 2030 thus calls for a new development model, based on the principles of environmental sustainability coupled with social inclusion. The 17 goals comprise different but related objectives, from decarbonisation to improvements in agriculture and from healthcare to transportation and urban environments, and request all levels of society and all global regions to contribute towards this new paradigm.

[1] Ferretti S., Feustel-Büechl J., Gibson R., Hulsroj, P., Papp A., Veit, E. (2016) ESPI Report 59 Space for Sustainable Development. ESPI, Vienna.

[2] Sachs J. (2015) The Age of Sustainable Development. Columbia University Press, New York, NY.

S. Ferretti (✉)
European Space Policy Institute (ESPI), Vienna, Austria
e-mail: Stefano.ferretti@esa.int

C. Al-Ekabi, S. Ferretti (eds.), *Yearbook on Space Policy 2016*,
Yearbook on Space Policy, https://doi.org/10.1007/978-3-319-72465-2_3

This call must also be heeded by the space sector, a vital contributor to our everyday lives and a potential game-changer in the efforts towards a more sustainable development in the years to come.[3] The Space for Sustainable Development project, carried out at the European Space Policy Institute (ESPI), provides impetus to these efforts and creates a framework within which the space sector can provide its assets and expertise towards the achievement of the SDGs. The Space for Sustainable Development theme has been addressed by ESPI along three main axes:

1. Policy, strategy and programmatic framework
2. Stakeholder mapping, engagement and partnerships
3. Capacity building, dialogue platforms and societal challenges

This section provides an overview of the above-mentioned framework including key recommendations emerging from high-level discussions carried out in the recent years, collecting inputs from a range of global space actors, from NGOs to space agencies, that have contributed to this vision.

3.1.2 Why Space for Sustainable Development

The space sector, as many others, is on the verge of a revolution due to digitalisation, big and open data, smallsats, private actors and the improvement of connectivity. Since the days of the space race, it has supported economic development, through the provision of innovative services and their technology spinoffs for terrestrial applications. ESPI believes the call to action accompanying the adoption of the SDGs should act as a catalyst for the space community, to reflect on how its activities can play a central role in satisfying end-user needs in sustainable development, with a focus on emerging economies. In particular, satellite infrastructure, data and telecommunications capabilities need to be transformed into relevant services for end users operating in the ambit of sustainable development. Indeed, the benefits of space technologies have been enjoyed both by developed and developing nations. Regarding developed nations, Earth Observation technologies provide functionalities that often only space-based assets can offer, for example, monitoring pollution over large urban areas through Earth Observation satellites or managing financial transactions through the precision of GPS-based timing.[4]

For developing countries in particular, space-based assets can largely mitigate the lack of terrestrial infrastructures, providing remote health management and education or meeting the need for in situ measurements to monitor pollution, coastal areas and land and water use. For example, India is a leader in space-enabled remote healthcare technologies, while African countries are accessing the benefits of Earth Observation for water through the Tiger Programme in cooperation with the

[3] Ferretti S., Feustel-Büechl J., Gibson R., Hulsroj, P., Papp A., Veit, E. (2016) ESPI Report 59 Space for Sustainable Development. ESPI, Vienna.

[4] Ibid.

European Space Agency (ESA). Private companies are also initiating interesting initiatives, for example, OneWeb is attempting to provide connectivity across all areas of the globe including underserved remote regions.[5]

Although developing countries can achieve great benefits using space technologies, current space programmes tend to draw on industrialised countries' needs, where the space industry is more prominent, and funding is readily available to promote the use of space services in everyday life. To have a wider applicability, however, space assets and services also need to be tailored for the needs of emerging economies. Moreover, space-based services need to be adopted at the grassroots level, through user communities that can develop relevant skills and be advocates for such technologies in the field. To achieve this, space actors need to be attuned to specific needs within such communities, which are complicated by the multitude of actors with different strategies and skills, spread out along a fragmented value chain.[6]

The space industry has the potential to become a pivotal force supporting sustainable development efforts, if it can efficiently engage relevant actors in the field to tune into their needs and aspirations, thereby transforming data and assets into value-adding services. There is still a large potential for growth in this area; however, only through a sustained dialogue and cooperation can Space for Sustainable Development truly reach its full potential in improving life on Earth, being recognised as a leading and essential asset available to humankind, thereby also creating new business opportunities for the industry at large.[7]

3.2 The Timeliness of the Further Involvement of Space

The transition from MDGs to SDGs has not only widened the scope of countries and actors involved in development efforts, but through its new holistic perspective, it has also paved the way for the involvement of industrial sectors that were previously distant and separate from sustainability considerations.[8] This was operationalized, for instance, through the inclusion of the outcomes of the UN World Conference on Disaster Risk Reduction and the 21st Conference of Parties (COP21) on Climate Change, within the SDGs, placing Agenda 2030 as a set of principles to be respected in all forthcoming global cooperation efforts across a variety of fields.[9] In encouraging new opportunities, a more holistic approach also presents the challenge of integrating new technologies, industries and mindsets to pave the way for solutions that respect the sustainability principles of Agenda 2030. The space sector, with its

[5] Ibid.

[6] Ibid.

[7] Ibid.

[8] Veit, Elisabeth. "ESPI Perspectives 74. From MDGs to SDGs: why now is the time to further integrate space into development." 22 Jan 2016. ESPI 23 Oct 2017 http://www.espi.or.at/images/stories/dokumente/studies/Perspective_74.pdf

[9] Ibid.

strategic assets that can serve a range of industries, as well as its long-standing capabilities to integrate technologies and systems to provide innovative solutions to complex problems that address social and economic challenges, is uniquely poised to benefit from and provide benefits to sustainable development solutions. Indeed, this was recognised during COP21, through the space agencies' resolution that was adopted with a broad consensus.[10] Two years into Agenda 2030, as sustainability efforts are being ramped up and an increasing number of actors are formulating their contributions, the space sector is challenged to consider their involvement from a strategic perspective. Although space presents clear benefits for many aspects of sustainable development, there is insufficient awareness of these benefits in the field, which, coupled with a lack of channels to communicate challenges and needs to space actors, is leading to an underutilisation of space technologies and services in the sustainable development space.[11]

3.2.1 Current Efforts by the Space Sector and Its Potential

To understand how space and sustainable development can come together, it is important to gain a deep understanding of the needs that are connected to the different SDGs, how they can be addressed by current technologies and which gap space could potentially contribute towards filling. These needs include those linked to societal evolution and those related to external factors such as climate change.[12]

An initial high-level mapping of the current contributions of space assets and services towards the SDGs was carried out at ESPI and is reported below (see Table 3.1). It clearly shows both the ubiquity of space solutions for the SDGs and their potential to affect development across many different domains. Especially due to this high potential, the space sector should seek increased integration with actors that can spread the use of space technologies and feed this information into future roadmaps for their full implementation of sustainable development beyond the present day.[13] The following paragraphs present a more in-depth analysis of current solutions and future needs.

[10] Ferretti S (2016) Space as an enabler of sustainable development. 4th ICSD, Columbia University, New York, NY.

[11] Ibid.

[12] Ferretti S., Feustel-Büechl J., Gibson R., Hulsroj, P., Papp A., Veit, E. (2016) ESPI Report 59 Space for Sustainable Development. ESPI, Vienna.

[13] Ibid.

Table 3.1 Actual or possible contributions of space to SDGs

SDG topic	Actual or possible contribution of space
SDG 1: No poverty	Improved communications and more environmental data as a driver of growth, better logistics management using satellite navigation
SDG 2: Zero hunger	Earth Observation data for optimised agriculture and livestock management, more efficient crop markets through better telecommunications, better emergency response enabled by Earth Observation data and telecoms, better delivery systems using satellite navigation
SDG 3: Good health and well-being	E-health, including telemedicine and medical tele-training and learning
SDG 4: Quality education	Tele-learning
SDG 5: Gender equality	Female empowerment through links to the Information Society, thanks to telecommunications, tele-learning, telecommunications enabling female entrepreneurship
SDG 6: Clean water and sanitation	Earth Observation data for water management, water detection and water pollution monitoring
SDG 7: Affordable and clean energy	Earth Observation data for renewable energy management and grid management
SDG 8: Decent work and economic growth	Space services as enablers of economic growth and higher-quality jobs in all economic sectors
SDG 9: Industry, innovation and infrastructure	Space as enabler of innovation both in its own sector and others, space-based data and communication abilities key for industrial processes, space telecoms compensating for lack of terrestrial networks, Earth Observation compensating for lack of in situ stations, satellite navigation for best use of transport infrastructure and banking systems
SDG 10: Reduced inequalities	Access to Information Society through telecommunications is a leveller, fosters transparency and hence helps fight against corruption, space services as enablers of employment opportunities
SDG 11: Sustainable cities and communities	Earth Observation data for pollution monitoring, energy management and land use planning, satellite navigation for traffic management, telecommunications for efficient information exchange
SDG 12: Responsible consumption and production	Earth Observation data for optimised supply management, energy management, satellite navigation for logistics management
SDG 13: Climate action	Earth Observation data is key for climate change monitoring and definition of mitigation strategies
SDG 14: Life below water	Earth Observation data is key for monitoring the health of oceans and other water systems, for fisheries management and policing
SDG 15: Life on land	Earth Observation data used for biodiversity monitoring, pollution monitoring, land use management and policing

(continued)

Table 3.1 (continued)

SDG topic	Actual or possible contribution of space
SDG 16: Peace and justice – strong institutions	Telecommunications empower civil society by providing connection to the Information Society, enabling also e-voting. Legal evidence, treaty compliance monitoring, security management provided through Earth Observation systems
SDG 17: Partnerships	Space community is part of an international fabric of partnerships. Possibilities of reinforcement of links with development actors

Source: ESPI Report 59

3.2.2 *Space and Data*

Agenda 2030 largely depends on the availability of data to determine the status quo, gauge progress and evaluate the effectiveness of different strategies.[14] An overall framework comprising 231 indicators[15] has been proposed by the UN Statistical Commission's Inter-agency and Expert Group (IAEG-SDGs); however, the responsibility for their measurement rests with individual states, who will also have the right to propose complementary measurements. Different sectors such as transportation and education, as well as NGOs, are also expected to identify specific indicators that can best measure their progress towards the SDGs. This approach outlines again the need for a concerted effort from governments, industry, NGOs and society at large.[16] The space sector is a key driver for the type of data that all the above-mentioned actors may need to inform their decisions. It can provide such data with a regular frequency and on a global scale without relying on in situ infrastructure, in a wide variety of fields such as climate change, emissions monitoring, urbanisation, etc.[17] For instance, the International Charter on Space and Major Disasters,[18] which uses space data to support governments and local actors in responding to local emergencies, was recently used during the Ebola public health crisis. Talks are underway to expand its use to respond to sustainable development needs as well. Moreover, private companies, such as Planet, have evaluated how Earth Observation data can contribute information towards 12 of the 17 SDGs, indicating how space assets can contribute to global monitoring efforts.[19]

---

[14] Ferretti S (2016) Space as an enabler of sustainable development. 4th ICSD, Columbia University, New York, NY.

[15] Agenda of the United Nations Statistical Commission of the Economic and Social Council, 47th Session. Report of the Inter-Agency and Expert Group on Sustainable Development Goal Indicators. UN Doc. E/CN.3/2016/2 of 17 December 2015. New York: United Nations.

[16] Ferretti S (2016) Space as an enabler of sustainable development. 4th ICSD, Columbia University, New York, NY.

[17] Ibid.

[18] "The International Charter". International Charter Space and Major Disasters 23 Oct 2017 https://disasterscharter.org/web/guest/home;jsessionid=D7D82D4BBCF5EAE9B8C4067708CB15C8.jvm1

[19] "Social Impact" Planet 23 Oct. 2017 https://www.planet.com/markets/impact/

In addition to serving the SDGs at a global level, space data has the potential to support a multitude of needs arising from local communities with differing strategies to tackle sustainable development. Indeed, overarching objectives, such as access to food and water or healthcare, are influenced at a local level by factors such as population size and distribution, geography, climate and institutional stability. For instance, the African continent has wide and scarcely distributed populations, which causes logistical and infrastructure challenges for remote communities, which then encourages migration towards large cities, which in turn experience logistical difficulties in supporting this growth. Moreover, the population is expanding rapidly, while resources such as water are being affected by climate change, placing further pressure on migration and increasing difficulties for citizens to access basic services. A similar population rise is occurring in Asia, with the addition of rapid industrialization that fuels demand for cheap energy from fossil fuels. Large areas of the continent already suffer the consequences of climate change – while the Middle East struggles with water scarcity, Southeast Asia is experiencing severe rainfall and flooding. Sea level rise due to climate change is the main issue faced by small island states in the Pacific and Caribbean. South American states are instead grappling with security challenges due to the illegal drugs market, which is fuelling corruption and organised crime.[20] The European continent is grappling with uncontrolled immigration both from Asia and Africa as the younger generations seek opportunities and stability.

These differing challenges are addressed at a macro level by a multitude of regional organisations, often created to foster economic development and encourage international cooperation, which are indispensable counterparts to any space actor wishing to contribute to sustainable development efforts.[21] In Africa this role is represented by the African Union, with its mandate to promote "Sustainable Development at the economic, social and cultural levels as well as the integration of African economies".[22] The South American situation is more fragmented, with many regional organisations that are operating independently from the overarching Community of Latin American and Caribbean States (CELAC).[23] The same situation is found in Asia, where countries pursue heterogeneous objectives and find themselves at different stages of development.[24] The two largest organisations promoting social well-being and economic development are the Association of Southeast Asian Nations (ASEAN) and the South Asian Association for Regional Development (SAARC).[25] Lastly, the Pacific Islands Forum strives to contribute to

---

[20] Ferretti S (2016) Space as an enabler of sustainable development. 4th ICSD, Columbia University, New York, NY.

[21] Ibid.

[22] "African Union Charter". African Union 23 Oct. 2017 https://au.int/sites/default/files/treaties/7759-sl-oau_charter_1963_0.pdf

[23] Ferretti S (2016) Space as an enabler of sustainable development. 4th ICSD, Columbia University, New York, NY.

[24] Ibid.

[25] Ferretti S (2016) Space as an enabler of sustainable development. 4th ICSD, Columbia University, New York, NY.

sustainable development and economic growth through capacity building and infrastructure development.[26]

At an even more local and fine-grained level, a multitude of NGOs operates on the field, eliciting and defining development needs, executing programmes and gathering feedback and lessons learnt on new technologies and sustainable development models. These organisations often operate in situations where governmental actors are unable to provide essential infrastructure and services such as healthcare and education, access to clean water and financing. In providing these services, NGOs often cooperate with the local population, providing training and assistance, thereby also contributing to capacity building. Through these activities, they build up a wealth of information which they can then relay at higher levels, for example, towards government organisations, thereby constituting a precious resource for policy-making and solutions development and filling an essential gap between citizens and institutions.[27]

Following from these considerations, one of the pillars of Space for Sustainable Development has been defined as a dialogue platform between different stakeholders in the field of sustainable development domain and those in charge of defining and developing future space programmes, to be able to identify the most relevant requirements, creating space services that can offer the most value towards concrete sustainable development policies and programmes.[28]

3.2.3 Space and Technologies

Besides the information potential that space technology offers, its services can also support developing countries in benefiting from new technologies without building costly infrastructure. This technological leapfrogging will most likely first manifest itself in the telecommunications sector, where satellites and drones may provide connectivity to underserved areas. This in turn could increase the availability of remote healthcare and education in areas where these services are poor and scarcely accessible. Moreover, improved access to the web and online services can also serve as a catalyst for the development of rural areas in terms of economic welfare and opportunities.[29]

Another sector of paramount importance to sustainable development, that is set to benefit from space technologies, is the energy sector. Demand for energy is constantly rising, and this will only accelerate as population inequalities are reduced. As global economies increasingly adopt renewable energy sources and localised energy production, space services can be employed to identify the optimal location

[26] Ibid.

[27] Ibid.

[28] Ibid.

[29] Ibid.

and type of energy generation, which can benefit developed nations as well.[30] Not only energy demands but also demand for raw materials is set to increase in future years, coupled with waste production and disposal. Recycling and reusability techniques will therefore have a prominent role to play, and manned space missions have been tackling this issue from their outset, creating opportunities for technology transfer to terrestrial applications.[31]

As urban centres increasingly concentrate most of the global population, current resource allocation and management will necessarily evolve to account for increased complexity and pressure on resources and services. New technologies such as the Internet of things (IoT) will enable a fine-grained collection of ground-based data, to control and improve resource utilisation and quality of life. These knowledge networks, stemming from buildings, transportation systems and energy and environmental control, will benefit from integration with EO and GNSS data to provide complete situational awareness. The management of this type of data should enable the production of smart dashboards, to inform decision-making, much like what is currently being done within space missions.[32]

Finally, additive layer manufacturing technologies are set to disrupt manufacturing and construction, and the space sector is developing them with a focus on extraterrestrial settlement, thus considering different types of in situ materials, extreme conditions and highly resilient robotic systems. These technology developments could in turn support urbanisation efforts in remote and underserved areas, eliminating the need to transport costly equipment and materials. Moreover, delocalising manufacturing could help provide necessary tools and equipment in a decentralised fashion, thereby also boosting local economies.[33]

3.2.4 Sustainable Development in Europe: History and Perspectives

The overall idea of creating a global agenda for sustainable development grew out of the work carried out by the United Nations, supported by the preparatory work of Prof. Sachs and his team, which is ongoing at the Earth Institute of Columbia University together with the Sustainable Development Solutions Network. From a historical perspective, though, the concept of sustainability was presented in 1713 by Hans Carl von Carlowitz,[34] the Chief Mining Administrator in the German region of Saxony, who held that forest timber sold to miners must be replaced by new trees,

[30] Ibid.

[31] Ibid.

[32] Ferretti S (2016) Space as an enabler of sustainable development. 4th ICSD, Columbia University, New York, NY.

[33] Ibid.

[34] Environment and Society Portal. "Hans Carl von Carlowitz and Sustainability" http://www.environmentandsociety.org/tools/keywords/hans-carl-von-carlowitz-and-sustainability

therefore ensuring a sustainable use of natural resources.[35] The much-cited definition of sustainable development that "meets the needs of the present without compromising the ability of future generations to meet their own needs" was also coined by a European, former Norwegian Prime Minister Gro Harlem Brundtland.

Europe is currently well engaged at various levels in sustainable development, including programmes and applications making use of space assets and data, such as Copernicus and Galileo. It is noteworthy that these programmes have been created by political actors and European institutions for the benefit of European citizens but provide valuable data globally, in line with an open free and full data policy, stemming from the best intentions of the Baveno Manifesto.[36]

Copernicus is the first operational programme in Earth Observation worldwide that provides critical information on land and sea, helping decision-makers to create better living conditions for citizens, while monitoring and protecting the planet.[37]

Europe is also investing in the economic growth and further development of partner countries, to extend its own space capabilities where they are most needed (e.g. the GMES and Africa programme), and is planning to soon provide operational services in the field of navigation, thanks to the full deployment of the Galileo GNSS system. Possible future initiatives include the provision of high-bandwidth connectivity by satellite in rural and underserved areas, to enable valuable services such as e-health and tele-education.[38]

Europe can also play a key role in the civil protection services of the future, including natural disasters and particularly earthquakes,[39] for which joint efforts by Italy and China have led to the addition of an Italian payload on board the China Seismo-Electromagnetic Satellite (CSES).[40]

European efforts are currently centred on government programmes. It is plausible, however, that downstream services stemming from open data will enable the private sector to also have an impact on global sustainable development.[41]

---

[35] Prof. Vittorio Prodi, Contribution to the ESPI 10th Autumn Conference on Space for Sustainable Development (2016). 26th Oct. 2017 https://www.espi.or.at/10th-autumn-conf

[36] Ibid.

[37] "Copernicus" European Commission 23 Oct 2017 https://www.eea.europa.eu/about-us/what/seis-initiatives/copernicus

[38] Prof. Vittorio Prodi, Contribution to the ESPI 10th Autumn Conference on Space for Sustainable Development (2016). 26th Oct. 2017 https://www.espi.or.at/10th-autumn-conf

[39] Prof. Vittorio Prodi, Contribution to the ESPI 10th Autumn Conference on Space for Sustainable Development (2016). 26th Oct. 2017 https://www.espi.or.at/10th-autumn-conf

[40] "CSES – CHINA SEISMO ELECTROMAGNETIC SATELLITE" Istituto Nazionale di Fisica Nucleare 23 Oct. 2017 http://cses.roma2.infn.it

[41] Prof. Vittorio Prodi, Contribution to the ESPI 10th Autumn Conference on Space for Sustainable Development (2016). 26th Oct. 2017 https://www.espi.or.at/10th-autumn-conf

3.3 The Key Actors

Having identified how space assets and technologies can support sustainable development efforts, it is now important to identify the relevant actors and stakeholders, both in the space sector (the supply side) and in sustainable development (the demand side).

3.3.1 The Supply Side

On the supply side, we find space agencies, government bodies, the satellite manufacturing industry, operators and value-added downstream service providers.

3.3.1.1 Space Agencies

The *European Space Agency*'s expertise is particularly suitable in the context of sustainable development, thanks to a proven capacity in space technology development, project management within international cooperation frameworks, Earth Observation missions' end-to-end management and the Integrated Applications Programme. Many of the *national space agencies* of Europe are involved in activities supporting developing countries in their efforts towards the full implementation of the UN Agenda 2030. Their role has relevance also because it builds upon traditional links of cooperation between a given donor and recipient countries and enhances the links at national level with general governmental aid institutions and other sectors.[42]

EUMETSAT has been an early mover, seeking to make meteorological and climate change data from space available and useful for sustainable development through its involvement, for example, in the Preparation for the Use of MSG in Africa (PUMA) programme of the early 2000s. This involvement was followed by strong participation in the African Monitoring of the Environment for Sustainable Development (AMESD) and the current Monitoring for Environment and Security in Africa (MESA) programmes. A characteristic of these programmes is that they bundle together the efforts and capabilities of many organisational actors receiving strong financial backing by the European Commission and placing the African Union and African regional economic communities centrally in the governance and implementation. These programmes can thus be pathfinders for inter-institutional cooperation on sustainable development also in other fields.[43]

[42] Ferretti S., Feustel-Büechl J., Gibson R., Hulsroj, P., Papp A., Veit, E. (2016). ESPI Report 59 Space for Sustainable Development. ESPI, Vienna.

[43] Ibid.

3.3.1.2 Government Bodies

It is not only space technology or service providers that are paramount in sustainable development efforts. Indeed, several essential coordination functions are performed by international and government organisations.[44]

The European Union, with its Copernicus and Galileo programmes, provides data and information of the greatest importance for sustainable development. The European Commission also created a dedicated activity, GMES and Africa, which has been aiding the more effective use of Copernicus data in the region. In addition, the Commission has funded several terrestrial equipment and training schemes relevant to the use of Earth Observation data. What is most important at present is that effective mechanisms are created that allow space-related information to flow freely between the various relevant Directorates-General of the Commission, so that, for instance, DEVCO and ECHO can even better integrate space services in their efforts.[45]

The *United Nations Office of Outer Space Affairs (UNOOSA) has a prominent role in facilitating and disseminating the adoption of space-based services within developing countries*. It serves as the secretariat for the United Nations Committee on the Peaceful Uses of Outer Space (COPUOS) and for the International Committee on Global Navigation Satellite Systems (ICG). UNOOSA also conducts activities that build human capacity in the fields of remote sensing, satellite navigation, meteorology and tele-education for the benefit of developing nations. Its role as the manager of the United Nations Platform for Space-based Information for Disaster Management and Emergency Response (UN-SPIDER) serves to support the use of space in disaster mitigation and relief.[46] The applications programme of UNOOSA has been very beneficial for developing countries, because it facilitates the application of space-based services to the local environment, eventually supporting also UNOCHA and other humanitarian assistance activities in the field.[47] UN-Space convenes annual sessions of the Inter-Agency Meeting on Outer Space Activities to discuss current and future activities, emergent technologies of interest and other related matters among UN system entities. For broader stakeholders' consultations, UN-Space organises informal sessions that are open to member states, the private sector, non-governmental institutions and academia.[48]

The *International Telecommunication Union* conducts similar activities in the field of telecommunications. Its mission to connect all the world's people entails

---

[44] Ibid.

[45] Ferretti S, Feustel-Büechl J., Gibson R., Hulsroj, P., Papp A., Veit, E. (2016) ESPI Report 59 Space for Sustainable Development. ESPI, Vienna.

[46] Ferretti S (2016) Space as an enabler of sustainable development. 4th ICSD, Columbia University, New York, NY.

[47] Ferretti S, Feustel-Büechl J., Gibson R., Hulsroj, P., Papp A., Veit, E. (2016) ESPI Report 59 Space for Sustainable Development. ESPI, Vienna.

[48] Ferretti S (2016) Space as an enabler of sustainable development. 4th ICSD, Columbia University, New York, NY.

allocating the global radio spectrum and satellite orbits as well as improving access to ICTs.[49]

Other important coordinating bodies, which have a mandate to extend use of, and access to, space technologies globally, include the *Group on Earth Observations* (GEO)[50] and the *Committee on Earth Observation Satellites* (CEOS), as well as the *International Charter on Space and Major Disasters*.

3.3.1.3 The Satellite Manufacturing Industry

The *satellite manufacturing industry* is important for sustainable development particularly in the planning of future space programmes, guiding customers towards innovative solutions and showcasing the possibilities of opening new services of relevance for developing countries.[51] Through R&D, also in cooperation with scientists and universities, the industry spearheads the development of new capabilities and new instrumentation, which will form the basis of future services that can be offered by satellites and constellations.

3.3.1.4 The Operators

The *operators* in the telecoms field have a crucial role to play as enablers in the development of new markets and business cases, thanks also to their ability to attract funding, including private investments.[52]

3.3.1.5 Value-Added Downstream Service Providers

Engaging the service providers of the *downstream segment* in filling the needs of development actors is key for making space ever more useful for sustainable development. Local entities and companies can collect local needs and requirements in much more efficient ways, and the associated development of the downstream segment can become a substantial economic activity for developing countries, also because entry barriers are relatively low in this segment.[53]

---

[49] "About ITU." International Telecommunications Union 10 Feb 2016 http://www.itu.int/en/about/Pages/default.aspx

[50] "GEO Vision" Group on Earth Observations 9 Feb 2016 https://www.earthobservations.org/vision.php

[51] Ferretti S, Feustel-Büechl J., Gibson R., Hulsroj, P., Papp A., Veit, E. (2016) ESPI Report 59 Space for Sustainable Development. ESPI, Vienna.

[52] Ferretti S., Feustel-Büechl J., Gibson R., Hulsroj, P., Papp A., Veit, E. (2016) ESPI Report 59 Space for Sustainable Development. ESPI, Vienna.

[53] Ibid.

Moreover, private or individual actors, such as notable billionaires, have a history of engaging in charity, and wealthy entrepreneurs have recently funded several space ventures. Altruism and technology are being combined by individuals like Zuckerberg with his internet.org campaign. However, the involvement of individual actors in development activities is not simple, requiring close coordination with local actors to avoid duplication of work and potential obstruction of activities without providing tangible benefits.[54]

3.3.2 *The Demand for Space Services in Sustainable Development*

The actors that can benefit from space services in the development field can be roughly grouped into seven categories: national governmental actors of developing countries, citizens in developing countries, non-governmental actors in developing countries, governmental aid organisations operating in developing countries, intergovernmental organisations in the field, the Office of the UN Special Representative for the SDGs and development consultancy firms.[55]

3.3.2.1 National Governments

National governments are central to any sustainable development effort as they have both policy-making and regulatory powers and provide essential services to their citizens. Development projects therefore need to be supported, integrated and coordinated with national governments and local authorities, as they are key enablers with their knowledge of local conditions and policies.[56]

3.3.2.2 Citizens

The ultimate beneficiaries of sustainable development efforts are the *citizens of a country*, and it is highly relevant to involve them beginning with the definition and planning of development activities. Nowadays useful information on societal trends and market demands can be extrapolated from big data sourced from social media platforms, as well as from space data.[57]

[54] Ferretti S. (2016) Space as an enabler of sustainable development. 4th ICSD, Columbia University, New York, NY.

[55] Ferretti S., Feustel-Büechl J., Gibson R., Hulsroj, P., Papp A., Veit, E. (2016) ESPI Report 59 Space for Sustainable Development. ESPI, Vienna.

[56] Ibid.

[57] Ferretti S., Feustel-Büechl J., Gibson R., Hulsroj, P., Papp A., Veit, E. (2016) ESPI Report 59 Space for Sustainable Development. ESPI, Vienna.

3.3.2.3 NGOs

NGOs play an extraordinarily important role in developing countries as they have realistic views on development challenges, particularly on those they themselves encounter in the field. NGOs are often stuck in the immediacy of trying to fill urgent needs and therefore might not always take the time to take the longer view or explore the utility of new approaches or tools, due also to some hesitancy in investing and adopting new technologies because donors tend to focus on short-term need alleviation.[58]

3.3.2.4 Governmental Aid Organisations

Governmental aid organisations also have a good view of conditions and challenges on the ground. However, they might be ready to entertain longer perspectives and innovation in tools and approaches, because of their links to general policy-making. These links might sometimes hamper actual operations but are also a good opportunity to try to make sure that new approaches fit in a broader context.

This is important for the deployment of new space services, the introduction of which might sometimes require dialogue between donor and recipient governments, and need to be carefully weighed as part of the prioritisation processes.[59]

3.3.2.5 Intergovernmental Organisations

From a top-level perspective, *international financial institutions* (IFIs), *regional organisations and international organisations* (IOs) also stand to benefit from the coordinating functions of space technology use. IFIs such as the World Bank; regional actors, such as the African Union and ASEAN; and IOs, such as the United Nations, all have a mission to help countries to develop by financing, advising or coordinating efforts.[60]

Intergovernmental organisations in the field are mostly part of the UN family. In many respects these organisations are indistinguishable from governmental aid organisations; yet, they always feature multicultural approaches and allow the leveraging of experiences over a wide spectrum of activities in many different settings. As has been recognised by the UN through its "Delivering as One" initiative, very good coordination is necessary to make sure that experiences of different actors in different environments are leveraged properly across the board and that ultimately efficiency of efforts is ensured.[61]

[58] Ibid.

[59] Ibid.

[60] Ferretti S. (2016) Space as an enabler of sustainable development. 4th ICSD, Columbia University, New York, NY.

[61] Ferretti S., Feustel-Büechl J., Gibson R., Hulsroj, P., Papp A., Veit, E. (2016) ESPI Report 59 Space for Sustainable Development. ESPI, Vienna.

Some UN organisations have expert knowledge and practical experience in leveraging space assets for their efforts, and the UN Office of Outer Space Affairs manages space applications programmes for sustainable development. In addition, the annual coordination meeting called UN-Space is a promising forum that should be expanded and used to become an even stronger entry point for the space community to seek dialogue with the UN system.[62]

Intergovernmental aid organisations not in the field, for example, the World Bank and the Directorates-General for Development and Cooperation and for Humanitarian Aid and Civil Protection of the European Commission, play a key role in funding development activities in developing countries.[63]

In this sense they are critical for the establishment of relevant programmes, and their knowledge of the potential of space is often decisive for whether space assets are utilised in the best manner in the overall development effort. The importance of raising such institutions' awareness of space as an enabler is further underscored by the advisory function that particularly the World Bank often fulfils. Like national aid authorities, this kind of institution is well suited to investing in targeted technology development for use in the field in developing countries and supporting in-house technology capacity building in NGOs and other sustainable development actors.[64]

3.3.2.6 The UN Special Representative for the SDGs

The holistic approach adopted by the SDGs puts a premium on good coordination at all stakeholder levels. The *UN Special Representative for the SDGs* plays a central role, and although it necessarily must take a bird's eye view and deal with a multitude of actors and specialisations, it would seem important[65] that space be understood as a substantial contributor to the achievement of the SDGs, so that also the Special Representative can raise relevant actors' awareness at all levels to assist consideration of whether needs might be best met by space assets in any situation. Thus, in some respects, the role of the Special Representative is to act as a conduit between the demand communities and the supply communities.[66]

[62] Ibid.

[63] Ferretti S., Feustel-Büechl J., Gibson R., Hulsroj, P., Papp A., Veit, E. (2016) ESPI Report 59 Space for Sustainable Development. ESPI, Vienna.

[64] Ibid.

[65] Ibid.

[66] Ibid.

3.3.3 The Importance of Dialogue

Space technologies can be relevant to the development projects of multiple SDG actors.[67] The key aspect to be underlined is the importance of the space community being aware of the great diversity of actors and tools, both within the space field and within sustainable development efforts at large, and focusing its efforts in optimising the mechanisms for dialogue and cooperation between this multitude of relevant contributors and the demand side,[68] with a view to ultimately proposing joint efforts based on creative programmatic thinking. This will be addressed in the following paragraphs.

3.4 Strategic Perspectives

This chapter began by outlining the three pillars on which the ESPI investigation rested: policy, strategy and programmatic framework; stakeholders mapping, engagement and partnerships; and, finally, capacity building, dialogue platforms and societal challenges.

Having examined how space programmes and policy are currently contributing to sustainable development efforts and the key actors involved both on the demand and supply sides, this paragraph focuses on stakeholder engagement through coordination and dialogue mechanisms and capacity building. These areas are ripe for a deeper involvement of space and can be useful to illustrate a future roadmap for space actors wishing to engage in sustainable development.

3.4.1 Current NGOs, Coordination and Dialogue Mechanisms

As we have previously seen, sustainable development spans across a multitude of disciplines, geographical boundaries and areas of responsibility. This creates a fragmented value chain, where interactions are required at different levels. Institutional dialogue mechanisms remain important to enjoy the support of national and local governments. But dialogue mechanisms with NGOs and organisations operating in the field are also a necessary complement to high-level coordination. The UN is a good example of this policy, hosting a special Non-Governmental Liaison Service,

[67] Ferretti S. (2016) Space as an enabler of sustainable development. 4th ICSD, Columbia University, New York, NY.

[68] Ferretti S., Feustel-Büechl J., Gibson R., Hulsroj, P., Papp A., Veit, E. (2016) ESPI Report 59 Space for Sustainable Development. ESPI, Vienna.

while UN offices also engage directly with prominent NGOs in their areas of operation.[69]

It is noteworthy that the Inter-Agency Standing Committee[70] (IASC) has created the Emergency Telecommunications Cluster (ETC), involving both service providers and end users, which is embedded within the UN.

Among the current dialogue mechanisms, involving both governmental and non-governmental actors, are the International Committee on Global Navigation Satellite Systems, the Global Earth Observation System of Systems, the International Charter on Space and Major Disasters and the Emergency Telecommunications Cluster.

3.4.2 Future Coordination

Optimising dialogue and cooperation mechanisms is central to allowing space to play its enabling role as effectively as possible vis-à-vis the sustainable development goals. At the political top level of the SDG management, there are the *UN Secretariat and the UN General Assembly*. Addressing the enabling functions of space in high-level political fora draws the attention of implementers, and being part of the dialogue means that space becomes part of the agenda.[71]

At this level, space technology is likely seen as a branch of hi tech, but it should be noted that the level of representation of space has been limited so far, since space agencies have been absent from the General Assembly discussions, except for the participation of UNOOSA. The *World Summit on the Information Society*, organised by the ITU and other UN specialised agencies, is also an important forum to reach sustainable development actors and be kept abreast of developments in hi-tech contributions to the SDGs. Clearly, space should become part of these discussions as well.[72]

The organisations and branches of the UN involved in the SDGs are many, and for many, space would be relevant. It would be highly advantageous if the space community, in whatever configuration, could become involved in the discussions centred on new programmes or the top-level monitoring of the implementation of existing programmes across the range of UN organisations. Within the UN system, the key figure is the *Special Representative for the SDGs*, appointed by the UN

[69] Ferretti S. (2016) Space as an enabler of sustainable development. 4th ICSD, Columbia University, New York, NY.

[70] The Inter-Agency Standing Committee (IASC) is the primary mechanism for inter-agency coordination of humanitarian assistance. It is a unique forum involving the key UN and non-UN humanitarian partners. The IASC was established in June 1992 in response to United Nations General Assembly Resolution 46/182 on the strengthening of humanitarian assistance. See "Welcome tot he IASC". Interagency Standing Committee 26 Oct 2017 https://interagencystandingcommittee.org/

[71] Ferretti S., Feustel-Büechl J., Gibson R., Hulsroj, P., Papp A., Veit, E. (2016) ESPI Report 59 Space for Sustainable Development. ESPI, Vienna.

[72] Ibid.

Secretary-General. Special efforts should be undertaken to organise joint initiatives, promoting information exchange at the highest level, together with central hi-tech representatives, such as Microsoft, Google and Facebook, and the space community, represented by the heads of the major space agencies and focused industrial actors in the telecommunications, navigation and Earth Observation domains.[73]

It is highly relevant that the space community creates synergies and develops partnerships also with the *World Bank* and other key *international financial institutions*, to address the issues related to financing of sustainable development initiatives.[74]

Within the European environment, a forum could be set up where once a year, high-level representatives of ESA and the national space agencies would meet with counterparts from the relevant Directorates-General of the Commission, e.g. GROW, DEVCO, ECHO, CONNECT, MOVE, ENVIRONMENT and CLIMATE ACTION, and with the External Action Service.[75]

The *International Astronautical Federation* has traditionally been a place where the issues and needs of developing countries are addressed. The annual International Astronautical Congresses (IACs) are the main events where space actors discuss the future evolution of their activities. These events could be the occasion to involve both high-level and working-level development actors. One possible initiative could be the creation of a Sustainable Development Networking Forum, dedicated to events bringing the space supply side together with sustainable development actors from the demand side. This would seem to chime with a desirable evolution of the IACs to bring together more those that could benefit more from space with space experts – a concrete manifestation of "Space for Earth".[76]

At the national level, it is recommended that space agencies and space industry become more directly involved with national aid agencies and aid communities, including especially NGOs, citizens and other actors working in the field. These actors could highly benefit from space services and therefore require attention, including participation in dedicated forums and events to showcase the potential of space technologies and applications. To this end each space agency could designate a "space ambassador" for outreach to sustainable development actors and for establishing two-way interaction that could help ingest the needs of the field into space agencies' future programmes.[77]

Finally, good stakeholder platforms are always promising keys for success, and therefore it is advisable to build on the successful work programme of the European Space Policy Institute on "Space for Sustainable Development", which has included an Advisory Board of senior advisors from both the supply and demand side, a high-level conference on the theme and the participation to political, social and

[73] Ibid.

[74] Ibid.

[75] Ibid.

[76] Ferretti S., Feustel-Büechl J., Gibson R., Hulsroj, P., Papp A., Veit, E. (2016) ESPI Report 59 Space for Sustainable Development. ESPI, Vienna.

[77] Ibid.

technical fora. This approach could be a good first step to improve dialogue on space and sustainable development across organisational boundaries to define more permanent dialogue mechanisms, either within existing structures or within newly created ones.[78]

3.4.3 New Tools for Cooperation?

Whereas dialogue mechanisms serve to channel needs and feedback on technological implementation, the use of space assets also requires diverse expertise and resources, which are not always readily available in developing nations. For this purpose, several types of actions could be implemented to join and share resources, to the advantage of all participants.[79]

Space agencies, for instance, could welcome third-party activities, by enabling the secondment of personnel from sustainable development agencies and being willing to carry out programmes with funding from developing countries, the private sector and NGOs, to channel their expertise into new fields and new regions.[80]

Thanks to its privileged position, UNOOSA could also co-ordinate the execution of joint programmes with the participation of different countries and global regions not covered by existing space agencies, to create new space-based services responding to unmet needs. Moreover, through its existing regional centres, local Space for Sustainable Development programmes could be initiated to further economic development and capacity building. Funding from aid agencies could be used in this context to elicit support from established space agencies on the more technical and programmatic aspects.[81]

Global student exchange programmes have supported generations of young and motivated students in gaining access to new knowledge while discovering new cultures. Establishing similar programmes between talented engineers from developing countries and developed space agencies, possibly including two-way exchanges, could thus support capacity building and promote future cooperation between the countries involved.[82]

Finally, existing frameworks such as the Charter on Space and Major Disasters, currently concentrating on disaster relief, would be extremely effective if extended to cover sustainable development efforts, supporting developing countries with specific requests to tackle their largest challenges.[83]

[78] Ibid.

[79] Ibid.

[80] Ibid.

[81] Ibid.

[82] Ferretti S., Feustel-Büechl J., Gibson R., Hulsroj, P., Papp A., Veit, E. (2016) ESPI Report 59 Space for Sustainable Development. ESPI, Vienna.

[83] Ibid.

3.4.4 Capacity Building

The space sector is currently entering a new era characterised by changes in technologies; social and environmental conditions, such as the advent of digitalisation; increased access to space data, thanks to free, full and open data policies; and new key players from start-ups, private companies and NGOs to entire spacefaring nations. In this context, new disciplines and methodological approaches are converging towards new ways of conducting business and academic research, which create opportunities for the emergence of a holistic capacity-building programme for all humankind.[84]

In the space sector, such a scenario could open the doors to achieving a seamless chain of innovation and exploitation of space, distributed across the world, thanks, for example, to digitalisation, innovative manufacturing technologies and miniaturisation of satellites. Access to space technology and services is becoming more open and can enable new solutions for sustainable development as well, by responding to the needs of the various communities involved. For example, the well-established United Nations/Japan Cooperation Programme on CubeSat Deployment from the International Space Station (ISS) Japanese Experiment Module (Kibo) "KiboCUBE" already provides developing countries the unique opportunity to deploy cube satellites (CubeSats), which they develop and manufacture.[85]

This enhanced networked and decentralised model, which includes concurrent design, advanced manufacturing techniques, innovative launch and operations concepts, could benefit from the Regional Centres for Space Science and Technology Education, which are affiliated to the United Nations. These centres are uniquely positioned to engage in synergistic partnerships with space agencies and industries to share design best practices, convey specific needs and requirements and identify existing matching space solutions.[86]

Improved connectivity in rural and underserved areas could also create new opportunities for both schools and universities, which could start with cocreating dedicated curricula to enhance STEM education opportunities for the younger generations. Thanks to their access to advanced space know-how, space agencies, intergovernmental organisations, academia and industry could sponsor and contribute to the creation of innovative interdisciplinary curricula, thereby preparing more citizens to fully exploit the benefits arising from innovation, space technologies and data.[87]

This capacity-building approach could further the potential for economic growth and sustainable development in the regions covered, which could benefit from technology spin-offs, a larger number of educated young people and the availability of

---

[84] Ferretti S. (2016) Space as an enabler of sustainable development. 4th ICSD, Columbia University, New York, NY.

[85] Ibid.

[86] Ibid.

[87] Ibid.

key services in the agricultural sector, in natural resources management and in e-health and tele-epidemiology domains, ensuring a more sustainable future for all on this planet.[88]

Such innovative capacity-building approaches are in line with the objectives of the UN Agenda 2030 and could contribute to the achievement of a substantial number of sustainable development goals, while also addressing thematic priority 7 of UNISPACE+50[89] for capacity building for the twenty-first century.[90]

3.5 Conclusions

The present chapter has outlined how space technology has distinctive advantages to contribute to the SDGs of Agenda 2030. It has identified relevant space solutions and technologies, as well as the space actors that need to be involved for new solutions to efficiently support sustainable development efforts. The opportunities that space technologies offer to developing countries, to leapfrog costly and resource-intensive technologies, are an indicator of the value they can and should provide in the immediate term. Moreover, space technology responds to essential data needs in the framework of the SGDs and provides access to advanced technologies that may be employed in disparate fields such as agriculture, smart cities, construction and renewable energies.

However, further actions are recommended to fully integrate space activities with the needs of the sustainable development community. First, actors in the field need to dispose of appropriate channels to be able to convey requirements towards space actors, which can then implement them within the appropriate programmes and technology roadmaps. These channels should be as direct as possible, fostering a two-way dialogue both in formalised settings, such as the UN, and more informal settings such as conferences or symposia. Moreover, an added benefit of cooperation between space actors and those involved in sustainable development could be the use of space as a platform for capacity building, sharing knowledge and ideas while encouraging new generations to pursue careers in STEM.

The year 2018 marks the 50th anniversary of the UNISPACE conference, and the celebratory UNISPACE+50 gathering will be an important discussion and decision point for global space activities. It is at this moment that additional efforts are required to ensure that the sustainable development community is included and heard from the point of view of the broad range of actors involved, so that their needs can inspire the key decision-makers as they look to the future.

[88] Ferretti S. (2016) Space as an enabler of sustainable development. 4th ICSD, Columbia University, New York, NY.

[89] "Unispace+50 Thematic Priorities". United Nations 26 Oct. 2017 http://www.unoosa.org/documents/pdf/unispace/plus50/thematic_priorities_booklet.pdf

[90] Ferretti S. (2016) Space as an enabler of sustainable development. 4th ICSD, Columbia University, New York, NY.

ESPI's activities have supported this vision, by initiating pilot dialogue platforms between space actors and civil society and moderating debates on how real needs may be met through the delivery of space-based services. These activities have also served to raise awareness of space technologies among development actors that were previously unaware of their potential, thus encouraging further uptake of space solutions.

Regarding future activities, there are several fields on which attention must be placed for these efforts to evolve from local initiatives towards holistic practices that are embraced by society at large. The recommendations outlined in the following chapters, which have been produced by the speakers at the ESPI 10th Autumn Conference "Space for Sustainable Development", provide additional insights along three dimensions: political, economic and governance.

At the political level, many governmental actors are engaged in sustainable development discourses, and they should be encouraged to include space communities within their communication and outreach fora, to create a constant and virtuous cycle of information that can be beneficial to all parties involved. Regional and local actors, for example, the European Union or the African Union, should participate in this process, to increase economic and social well-being by adapting solutions to pressing needs.

Sustainable development initiatives involving space technologies are not exempt from funding requirements, which need to be addressed with the appropriate providers. These could come from both the space sector and from the sustainable development domain, from UN development agencies, international financial institutions, local governments including rising economic powers, through to private actors and new space companies.

As the space sector evolves in response to new technologies and geopolitical shifts, space governance aspects will need to address sustainable development as a key element of future policies.

We conclude with the recommendation that the approach outlined above should be taken into serious consideration by all the relevant actors, for space activities to be fully utilised towards a more sustainable planet.

Acknowledgements The author gratefully acknowledges the valuable insights and contributions of Jörg Feustel-Büechl, former Director at ESA; Roy Gibson, former Director General of ESA; Peter Hulsroj, former Director of ESPI; Andreas Papp, International Director at SOS Children's Villages; and Elisabeth Veit, former Junior Researcher at ESPI, and Ambassador Thomas Stelzer in the elaboration of this contribution.

The author would like also to recognise the significant contribution to the ESPI Space for Sustainable Development programme of the participants to the dialogue platform with NGOs and of the speakers of the 10th ESPI Autumn Conference "Space for Sustainable Development".

Sincere gratitude is extended to the European Space Agency, the Italian Space Agency and the United Nation Office for Outer Space Affairs for their support and collaboration in the framework of the many activities undertaken in the last years. Finally, the author would like to thank Jean-Jacques Tortora, Director of ESPI; Alessandra Vernile, former ESPI Resident Fellow; as well as the ESPI team.

References

Ferretti S, Veit E (2016) "Space for sustainable development" (IAC-16-E3.1.15). Proceedings of the 67th International Astronautical Congress (IAC), Guadalajara, Mexico

Ferretti S, Feustel-Büechl J, Gibson R, Hulsroj P, Papp A, Veit E (2016a) ESPI Report 59 space for sustainable development. ESPI, Vienna

Ferretti S, Tortora JJ, Veit E, Vernile A (2016b) ESPI Report 60 engaging with stakeholders in preparation for UNISPACE+50. ESPI, Vienna

Ferretti S, Imhof B, Balogh W (2016c) "Future space technologies for sustainability on earth" (IAC-16-D4.2.1). Proceedings of the 67th International Astronautical Congress (IAC), Guadalajara, Mexico

Veit E (2015) From MDGs to SDGs: why now is the time to further integrate space into development. ESPI perspectives 74. ESPI, Vienna

Chapter 4
Challenges of Development and the Role of Space

Justin Loiseau

Abstract What is the status of global development? What role can space community stakeholders play in contributing to poverty alleviation efforts? This essay describes progress achieved to date by the international development community and challenges ahead, suggesting a complementary framework and two specific contribution opportunities for the space community's consideration.

What is the status of global development? What role can space community stakeholders play in contributing to poverty alleviation efforts? This essay describes progress achieved to date by the international development community and challenges ahead, suggesting a complementary framework and two specific contribution opportunities for the space community's consideration.

4.1 The Status of Global Development

4.1.1 Recent Progress

Over the past 25 years, the world has celebrated marked improvement on important development indicators. The United Nations Millennium Development Goals (MDGs) provide a clear benchmark for success. In many cases, there is reason to celebrate international progress. In developing countries, the extreme poverty rate (defined as those living on less than US$1.25 per day) dropped 33% points, from 47% in 1990 to 14% in 2015 (Fig. 4.1).[1]

[1] United Nations. Department of Economic, and United Nations. Department of Public Information. The Millennium Development Goals Report 2015. United Nations Publications, 2015. 3 http://www.un.org/millenniumgoals/2015_MDG_Report/pdf/MDG%202015%20Summary%20web_english.pdf

J. Loiseau (✉)
Abdul Latif Jameel Poverty Action Lab, Massachusetts Institute of Technology (MIT), Boston, MA, USA
e-mail: jloiseau@povertyactionlab.org

C. Al-Ekabi, S. Ferretti (eds.), *Yearbook on Space Policy 2016*, Yearbook on Space Policy, https://doi.org/10.1007/978-3-319-72465-2_4

Fig. 4.1 Extreme poverty rate in developing countries (Source: United Nations)

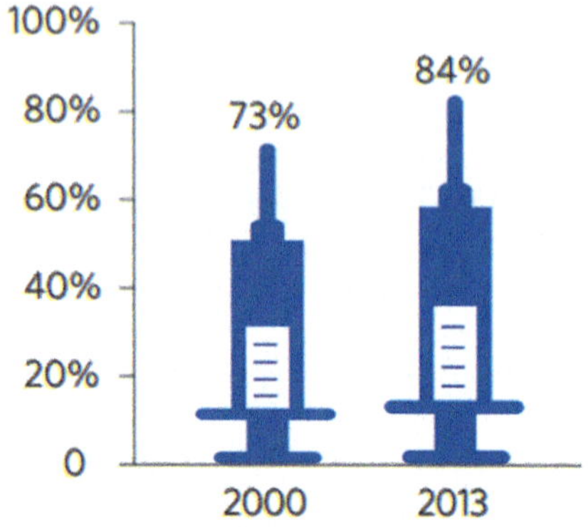

Fig. 4.2 Global measles vaccine coverage (Source: United Nations)

Worldwide, people are not only exiting extreme poverty, but important health outcomes are also improving. The global number of deaths of children under 5 has more than halved, dropping from 12.7 million children in 1990 to 6 million in 2015. Measles vaccinations are credited with having helped to prevent 15.6 million deaths between 2000 and 2013. During that period, global measles vaccine coverage increased 11 points to reach 84% of all children (Fig. 4.2).[2]

To improve opportunities for the world's youth, it is additionally essential to think beyond health and wealth, alone. Access to education is an important step in improving development opportunities for children and communities. In 1990 in sub-Saharan Africa, around half of all children were not enrolled in school. Over the past 25 years, however, countries in this region have made marked improvements in increasing access to education (Fig. 4.3). By 2015, 80% of children were enrolled in primary school.[3]

4.1.2 Challenges Ahead

These advances in development should be lauded. However, alongside advances in well-being, there is an increasing awareness in the international development community that there are still many problems for which no solutions exist—and even more troublesome, many things we believed to be solutions that haven't actually

[2] Ibid., 4.

[3] Ibid., 3.

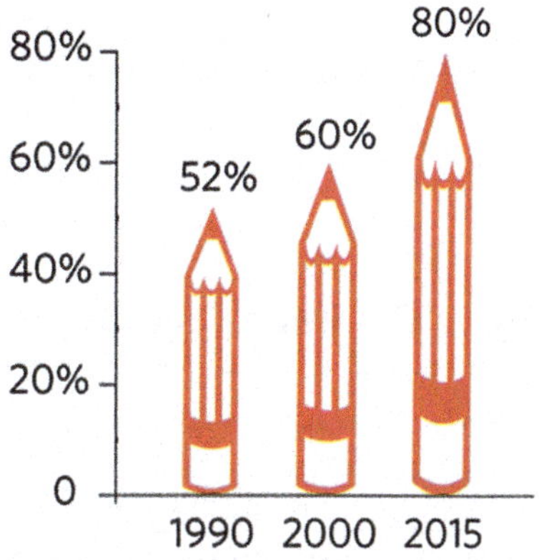

Fig. 4.3 Primary school net enrollment rate in sub-Saharan Africa (Source: United Nations)

Fig. 4.4 Evolution of the goal of achieving universal primary education leading to relevant and effective learning outcomes (Source: United Nations)

solved any problems. In general, much of the progress that has been made has been on first-level issues, and much of that progress has not been applied equitably across and within countries.

Consider the previously mentioned advances in enrollment. While ensuring that children are enrolled in school is an important goal, it is unfortunately not an indicator that will, in and of itself, improve a child's well-being. Instead, as the United Nations Sustainable Development Goals (SDGs) make clear, the international development community's objectives have evolved to include important outcomes. Rather than focusing on achieving universal primary education, the education-focused SDG lays out an objective of "free, equitable, and quality…education leading to relevant and effective learning outcomes" (Fig. 4.4).

Equally important to ensuring that education leads to learning outcomes is ensuring that education leads to equitable learning outcomes for all children. While 80% of children in sub-Saharan Africa are now enrolled in primary school, the 20% that are not enrolled likely suffer from greater marginalization. They may be girls, orphans, those forced to work, or live far from public amenities. These children, as well as others who the international development community has thus far failed to help achieve their full potential, will, in some ways, be even more difficult to reach than any of the 80% who are already enrolled in primary school.

4.2 The Role of Space

Having provided an overview of the status of global development, as well as some of the challenges ahead, it is clear that there is ample opportunity for various communities to collaborate further to address these challenges. The European Space Policy Institute (ESPI) recognizes that there is a key role that the space community already plays in international development, as well as some exciting opportunities for further collaboration. The idea of inter-sector collaboration is especially salient to J-PAL, as it is a core component of our own mission. J-PAL works to reduce poverty by ensuring that policy is informed by scientific evidence. We achieve this by conducting research through rigorous impact evaluations, disseminating evidence on what works through policy outreach, and building the capacity of policymakers and practitioners to accurately consume and generate evidence on what works.

There exist myriad overlap opportunities between space and international development sectors. Some are de facto overlaps, such as the natural spillover of space technology to earth-based applications. At J-PAL, we reflect often on our comparative advantage in a community comprised of many different types of development stakeholders. The space community, as well, has a comparative advantage in several contribution opportunities, two of which are briefly summarized below.

4.2.1 Institutionalizing Overlap

Space technology is now nearly inseparable from the way in which we live our lives on earth, whether it is the food we eat or the phone we use. Some of these technologies benefit from clear and unobstructed opportunities to directly apply themselves to a real-world opportunity. For others, applications beyond their original design may be less clear, or there may be other obstacles such as regulation or cost that prohibit simple replication. Fortunately, there are institutes and divisions that make this search for overlap an express mandate of the space community.

To date, NASA's Technology Transfer Program has amassed a database of nearly 2000 spinoff technologies and over 1000 products available for licensing.[4] The European Space Association (ESA) has seven business incubation centers that support more than 60 start-ups every year dedicated to ensuring space technology has a

[4]"NASA Spinoff". 2016. NASA 20 Dec. 2016 https://spinoff.nasa.gov/

place on Earth.[5] There are also international charters that exist outside the scope of any one organization. For example, the International Charter on Space and Major Disasters makes satellite-derived information available to support disaster response efforts.[6]

The opportunities for overlap are especially compelling when considering the focus of each community's work. Space agencies are forced to innovate to deal with some of the most extreme and resource-scarce corners of the universe and, on Earth, many of those same innovations can be used in similarly extreme and resource-scarce regions of the world. Supporting, proactively publicizing, and making use of these institutional overlaps is key to ensuring that the international development community and the space community do not operate in parallel worlds. However, the development community has continued to learn that technology and innovations are not solutions, in and of themselves. Many space technologies are ultimately not applicable to international development, and it is essential to continue to rigorously evaluate any such attempted adaptation to determine the true impact of capitalizing on this overlap.

For example, J-PAL's Africa office recently conducted some exploratory scoping work in Sierra Leone to learn more about the country's health systems and usage following the 2014 Ebola outbreak. Sierra Leone ranks seventh from last on the United Nations Development Programme's (UNDP) 2014 Human Development Index[7] and has an average life expectancy of 51, the third lowest in the world.[8] Electricity reaches less than 10% of the population,[9] and, historically, many families have been forced to travel long distances to village centers to receive immunizations for their children. But solar-powered battery-free heavily insulated refrigeration units, spun-off from NASA in 1999, have changed immunization storage forever.[10] Since 2002, nurses in Sierra Leone have been able to store vaccines in individual refrigeration units at local health clinics (Fig. 4.5).

This space technology has certainly benefited Sierra Leoneans. However, it does not imply that Sierra Leone subsequently achieved universal immunization. While this technology has expanded to many health centers, 60% remain without refrigeration units. Where units do exist, a 2015 assessment found that 30% are

[5] "Spin-Off". 2016. European Space Agency 20 Dec. 2016 http://www.esa.int/Our_Activities/Space_Engineering_Technology/TTP2/Highlights/Spin-off

[6] "International Charter Space And Major Disasters; UN-SPIDER Knowledge Portal". 2016. *Un-Spider.Org*. 20 Dec. 2016 http://www.un-spider.org/space-application/emergency-mechanisms/international-charter-space-and-major-disasters

[7] "Human Development Report 2015". 2015. United Nations Development Programme 20 Dec. 2016 http://hdr.undp.org/sites/default/files/2015_human_development_report.pdf

[8] "Life Expectancy At Birth, Total (Years) | Data". 2016. Data.Worldbank.Org. 20 Dec 2016 http://data.worldbank.org/indicator/SP.DYN.LE00.IN?year_high_desc=false

[9] "National Energy Profile of Sierra Leone". 2012. United Nations Development Programme 20 Dec. 2016. http://www.undp.org/content/dam/sierraleone/docs/focusareadocs/undp_sle_energy-profile.pdf

[10] "Solar Refrigerators Store Life-Saving Vaccines". 2016. NASA Spinoff 20 Dec. 2016 https://spinoff.nasa.gov/Spinoff2013/hm_4.html

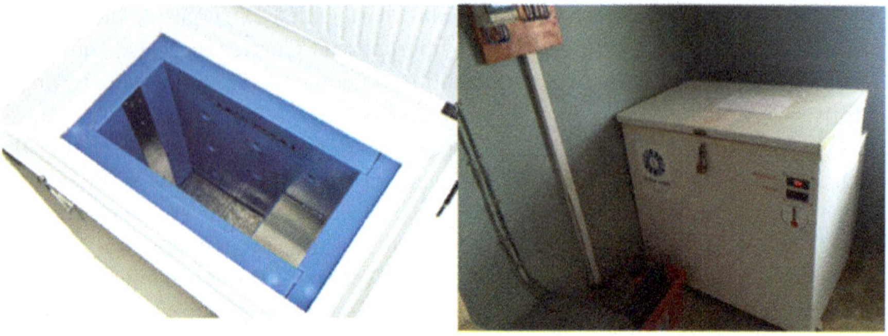

Fig. 4.5 SunDanzer Refrigeration's Solar Direct-Drive Vaccine Refrigerator (Source: NASA (left), Loiseau (right))

dysfunctional, and a further 10% may function but are disabled due to unreliable power supplies.[11] Perhaps most importantly, however, there may be reasons beyond the control of any refrigeration technology that prevent full immunization from being achieved. People in Sierra Leone, as everywhere, are often busy, cannot afford to take time off work, and may make informed decisions to delay immunizing their children when it does not provide immediate benefit to them. For all these reasons, the space community can contribute positively to international development by not just institutionalizing overlap to identify innovative collaboration opportunities but by also assisting the international development community in testing these and other innovations.

4.2.2 Testing Innovation

In addition to innovating for decades, the space community has also been testing innovations. Its methodical trial-and-error approach has resulted in some difficult and tragic lessons, and it continues to tweak and hone its analyses to be increasingly accurate, efficient, and effective. It is only through such unswerving analysis that the space community can ensure that its innovations are performing as intended.

The international development community, in general, and impact evaluation organizations such as J-PAL, in particular, have benefited from innovations in other disciplines, including space, to conduct better research. Less than a decade ago, it was a nearly universal norm to conduct household-level surveys using pen and paper. For one such evaluation in Uganda, surveyors completed over 5000 20-page surveys at 250 locations around the country. This resulted in around 50,000 survey pages that ultimately required manual data entry in order to be statistically summarized and analyzed. In 2016, the norm shifted to electronic surveying.

[11] Sierra Leone Health Facility Assessment March 2015 – MoHS Validation Workshop2015. Sierra Leone Ministry of Health & Sanitation and UNICEF.

Instead of printing 50,000 pages, research teams can simply use 50 smartphones, an innovation with innumerable ties to space technology advancements. Electronic surveying reduces human error and additionally enables valuable new data types to be gathered, such as GPS coordinates to "geo-tag" individual households in remote villages.

A recent evaluation by Northwestern University Associate Professor and J-PAL affiliate Seema Jayachandran,[12] and co-authors, utilized a different type of space technology to ask a previously unanswerable question. While satellite imagery is an important component of much descriptive geographical analysis, Dr. Jayachandran partnered with Uganda's local forest authority to use satellites to not only map tree coverage in rural Uganda but to also evaluate the effectiveness of a payment for ecosystem services (PES) program.[13]

PES is a policy approach to curb deforestation where individuals are paid to refrain from cutting down trees on their land. PES programs are increasingly popular, especially in developing countries, because they are voluntary and so do not force people to protect the environment or impoverish them by taking away a key source of income. However, it is not clear what portion of payments goes to forest owners who were not intending to cut down trees anyway. Furthermore, individuals could comply with the contracts for covered land to receive payment but shift deforestation activities to other land. To answer these and other questions, Dr. Jayachandran and co-authors conducted the first ever randomized evaluation of PES across 121 Ugandan villages[14] from 2010 to 2013.

To determine the program's effectiveness, it was necessary to gather information on changes in tree coverage over time. Researchers determined that on-the-ground tree coverage assessments such as tree counting were prohibitively difficult and expensive and so instead had surveyors calculate the GPS coordinates of landowners' homes. This allowed them to independently confirm how many trees were around each landowner's property using commercial satellites. A computer algorithm determined if each pixel in the satellite image (measuring 2.4 by 2.4 m) contained a tree at the beginning and end of the 1.5-year study period. Dr. Jayachandran and co-authors found that, in fact, farmers who were randomly selected to be offered these payments cut down around half as many trees as those that were not offered PES. In this case, satellite imagery not only served as a tool for better research evaluation but was also the foundational innovation upon which a PES program like this could be cost-effectively scaled up across the country. This is an especially promising policy solution for Uganda, which has the third highest deforestation rate in the world.[15]

---

[12] "Seema Jayachandran". J-Pal 20 Dec. 2016 https://www.povertyactionlab.org/jayachandran

[13] Jayachandran, Seema, De Laat, J., Lambin, E., Stanton, C. "Cash for Carbon: A Randomized Controlled Trial of Payments for Ecosystem Services to Reduce Deforestation." Working Paper Oct. 2016 https://www.povertyactionlab.org/sites/default/files/publications/465_Cash-for-Carbon_Seema_Oct2016.pdf

[14] Ibid.

[15] Ibid.

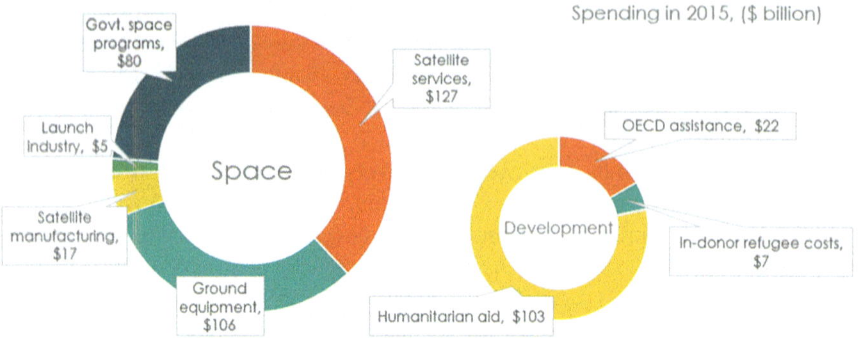

Fig. 4.6 Space industry spending compared with development aid (Source: Tauri Group via Economist, OECD)

4.3 A Case for Collaboration

In 2015, space industry spending amounted to $303 billion.[16] In the same year, around $130 billion was spent on development aid, worldwide[17] (Fig. 4.6).

Those who believe that the space community and the international development community have no opportunity to overlap might, then, debate the optimal distribution of resources for these two purposes. However, as discussed here, there are real and actionable ways in which each community can complement the other. By institutionalizing such overlap and further contributing to the testing of its own innovations and those of other development stakeholders, the space community can play an increasingly important role in ensuring that we continue to reduce poverty around the world.

About J-PAL

The Abdul Latif Jameel Poverty Action Lab (J-PAL) is a network of more than 140 affiliated researchers from over 40 universities. Our mission is to reduce poverty by ensuring that policy is informed by scientific evidence. We engage with hundreds of partners around the world to conduct rigorous research, build capacity, share policy lessons, and scale up effective programs. J-PAL was launched at the Massachusetts Institute of Technology (MIT) and now has regional offices in Africa, Europe, Latin America and the Caribbean, North America, South Asia, and Southeast Asia. For more information, visit povertyactionlab.org.

[16] "A Sudden Light". 2016. The Economist 20 Dec. 2016 http://www.economist.com/technology-quarterly/2016-25-08/space-2016

[17] "Development Aid Rises Again In 2015, Spending On Refugees Doubles – OECD". 2016. Oecd.org. 20 Dec. 2016 http://www.oecd.org/dac/development-aid-rises-again-in-2015-spending-on-refugees-doubles.htm

References

A Sudden Light (2016) The Economist. 20 Dec. 2016. http://www.economist.com/technology-quarterly/2016-25-08/space-2016

Development Aid Rises Again in 2015, Spending on Refugees Doubles – OECD (2016) Oecd. Org. 20 Dec. 2016. http://www.oecd.org/dac/development-aid-rises-again-in-2015-spending-on-refugees-doubles.htm

Human Development Report (2015) United Nations Development Programme. 20 Dec. 2016. http://hdr.undp.org/sites/default/files/2015_human_development_report.pdf

International Charter Space and Major Disasters; UN-SPIDER knowledge portal (2016) Un-Spider. Org. 20 Dec. 2016. http://www.un-spider.org/space-application/emergency-mechanisms/international-charter-space-and-major-disasters

Jayachandran S, De Laat J, Lambin E, Stanton C (2016) Cash for carbon: a randomized controlled trial of payments for ecosystem services to reduce deforestation. Working Paper Oct. 2016. https://www.povertyactionlab.org/sites/default/files/publications/465_Cash-for-Carbon_Seema_Oct2016.pdf

Life Expectancy at Birth, Total (Years)|Data (2016) Data.Worldbank.Org. 20 Dec. 2016. http://data.worldbank.org/indicator/SP.DYN.LE00.IN?year_high_desc=false

NASA Spinoff (2016) NASA. 20 Dec. 2016. https://www.spinoff.nasa.gov/

National Energy Profile of Sierra Leone (2012) United Nations Development Programme. 20 Dec. 2016. http://www.undp.org/content/dam/sierraleone/docs/focusareadocs/undp_sle_energyprofile.pdf

Sierra Leone Health Facility Assessment March 2015 – MoHS Validation Workshop (2015) Sierra Leone Ministry of Health & Sanitation and UNICEF

Solar Refrigerators Store Life-Saving Vaccines (2016) NASA Spinoff 20 Dec. 2016. https://spinoff.nasa.gov/Spinoff2013/hm_4.html

Spin-Off (2016) European Space Agency. 20 Dec. 2016. http://www.esa.int/Our_Activities/Space_Engineering_Technology/TTP2/Highlights/Spin-off

Sustainable Development Goals – United Nations (2016) United Nations Sustainable Development. 20 Dec. 2016. http://www.un.org/sustainabledevelopment/sustainable-development-goals/

United Nations Department of Economic, and United Nations. Department of Public Information. "The Millennium Development Goals Report 2015". (2015). United Nations Publications. 20 Dec. 2016 http://www.un.org/millenniumgoals/2015_MDG_Report/pdf/MDG%202015%20Summary%20web_english.pdf

Chapter 5
Reflecting on a Decade of Collaboration Between NASA and USAID: Deriving Value from Space for International Development

Jennifer Frankel-Reed

5.1 "Connecting Space to Village" Through the SERVIR Collaboration

5.1.1 Origins of the SERVIR Collaboration

The US Agency for International Development (USAID) and the National Aeronautics and Space Administration (NASA) are US government agencies that have a lot in common despite their very different purposes. Among NASA's goals is improving life on Earth. It advances this goal through a constellation of satellites that generate Earth observation data, by making that data available for free to the public, as well as through an extensive applied research portfolio aimed at discovering and demonstrating innovative and practical uses of Earth science data and knowledge. NASA funds researchers at US institutions conducting activities all around the world but has a limited ability to fund international partners to advance the application of data in their countries. Within USAID's mission is the objective of ending extreme poverty and promoting resilience in developing countries, which it does through an extensive network of in-country assistance programmes and local staff and offices that work with local partners. USAID partners with local authorities and communities to address issues such as food security, natural resources management, health, and humanitarian assistance that require good data in places that are often very data-limited.

It is against this backdrop that NASA scientists and USAID regional development and geospatial experts looked at each other's agencies in 2004 and saw the potential for powerful collaboration. Through SERVIR, NASA's science and technology and USAID's development investments are brought together with the aim of

J. Frankel-Reed (✉)
U.S. Agency for International Development (USAID), Washington DC, USA
e-mail: jfrankel-reed@usaid.gov

C. Al-Ekabi, S. Ferretti (eds.), *Yearbook on Space Policy 2016*,
Yearbook on Space Policy, https://doi.org/10.1007/978-3-319-72465-2_5

empowering developing countries to better monitor, predict, and respond to environmental challenges through the use of Earth observation satellite data. The two agencies now see their partnership as essential to helping each one fully deliver on its mission. This is how the SERVIR programme motto "connecting space to village" was born.

Today, the SERVIR network of experts around the world possesses a wealth of experience in creating user-driven information products and services with Earth observation data. Many of these efforts have succeeded over the years but certainly not all of them. Based on that learning, and supported by a spirit of adaptive management, the programme is capturing good practices that are relevant for a much broader community of practice. That process continues, and SERVIR aims to continue to make new connections and evolve. This chapter describes how the programme works, how it takes advantage of different partners' strengths, and shares some lessons that may be relevant to others working on similar efforts.

5.1.2 History

The first activities of the SERVIR programme, SERVIR-Mesoamerica, were launched in 2005 through a partnership with the Water Center for the Humid Tropics of Latin America and the Caribbean (CATHALAC) headquartered in Panama. SERVIR-Mesoamerica served Central America and the Dominican Republic. Due to the breadth of SERVIR's activities in the region, the 2007 Earth Observation Summit recognized SERVIR in the First 100 Steps towards the implementation of the Global Earth Observation System of Systems (GEOSS).[1] It was in Mesoamerica that the programme name SERVIR, meaning "to serve" in Spanish, was started. Based on the success of this first effort, the programme continued the model of supporting regional "hub" institutions (see Sect. 5.2.3 for more about hub requirements). The next region for expansion was Eastern and Southern Africa, which launched in 2008. The SERVIR-Eastern and Southern Africa hub was established through a partnership with the Regional Centre for Mapping of Resources for Development (RCMRD), a leading intergovernmental organization headquartered in Kenya with 20 contracting member states. SERVIR-Eastern and Southern Africa focuses on six priority countries in the region.

In 2010, the next hub was established at the International Centre for Integrated Mountain Development (ICIMOD), a regional intergovernmental learning and knowledge-sharing centre in Nepal serving its eight regional member countries of the Hindu Kush-Himalaya region. SERVIR-Hindu Kush Himalaya focuses on five priority countries in the region. In 2015, SERVIR-Mekong was launched in Southeast Asia, implemented by the Asian Disaster Preparedness Center (ADPC) and its consortium partners Spatial Informatics Group (SIG), Stockholm

[1] "The First 100 Steps to GEOSS." 30 Nov. 2007. Group on Earth Observations Secretariat 20 Dec. 2016 https://www.earthobservations.org/documents/the_first_100_steps_to_geoss.pdf

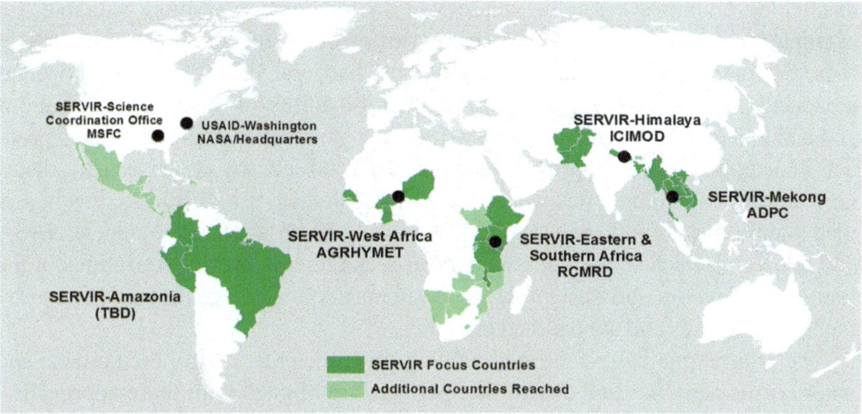

Fig. 5.1 SERVIR network map

Environment Institute (SEI), and Deltares. ADPC, based in Thailand, is a regional resource centre on disaster risk reduction and climate change adaptation in Asia. SERVIR-Mekong serves five countries in the Lower Mekong Region.

The fifth hub, SERVIR-West Africa, launched in 2016. This hub is coordinated by the Agriculture, Hydrology and Meteorology (AGRHYMET) Regional Centre, a subsidiary of the Comité Permanent Inter-Etats de Lutte contre la Sécheresse dans le Sahel (CILSS). AGHRYMET is based in Niger, and serves West African countries. SERVIR-West Africa focuses on four priority countries in the region. A number of West African organizations are consortium partners in SERVIR-West Africa, including the Centre for Remote Sensing and Geographic Information Services (CERSGIS) in Ghana and Centre de Suivi Ecologique (CSE) in Senegal.

SERVIR-Amazonia is planned for 2017. In total, the SERVIR network now reaches more than 40 countries (Fig. 5.1).

5.1.3 *Purpose and Principles*

The value of Earth observation data to support evidence-based decision-making is well established in a variety of areas.[2] That value is relevant around the world, but the use of Earth observation data is particularly limited in developing countries. Developing countries often lack the necessary skilled analysts and experts, research and monitoring budgets to support data collection and validation, computing infrastructure and bandwidth, budgets for software licences, and enabling environments that support accountability and data sharing. Moreover, in countries where ground-based observations are lacking, sporadic, or incomplete – as is the case in many

[2]"Our Mandate." Group on Earth Observations 8 Jul. 2017. https://www.earthobservations.org/vision.php

developing countries – satellite sources of data become even more valuable, for example, for monitoring and enforcement of resource use or disaster impacts in remote or hard-to-access areas. In light of those barriers, SERVIR was developed as a capacity building programme to strengthen the ability of regional organizations and national decision-makers to work with Earth observation data and derived products and applications to improve their countries' food security, water security, disaster management, climate resilience, and sustainable land use. SERVIR aims higher than simply developing analytical products using Earth observation data at NASA and delivering them to partner countries. It seeks to empower organizations in the regions to develop and sustain decision support solutions themselves for use by member countries and subnational entities.

The programme has established guiding principles that clarify the purpose and nature of its activities and ensure that partners are aligned with these approaches. They reflect that SERVIR is not primarily a research programme, and it is not primarily a training programme, although elements of applied research and training are involved. It is about capacity development and applying science to policy and practice at its core. SERVIR is focused on the following principles:

- Delivering solutions that respond to regional development needs
- Connecting the most appropriate Earth observation data and high-quality applied science
- Working through co-development on integrated, user-tailored services
- Building capacity to strengthen service providers and promote sustainability
- Promoting free and open data
- Sharing knowledge and promoting exchange across hub and partner networks

In line with these principles, SERVIR activities must be needs-driven rather than supply-driven, use the right data and science (not necessarily the state-of-the-art), and be developed through collaboration, with sustainability built in from the start. All of the products, applications, and source code developed by the programme are available online, in keeping with the principles of open data and exchange (with accommodations made to safeguard certain sensitive input data if required by local partners).

Priority "service areas" or themes have been selected to increase technical focus. SERVIR prioritizes the areas of greatest demand: agriculture and food security; water resources and hydrological disasters; land cover, land use, and ecosystems; and weather and climate.

5.1.4 The SERVIR Theory of Change

Theories of change are used in the development community for programme planning and evaluation. They are useful because they define a long-term goal or objective and work backward to identify necessary preconditions to achieving that goal. They are similar to a scientific hypothesis. SERVIR has used a theory of change to

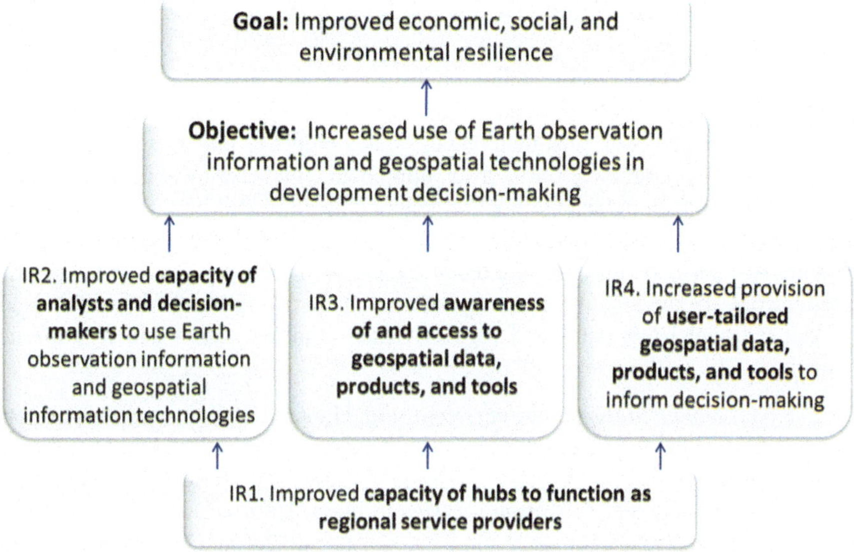

Fig. 5.2 SERVIR theory of change

structure the programme simply, with the overall objective of increasing the use of information in decision-making. Scientists and researchers who have worked on applications of their research and analysis understand that this is a humbling endeavour and easier said than done. The theory of change, or hypothesis driving the programme, is depicted in Fig. 5.2 and can be described as follows, from bottom to top.

If SERVIR *works with regional institutions to improve their capacity* to provide services to actors in their regions (Intermediate Result or IR 1), and as a result (a) *analysts and decision-makers have improved capacity to use Earth observation information* (IR2), (b) there is *improved awareness of and access to relevant information* (IR3), and (c) *user-tailored information services are provided to decision-makers* (IR4), then we will achieve *increased use of Earth observation information in development decision-making*. Ultimately, the use of improved information supports improved economic, social, and environmental resilience in developing countries.

The theory of change reflects what SERVIR has learned: that if any of the results are missing, the use of information in decision-making is not achieved. For instance, if the authorities responsible for local flood preparedness know how to work with information services providing flood early warnings (IR2), and they are able to access it (IR3), but it does not provide properly tailored and contextualized information to answer the question they need to act upon (IR4), it will not be used. Perhaps the warning does not come soon enough, or it is not accurate enough, or the variable that local authorities care about is not provided. In another case, if the flood prediction is well-tailored to the authorities' needs (IR4), and they are able to access the information (IR3), but they do not have the software or know-how to integrate the warning into their existing disaster management system (IR2), it will not be used.

5.1.5 How SERVIR Works

Making the theory of change operational means that USAID and NASA provide support to strengthen the skills, data, science, and infrastructure at regional organization and that regional hubs then engage stakeholders to understand their needs; help to access and integrate Earth observation data with other local data; work with models and other analytical methods to generate user-tailored information services; develop products, training, and tools to package and deliver information effectively; and manage partnerships to ensure feedback and improvements over time. The programme has six key institutional components that are carefully coordinated through annual regional work plans and regular virtual communication and face-to-face meetings:

- *Regional hub institutions, partners, and staff.* Hubs and their partners are responsible for conducting needs assessments in focus countries, identifying opportunities to work with decision-makers to design information services that apply Earth observation data and delivering services to users. Hubs are staffed by teams of geospatial analysts, remote sensing specialists, subject matter experts, training experts, stakeholder engagement experts, communications professionals, and more.
- *NASA's SERVIR Science Coordination Office.* The SERVIR Science Coordination Office is staffed with NASA experts responsible for supporting scientific activities across the network, including training for hub staff and scientific assistance for products and applications; ensuring thematic exchange across regions through the facilitation of thematic working groups; providing support for geospatial information technology, data sharing, and best practices; and identifying subject matter experts that can assist hubs as needed.
- *SERVIR Support Team.* The support team is a small contracted team responsible for organizing technical exchanges to encourage sharing and capacity building across the network, providing capacity building support to hubs on organizational management topics, maintaining a global website, and managing communications.
- *USAID's field offices and partners.* USAID field office and headquarters staff facilitate relationships with partners implementing other development programmes to strengthen buy-in and uptake of SERVIR services. USAID also oversees hub funding agreements.
- *US science collaborators.* NASA competitively selects additional SERVIR applied science researchers on a periodic basis, typically via 3-year grants. Through this process, principal investigators at US-based research institutions work with SERVIR hubs on joint proposals in areas of need identified by the hubs. Activities include applied research, product and tool co-development, dissemination, and training.
- *Private sector partners.* Private sector SERVIR partners include Esri, Google, Amazon, and DigitalGlobe. Esri provides hubs with GIS software solutions at discounted prices. Google facilitates access and use of its Google Earth Engine

by the hubs, which then provide training to users in their regions on the platform. Amazon and Google are working with SERVIR on cloud solutions at discounted prices. DigitalGlobe is providing hubs with reduced rates for high-resolution satellite imagery and value-added services.

The process for developing SERVIR activities has three steps that ensure that the programme is not pushing science and data out but rather understanding needs and establishing relationships in order to increase the chances of sustained use of science and data (Fig. 5.3). In the first step, Consultation and Needs Assessment, hub teams engage local stakeholders broadly, including USAID field offices and their partners, to identify and prioritize development challenges and identify SERVIR's specific niche and likely user groups to work with to address priority challenges. In the second step, Service Design, hub teams engage a smaller set of decision-makers and experts from the NASA Science Coordination Office and collaborating scientists to design a specific service and develop plans for the component products, tools, data sets, trainings, and outreach activities involved. In the third step, Service Delivery, hub teams work with decision-makers and collaborating scientists to pilot, disseminate, and evaluate the service.

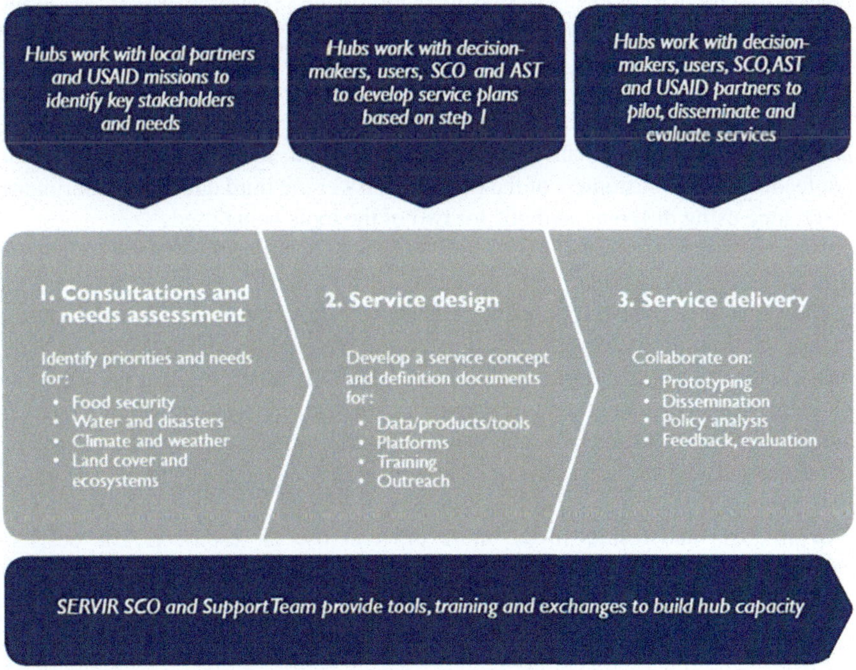

Fig. 5.3 Service planning and delivery process

5.2 Results, Evaluating Progress, and Lessons Learned

5.2.1 Results

Over the years, SERVIR has developed more than 70 products and applications to meet the needs of decision-makers. The programme uses data from 28 Earth observing satellites and sensors, has trained more than 3500 people to use its tools and information products, partnered with 250 institutions globally, and enabled more than 400 decision-makers and scientists to participate in technical exchanges. The programme has also engaged 27 different US-based research collaborators to bring cutting edge techniques to help solve regional development challenges. The results cannot be fully captured by facts and figures alone, however. There are many dimensions of positive benefit that are more difficult to quantify, such as science diplomacy, promoting data sharing, engaging youth in science and technology, empowering female scientists, and promoting South-South exchange across hubs.

5.2.2 Evaluation

About a decade into the programme, USAID funded a major, third-party performance evaluation designed to answer three questions: (1) How are SERVIR's applications being used and with what results? (2) Are SERVIR hubs becoming stronger regional service providers? (3) What is the value of SERVIR in terms of the monetary value of its effects? To answer the first question, case studies were carried out for nine of SERVIR's longer-running products to ensure time for adoption and use and enable evaluation. To answer the second question, surveys were carried out to get feedback from customers of the hubs. To answer the third question, experiments were done using different methods for two of the tools being used.

The evaluation generated many useful findings, including the general conclusion that SERVIR products are part of larger decision-making systems in which the product may increase data confidence by filling a gap, complementing other sources of information, or enabling more efficient and targeted use of resources for local data collection. For example, in El Salvador, SERVIR maps of algal blooms complement on-the-ground testing of water and shellfish tissue samples. In Nepal, an agricultural monitoring tool is one way to monitor food insecurity that contributes to food assistance targeting. So-called administrative efficiency gains accrue due to better access to data than would be otherwise available without the Earth observation products and results in the more efficient use of resources. The costs of sampling more comprehensively or flying planes or sending out data collectors is much higher than the costs of analysing satellite datasets and targeting local data collection more narrowly.

With respect to the second question, the evaluation found that the hubs were well-regarded as centres of excellence for data analysis and training. For the third question, two products were studied that both showed monetary value to their users – forest fire hotspot monitoring in Guatemala and frost mapping in Kenya.

5.2.2.1 Forest Fire Hotspot Monitoring Tool in Guatemala

The Petén region in Northern Guatemala is significant for its history, ecology, and tourism, and 60% of the region is susceptible to forest fires. Each year, government authorities work to manage fires. A hotspot monitoring tool developed with SERVIR support in Guatemala enables the government to be more cost efficient in its forest fire efforts by using satellite-derived hotspot data, saving scarce public resources for other needs. The evaluation found that communities using the hotspot monitoring tool had fewer to no uncontrolled forest fires and that local farmers were more likely to follow local rules permitting agricultural burning because they knew that fire activity was being monitored daily by the system. The use of the hotspot maps appears to be linked to a change in behaviour, potentially decreasing the amount of irresponsible agriculture burning, informing the deployment of resources to fight fires, and improving forest fire management. A willingness-to-pay survey found that users of the forest fire hotspot monitoring tool were willing to pay an average of $78 per year for daily access to the hotspot maps, because of the tool's frequency of reporting and its reliability. The information is made available for free and will continue to be provided free of charge.

5.2.2.2 Frost Mapping and Forecasting in Kenya

Kenya's tea industry supports 10 per cent of the country's population, around 3 million families. The industry is prone to damage from frost, which can damage leaves and bushes, reducing production. SERVIR developed a tool to map frost potential with up to 3 days' warning and enable farmers to take actions to reduce the loss of their crop. Based on average reported losses for current and future harvestable leaf, and tea bush death, an average smallholder tea farmer in the frost-vulnerable regions of highland Kenya loses approximately $212 each year in potential income due to frost damage. The tool uses satellite-derived elevation data and MODIS land surface temperatures to generate frost alerts that can be modelled using weather forecasts from the Kenya Meteorological Department. A 3-day warning enables an annual average reduction of $80.47 in frost damage losses each year, which is equivalent to 25 days of household food spending or almost a full year of a child's school tuition. Frost information is not yet fully accessible to individual farmers, but the potential value of the information is promising.

5.2.3 Lessons Learned

SERVIR has learned a number of lessons that complement the findings of the evaluation. One lesson was to shift from one-off products to sustainable services. The Earth observation community uses the term "products" to describe the outputs of analytical efforts that add value to raw data, whether processed datasets, maps, bulletins, or other formats of analysis. These products are valuable in and of themselves, but they are more valuable as part of a sustained service that packages information in the most useful way for decision-makers and combines it with training, outreach, monitoring, and a dissemination strategy. A service requires a client orientation: understanding users' needs, tailoring services to meet those needs, and being responsive to feedback and updating the service components over time.

The programme has found that it requires more than technical capacity to inform decision-making. Seven dimensions of capacity have been identified that make a strong SERVIR hub:

Relationships

- *Political mandate* to convene and advise member states to influence decision-makers
- *Network of partners*, including an ability to connect with others generating, consuming, or translating/disseminating data, products, and tools

Technical

- *Technical ability* to work with remote sensing information, models, and GIS and provide expert training
- *Sector expertise* in agriculture, water, weather and climate, disasters, land use, and ecosystems

Organizational

- *Organizational and financial capacity* to manage projects and staff
- *Engagement and outreach capability* to reach and communicate effectively with users
- *IT infrastructure*, including the necessary hardware, software, and internet connectivity

SERVIR has also identified some general challenges to sustaining the use of Earth observation data products and services in developing countries that others should consider when pursuing similar approaches:

- Technical: geospatial information technology infrastructure, connectivity, power, and licences are difficult to maintain.
- Data sharing: lack of incentives in many countries for data sharing and integration.

- Usability and dissemination: lack of interdisciplinary staff and skills required to engage users, co-create user-friendly products, and disseminate them effectively.
- Awareness and demand: mixed levels of awareness and demand for satellite data and often unrealistic expectations for what satellite data can do.
- Sustainability: limited ability to maintain services over time and few opportunities for revenue-generating services.

Finally, SERVIR has learned that one type of organization cannot do it all to connect Earth observation data to decision-making. It takes diverse expertise and varied levels of investment at different entry points, depending on the context. SERVIR has developed a simple framework to help the programme think holistically about the context for Earth observation information. We used to talk about "supply and demand" and "providers and users". Now, however, we have a more nuanced understanding and use a "chain" analogy (Fig. 5.4) that reflects the different functions and types of organizations that are involved in connecting data to decisions. There is a tendency to focus on one point in the chain, depending on one's bias, for instance, assuming that data is the biggest gap and that if we improve the data or make it more accessible, it will naturally lead to the desired use and development results. Or sometimes focus is placed on the product itself or on the dissemination method, without spending sufficient time thinking about what comes before and after or how information will flow from one actor to another across the chain or whether users even have the ability to take action on the information they receive. The capacities and gaps that exist all the way through this chain must be examined, including enabling policies, in order to optimize investments to increase the use of Earth observation data. This framework has a lot in common with the programme-level theory of change in Fig. 5.2.

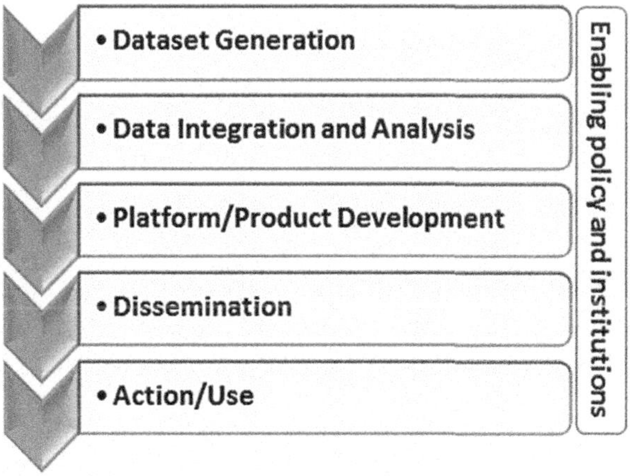

Fig. 5.4 The decision support investment chain

5.3 Conclusion

The SERVIR programme is just one of the many efforts globally to derive value from space to support economic growth and environmental sustainability in developing countries, but it is unique in its longevity and global breadth. The original aims and objectives of the programme from 2005 persist, but a culture of adaptive management and learning has enabled many adjustments over the years. Those adjustments have helped the programme get closer to truly putting Earth observation data and science into the service of society by finding ways to work across organizations and disciplines and more effectively capitalize on the different skill-sets required to connect space to village.

Chapter 6
The Sustainable Development Goals: A New Space for Action?

Petra Dannecker

Abstract The article trace the history and trajectory of the Sustainable Development Goals (SDGs). The guiding questions are whether the SDGs constitute a possible space or agenda for a profound paradigm shift in 'development' thinking and practice or, to put it differently, whether the SDGs have the potential to rethink the narrow, economic and outdated 'western' concepts of 'development' which are still embedded in the idea of continous growth. For understanding the new agenda a brief debate of the history of 'development' and the Millennium Development Goals (MDGs) is necessary before discussing the weaknesses and strengths of the SDGs and thus the guiding questions.

6.1 Introduction

In September 2015, the world's leaders adopted the '2030 Agenda for Sustainable Development', a new framework for global development. The resolution states that 'never before have world leaders pledged common action and endeavour across such a broad and universal policy agenda'.[1] The Sustainable Development Goals (SDGs), which succeeded the Millennium Development Goals (MDGs), are intended to guide policy makers worldwide for the next decade and a half. The SDGs were presented as the world's consensual basis and framework for any kind of cooperation and, at first glance, are breath-taking in scope and ambition.[2]

In this article, the potential impact as well as the transformative vision of the SDGs will be analysed and discussed, especially given the growing global challenges such as food insecurity, public debt, persistent poverty, climate change, rising inequality and international as well as national wars. These global challenges,

[1] Sengupta, Muti. "The Sustainable Development Goals: An Assessment of Ambition". 18 Jan. 2016. E-International Relations 9 Jul. 2017 http://www.e ir.info/2016/01/18/the-sustainable-development-goals-an-assessment-of-ambition/

[2] Langford, Malcolm. "Lost in transformation? The politics of the sustainable development goals." *Ethics & International Affairs* 30.2 (2016): 167–176.

P. Dannecker (✉)
Department of Development Studies, University of Vienna, Vienna, Austria
e-mail: petra.dannecker@univie.ac.at

C. Al-Ekabi, S. Ferretti (eds.), *Yearbook on Space Policy 2016*,
Yearbook on Space Policy, https://doi.org/10.1007/978-3-319-72465-2_6

which can be observed worldwide, are to a great extent the result of orthodox growth models, which have defined and promoted development nearly exclusively as economic development and the growth of material wealth.[3] Thus, the question that will guide this article is whether the SDGs constitute a possible space or framework for action leading to a profound paradigm shift and re-conceptualisation of responses to environmental and social challenges. Can and do the SDGs have the potential to stimulate rethinking of the narrow, economic and outmoded concepts that are based on continuous growth, and will they provide the necessary social innovations?

Before discussing the SDGs, however, it is necessary to start with a brief discussion of the history of 'development' and the Millennium Development Goals as the precursor of the post-2015 agenda and the Sustainable Development Goals. Subsequently, the SDGs will be introduced with highlights on their weaknesses and strengths on different levels before discussing the questions raised.

6.2 Development

Whereas the negotiations over the post-2015 agenda were accompanied by uncertainty and, as Langford (2016) argues, a level of frustration with the reductionism of the MDGs, during the first development decade in the 1960s, the UN and governments were driven by a positive vision and commitment about and to 'development'. 'Development' was conceptualised as economic growth, and the key elements to reach this goal were believed to be industrialisation, productivity and technology (including technology transfer), which would, so the vision and policy held, help so-called developing countries to achieve 'development' and 'catch up' with so-called developed countries within one decade.[4]

Since that time, the idea that economic growth will lead to development or, to put it differently, to human flourishing has prevailed. Despite the fact that in the second development decade, social goals such as employment, education, health, nutrition and participation were included and an explicit call for the eradication of poverty and hunger was made in the third development decade, as well as the fact that in the final development decade starting in the 1990s poverty and sustainability became more central, economic growth remained the key means to achieve development and the instrument with which to measure it.[5]

Thus, a single route to prosperity, as well as a single economic model, dominated the discourses and the politics from the end of the Second World War. However, critiques of this development vision, discourse and practice have also accompanied these processes over the past 20 years, challenging and questioning the transfer of

[3] Moore, Henrietta L. "Global Prosperity and Sustainable Development Goals: Global Prosperity and SDGs." *Journal of International Development* 27.6 (2015): 801.

[4] Koehler, Gabriele. "Seven Decades of 'Development', and Now What? The SDGs and Transformational Change 2015." *Journal of International Development* 27.6 (2015): 737.

[5] Ibid. 738.

economic models and their standards from the global North to the global South,[6] as well as the belief that development, as a problem and as a goal, seems an issue only relevant and applicable to countries in the global South and that '... development obscures (...) their power-political and historical dimension'.[7] But these critiques had little impact, perhaps, as Moore argues, because alternatives were hardly ever offered and '... political and theoretical poverty, as well as economic poverty' dominated.[8] Whether the SDGs are or can become such an alternative will be discussed after a brief introduction to, and contextualisation of, the MDGs.

6.3 The Millennium Development Goals

During the 1990s, a critical scientific discourse about the effectiveness of development projects and programmes emerged, including public debates in so-called developed countries. Whereas on a more practical level discussion focused on whether aid was the right tool to foster development, on a political and academic level, the transference of the economic liberal model that focused on privatisation and reduced government came under pressure, especially since empirical studies revealed that the transfer of this model did not lead to growth and prosperity but instead damaged education, health and essential services in many countries in the global South.[9] The various United Nations (UN) summits that took place in the 1990s reflect the political space that began to emerge as a result of these discussions and debates. Of these summits, the Copenhagen Social Summit 1995 can be seen as a stepping stone for more explicit calls for poverty eradication, social justice and human rights.[10]

These discussions and summits brought the UN, in particular, to the realisation that it could not continue with another development decade but rather needed a common framework to make aid more effective and focus more on poverty alleviation. The MDGs, which describe the focus and the tasks for international politics in the twenty-first century, are the result of this process. They were formulated by a group of high-level experts that was, from the beginning, a target of criticism.[11] However, in 2000, following the Millennium Summit of the United Nations, the UN

---

[6] Escobar, Arturo. Encountering Development: The Making and Unmaking of the Third World. Princeton, N.J: Princeton University Press, 2011; Rist, Gilbert. "'Development' as a Part of the Modern Myth: The Western 'socio-cultural Dimension of 'development'." The European Journal of Development Research 2.1 (1990); Ziai, Aram. Exploring Post-Development: Theory and Practice, Problems and Perspectives. London ; New York: Routledge Chapman & Hall, 2007.

[7] Tandon, Yash. "Aid without Dependence: An Alternative Conceptual Model for Development Cooperation." *Development* 52.3 (2009): 356.

[8] Moore 802.

[9] Hulme, David. The Millennium Development Goals (MDGs): A Short History of the World's Biggest Promise. Manchester: Brooks World Poverty Institute, (2009): 8.

[10] Koehler 741.

[11] Sengupta.

Fig. 6.1 Millennium Development Goals (Source: United Nations)

organisations as well as national governments decided to strive individually and collectively towards the developed eight goals to make aid more effective, focusing more on poverty reduction.[12]

The MDGs thus confirmed a shift from employment, productivity and redistribution to poverty alleviation and symptoms of poverty. Poverty, now defined as a multidimensional phenomenon, should be eradicated by the expansion of productive employment and the reduction of unemployment, accompanied by a strong focus on social development, gender equality and environmental concerns as well as social integration (Fig. 6.1).[13]

The first seven goals were exclusively directed to developing countries, especially when viewed at their target and indicator level.[14] Goal number eight – to 'develop a global partnership for development' – was only added following pressure from the so-called global south and the G77, but without a date set for reaching the goal or any clear targets,[15] which strengthened the impression that the MDGs were particular instructions for developing countries. Despite the focus on poverty reduction, a market-oriented neoliberal approach of development framed the MDGs, with the private sector as a new player.[16]

It can be argued that one of the strengths of the MDGs was that they were capable of directing the world's attention to the plight of the poorest by means of a constituted common framework with a manageable number of straightforward goals for promoting global development, which permitted the international community to engage in certain long-neglected social issues.[17] The new framework

---

[12] Hulme 19.

[13] United Nations. "Millennium Development Goals." UN City Copenhagen20 June 2017 http://un.dk/about-the-un/the-mdgs

[14] Sengupta.

[15] Hulme 41.

[16] Koehler 714.

[17] Langford 16.

helped to reverse the downward trend in official development assistance (ODA) in the 1990s, when many rich countries were about to cut their foreign aid budgets, turning their focus primarily to national issues, and it provided guidelines for countries in the South to plan their social and economic development.[18] The idea, however, that growth is the only solution to poverty eradication was maintained and structured the new framework. Thus, many analysts criticised the MDG design as neoliberal, however, under a 'pretty cover'.[19]

In the Millennium Goals Report 2015, the United Nations heralded the achievements of the MDG decade as a partial success. It was stated that 'the MDGs have produced the most successful anti-poverty movement in history' (see foreword by Ban Ki-Moon). Especially the facts that nearly half of the population in the developing world, who lived on less than $1.25 a day, dropped to 14% in 2015 and that the number of people living in extreme poverty declined by more than 50%, decreasing from 1.9 billion in 1990 to 836 million in 2015, were presented as the main indicators for the success of the MDGs.[20] Furthermore, the reduction of child mortality by half and the increased access to drinking water by 2.6 billion people, which means that the target of halving the proportion of people without drinking water was achieved, were presented worldwide as indicators of the effectiveness of the MDGs, together with primary school enrolment, which rose from 83% in 2000 to 91% in 2015 for boys and girls in equal numbers.

In fact, the meaningfulness of the measured quantitative improvement is highly questionable since the statistics about the enrolment of children in primary schools, for example, neither gives information about the quality of the education nor about the duration of the increased school attendance.

More interesting, however, is a closer analysis of the presented data. This shows, for example, that the reduction of poverty was achieved mainly through the so-called overachievers in Asia, especially China, India and Vietnam. However, within countries as well as between countries and regions, development was extremely uneven. Pogge (2014) calls the MDG outcome a 'scandal' primarily because of the huge surge in income and wealth inequalities during the MDG era.[21] Other authors have questioned the 'existence' of a correlation between the goals and individual policies since, for example, some countries, especially in Asia, had modified and adapted their more ambitious national goals and, hence, could report progress long before 2015. Fukuda-Parr and Greenstein go even further, arguing that the

[18] Obrosky, Michael. "Keynote." Konferenz: Universalität der SDGs und die Folgen für Österreich Politik und Zivilgesellschaft im Dialog. Karl-Renner-Institut, Vienna. 2015: 1.

[19] Fukuda Parr, Sakiko, Joshua Greenstein, and David Stewart. "How Should MDG Success and Failure Be Judged: Faster Progress or Achieving the Targets?" World Development 41 (2013): 19–30.

[20] United Nations. The Millennium Development Goals Report 2015. New York: United Nations, (2015): 2.

[21] Pogge, Thomas. "Die MDGs sind moralisch ein Skandal". 2014. Vereinte Nationen 7 July 2017 http://www.dgvn.de/fileadmin/publications/PDFs/Zeitschrift_VN/VN_2014/Heft_6_2014/04_Standpunkt_Pogge_VN_6-14_25-11-2014.pdf

achievements would have been likely reached without the framework of the MDGs.[22] Moreover, many African countries started at a lower point and were labelled as 'failures' for not meeting the goals, goals that were set up in a way that portray universal failure in Africa when actually there were important successes, as Easterly argued.[23] For him the 'Africa as failure image' follows a well-known discourse, namely, to exaggerate the role of 'the West as Saviour for Africa'.

On the political level, the MDGs contained neither a specific plan nor an increased budget and did not have a specific concept of responsibilities. For example, the eighth goal 'global partnership for development', which was meant to guide North-South cooperation, was rather vague from the beginning instead of providing clear guidelines.[24] Policy was absent and, thus, paved the way for a neoliberal agenda with policy prescriptions by international finance institutions. Against this backdrop, the achievements of the MDGs can at best be interpreted as attenuating the social consequences of the economisation of thinking underlying the MDGs, through a unidirectional transfer of resources and aid from North to South. The data presented, as well as the alleged achievements, disguise the widespread recognition that the dominant economic model is neither pulling large enough numbers of people out of poverty nor guaranteeing sustainable growth and equality. The understanding of sustainability, therefore, was very reductionist, especially since central concerns of sustainable policies – chiefly environmental objectives – were not sufficiently reflected.[25] To wrap up, it is useful to refer to William Easterly who stated that the 'MDGs gave far too much attention to middle-aged white male experts in the West debating what should be done for the rest of the world'[26]; experts who are still defining for all what constitutes and drives an economy.[27]

6.4 The Sustainable Development Goals

Taking the history of 'development' into consideration, the discourse and practice of the idea of development, which is conveyed through the Sustainable Development Goals, is – at least on a conceptual level – challenging the transfer of models or standards from North to South and doubtlessly presents an ambitious global vision. When it became clear that the MDGs would not be achieved by the deadline, the

[22] Fukuda-Parr et al.

[23] Easterly, William. "How the Millennium Development Goals Are Unfair to Africa." World Development 37.1 (2009): 35.

[24] Koehler 743.

[25] Loewe, Markus. "Post 2015: How to Reconcile the Millennium Development Goals (MDGs) and the Sustainable Development Goals (SDGs)?" Bonn: German Development Institute, (2012): 3.

[26] Easterly, William. "The SDGs Should Stand for Senseless, Dreamy, Garbled." 28 Sept. 2015. Foreign Policy, 24 June 2017 http://foreignpolicy.com/2015/09/28/the-sdgs-are-utopian-and-worthless-mdgs-development-rise-of-the-rest/

[27] Moore 808.

international community agreed on negotiating a second 15-year agenda – the Sustainable Development Goals – within an Open Working Group. A series of global consultations was conducted, both online and offline, and civil society organisations, citizens, scientists, academics and the private sectors from around the world were all actively engaged in the process. The extent of people's potential participation in designing a global framework was unique.

But this was by far not the only reason for the agreement on a new programmatic phase. The power structures had changed further from a bipolar to a multipolar system; the BRIC countries had emerged as new development actors; and 'classic' development aid was increasingly losing importance. With climate change as a global problem, the international community came to the realisation that far more attention had been given to socio-economic development than to environmental sustainability and that the model based on continuous growth had put a number of essential earth system services at risk, and with them poverty-reduction gains and the livelihood of future generations.[28]

Thus, there are numerous transformative shifts due to the new agenda. One central pillar of the SDGs is the guiding principle 'Leave No One Behind', which is defined as 'the goals and targets met for all nations and peoples and for all segments of society'.[29] The SDGs, as is already visible in the name, also put sustainable development at the core – not to define stand-alone goals but to negotiate in a cross-cutting manner a framework that is able to capture the interconnectedness of sustainable development concerns.[30] The aim of the new agenda is further to forge a new global partnership with a new spirit of solidarity, cooperation and mutual accountability.

After 3 years of consultation and debate, the commonly developed and elaborated 17 goals and 169 targets were first proposed in mid-2014 (Fig. 6.2[31]).[32]

The SDGs are broader than the MDGs – not just in terms of including goals, which require reforms, as social dimensions, but also considering goals, which focus or imply a reversal of current trends with regard to income inequality, the modes of sustaining and producing, climate and environment, inclusiveness as well as, for example, focusing not only on economic but also on social development.[33] Until the SDGs, human rights, governance, the environment and economic as well as social development were not seen as interconnected.

---

[28] Ibid. 805.

[29] United Nations General Assembly. Resolution adopted by the General Assembly on 25 September 2015. Transforming Our World: The 2030 Agenda for Sustainable Development. UN Doc A/RES/70/1 of 25 Sept. 2015.

[30] Sengupta.

[31] United Nations. "Sustainable Development Goals." United Nations Sustainable Development Goals 21 June 2017 http://www.un.org/sustainabledevelopment/sustainable-development-goals/

[32] Koehler 744; Scott, Andrew, and Paula Lucci. "Universality and Ambition in the Post-2015 Development Agenda: A Comparison of Global and National Targets: Universality and Ambition in the SDGs." Journal of International Development 27.6 (2015): 753.

[33] Koehler 744–45.

Fig. 6.2 Sustainable Development Goals (Source: United Nations)

Thus, there is doubtless progress on the conceptual level. The social dimensions are cast in a more rights-oriented way by fostering the notion of universalising access. There is an explicit call for equality in and among countries throughout the goals. Sustainable production and consumption is also a target. While the MDGs merely focused on the so-called developing countries, the SDGs apply for all countries. Some argue that this approach goes beyond international development cooperation and formulates the goal of global sustainable development in the economy, environment and society and '...disappointed those who wanted to see a more pragmatic, evidence-based approach to development'.[34]

Yet, on the political level, the SDG framework fails to address the synergies and trade-offs between different goals as well as the contradictions among goals and continues the tradition of former development decades as well as the MDGs, namely, the lack of an underlying policy framework.[35,36] Hence, the possibility of citizens' participation in global policy making is not specified. Furthermore, the responsibilities are not clear. There is no guidance on how the SDGs can be connected and embedded within national political systems.

The SDGs 'are defined as aspirational and global, with each government setting its own national targets guided by the global level of ambition but taking into account national circumstances'.[37] Each government has the authority to decide to what extent these 'aspirational' and global targets should be incorporated into national policies and strategies. Scott and Lucci (2015) conducted a study and analysed whether existing national targets had been taken into consideration during

[34] Langford 169.

[35] Koehler 745.

[36] Scott and Lucci 752.

[37] United Nations General Assembly.

the developing process of the SDGs. They identified notable differences between targets set by different countries, and also between national and global targets, thus concluding that the SDGs are weak in terms of articulating national and local priorities.[38]

But it is not only the missing political will with regard to transfer and translation of the global targets nationally, it is already obvious that there will be no funding available for the implementation of the SDGs framework. The UN estimated that US$ 3.9 trillion would be needed annually. However, the current level of public and private investment in the relevant areas account for just around US$ 1.4 trillion, which means that annual investments must increase dramatically. The question is whether increased foreign assistance accompanied by augmented private capital flows, both domestic and cross-border, and by an increase in domestic revenue mobilisation, as currently discussed by different actors and on different levels, is not just a financial issue but first of all a programmatic one. What are the implications for the SDG agenda and its possible implementation, if private capital flows and foreign investments are the main financial resources? As is argued in this paper, this would mean that, through the backdoor, global development will once again be exclusively equated with economic development and growth despite the unique possibilities that the SDGs provide on a conceptual level.

6.5 Conclusion: Can the SDGs Become a New Paradigm?

In conclusion, the initial question can be answered positively. The SDGs have potential to become a new paradigm since a 'safe operating space' has been constructed for searching and developing a new development paradigm.[39] Importantly, the discourse about post-2015 brought together an unusual array of people, public officials, economists, NGOs, local activists and billionaires. 'Sustainable development', however, is no innovation of the post-2015 era but rather a revival and a normative consensus on development based on the realisation of the need to bring together three dimensions, namely, economy, society and environment.

But since the SDGs – like the former development decades and the MDGs – are not binding, the question arises, whether a transformative agenda matters at all. It is not specified by which policies 'sustainable development' should be achieved, and, hence, current power structures are difficult to overcome without guidance. The lack of guidance goes hand in hand with the lack of political will. Currently, there is not much interest on the different political levels to operationalize the SDGs – on the contrary. Furthermore, financing still remains a challenge, especially since the private sector as well as private foundations seem not much interested in the global agenda settings but instead have their own rationales and visions of development.

---

[38] Scott and Lucci 758.

[39] Moore 811.

It must therefore be questioned whether the new actors are willing to subsume their interests under the SDGs.

This can also be said for many national actors, who seem to define and follow 'classic' visions of development instead of taking this momentum as an opportunity to advance more radical conceptual and practical approaches that challenge the current reductionist understanding of 'sustainability'. Furthermore, SDGs have not shifted public opinion towards the need for sustainable and global development and thus influenced formal politics as, for example, optimistically expressed as a possibility by Varun Gauri.[40] Despite the negotiations and the rewriting of goals and targets in the framework of the SDGs, it seems that the same economic and social models and development visions will persist without taking the planet's limited resources and their sustainability into account. If the scope and the normative gains that the SDGs provide are not taken further, the SDGs will most likely not make a difference.

[40] Cited by Langford 174.

Chapter 7
Space Applications for Development: The Indian Approach

Bhupendra Singh Bhatia

7.1 Introduction

Since its inception in the mid-1960s, the Indian space programme has aimed at achieving self-reliance in space technology and in demonstrating the applications of space technology for the development and betterment of society.

To work towards developing a self-reliant space programme, the Indian Space Research Organisation (ISRO) established a Launch Vehicles Programme and a Satellite Building Programme. Over just a few decades, it has successfully developed operational launch vehicles such as the Polar Satellite Launch Vehicle (PSLV) and Geosynchronous Satellite Launch Vehicle (GSLV) and is working towards the further enhancement of these capabilities. In the field of satellites and spacecraft, it started with small satellites such as APPLE, Aryabhata and Bhaskara and moved on to build a series of operational communications satellites (INSAT), a constellation of navigation satellites (IRNSS) and a series of remote sensing (Earth observation) satellites (IRS), including microwave remote sensing satellites (RISAT).

However the driving force of the space programmes has been the applications of space technology to support and accelerate the pace of national development. It has almost been an article of faith since its founder Dr. Vikram Sarabhai said:

> There are some who question the relevance of space activities in a developing nation. To us there is no ambiguity of purpose…We are convinced that if we are to play a meaningful role nationally and in the comity of nations, we must be second to none in the applications of advanced technologies to the real problems of man and society.

The views expressed are the author's own.

B. S. Bhatia (✉)
Vikram A Sarabhai Community Science Centre, Ahmedabad, India

Development and Educational Communication Unit (DECU), Indian Space Research Organisation (ISRO), Ahmedabad, India
e-mail: bsbhatia44@yahoo.com

C. Al-Ekabi, S. Ferretti (eds.), *Yearbook on Space Policy 2016*,
Yearbook on Space Policy, https://doi.org/10.1007/978-3-319-72465-2_7

The applications of space technology for development started in the early 1970s with experimentation in satellite-based television broadcasting to reach out to rural areas under the Satellite Instructional Television Experiment (SITE) and went on to grow into one of the largest satellite-based infrastructures to support education.

Similarly, in the field of remote sensing, it started as a small experiment in the use of aerial photography to detect and study coconut wilt disease and went on to conduct joint experiments leading to the operationalization of several applications and their integration and institutionalization in user agencies with the continuous involvement and support of the ISRO.

7.2 The Approach

The approach adopted by ISRO in India to implement space-based applications has had the following features:

1. Joint experimentation
2. Technology/technique development and transfer
3. Integration and institutionalization

7.2.1 Joint Experimentation

The first step in the implementation of any application has been joint experimentation with the user agency. The unique feature of such joint experimentations has been that they are conducted jointly from "end to end". This means that the space agency does not limit its role to providing the space segment, but it gets involved in each and every step of utilizing and evaluating the technology to see how the user can apply it most effectively. This requires that the space agency moves beyond providing space-related infrastructure and actually gets involved in every step of the user including data collection, processing, techniques development for specific applications, evaluating effectiveness and providing feedback to both the space segment and the user agency. Such an intensive and involved process of joint experimentation and evaluation helps in *developing techniques/methods* that are acceptable to the user and in demonstrating the strengths and advantages of the space inputs in the task of the user agency. This helps in *integrating* the space inputs into the routine working of the user agency. If it is considered necessary by the user agency, especially if multiple agencies are involved, a separate institution is created for effective utilization of the space inputs. Thus *new institutions* get created in the user agency for effective use of space inputs and their future growth. The space organization continues its involvement in a support role for the smooth functioning of the new institution.

The results of such joint working are presented periodically to decision makers and funding agencies. This enables the user agency to receive funding to accept, adapt and internalize the space inputs.

To understand this process of technology transfer and institutionalization, a few cases are presented here in brief.

7.2.2 Educational Communications

Long before the space programme in India had taken off, an experiment in satellite broadcasting for instructional television was planned and conducted. In the mid-1960s, India signed an agreement with NASA to borrow its ATS-6 satellite for 1 year to conduct a huge experiment in broadcasting instructional programmes to 2400 remote villages spread out in the backward regions of six states of the country. It was proposed to transmit programmes for school children in the mornings and adult rural audiences in the evening. As the experiment required broadcasting of television programmes, the Ministry of Information and Broadcasting (I&B) was approached and actively partnered in the conduct of the experiment by taking on the responsibility of production of programmes required for transmission. Similarly the Ministry of Education was involved for production and transmission of school educational programmes.

In the mid-1960s, the national electronic and media picture was such that there was almost no manufacture of ground electronics such as earth stations, receive terminals, television receivers, etc. in India. There was one broadcasting station using old, heavy and expensive equipment.

As part of its *end-to-end* approach, ISRO took on the responsibility of indigenous design, development and manufacture of the ground hardware required. It had to further take responsibility for deployment of the ground receive terminals in remote villages and arrange for their operation, upkeep and maintenance. This was probably much beyond the normal scope of a space agency, but ISRO took on the responsibility because the experiment would just not have happened without this, and the adoption of satellite broadcasting in India would not have been that rapid.

As earth stations were available only with ISRO, it had to take on the responsibility of broadcasting the morning and evening programmes. Broadcasting studios had to be set up. ISRO therefore became involved in the design of low-cost educational studios. It helped in setting up these studios for the Ministry of Education at the Centre for Education Technology (CET) of the National Council of Educational Research and Training (NCERT). It also worked with the Ministry of I&B to design similar, compatible studios at the base production units set up for SITE programme production.

Acknowledging that the effectiveness of satellite instructional television was greatly dependant on the quality of programmes as well as the functioning of the hardware, ISRO took upon itself the responsibility of producing science programmes for school telecasts and to produce rural adult-oriented programmes for the low-power transmitter of PIJ as part of SITE.

Table 7.1 Phases of educational communications development in India

Phase	Satellite	Partner	Sector
Experiment	ATS 6	CET/NCERT	Primary school
Technology transfer	INSAT	UGC	Graduate
		IGNOU	ODL
		VIGYAAN PRASAR	Science popularization
Operational	EDUSAT	All universities and state governments	All sectors
	OFFLINE	Sakshat	Primary school
		NPTEL	Engineering

Table 7.2 Phases of the crop forecasting programme

Phase	Partners	Sector
Experiment	IARI	CAPE
Technology transfer	IARI	FASAL
Operational	IARI	National Forecasting Centre

ISRO also took upon itself the evaluation of all aspects of the SITE experiment. The performance of hardware systems was evaluated, and the impact of the programmes on school children as well as adult audiences was evaluated employing all-round interdisciplinary evaluation methods. The results clearly indicated the success and effectiveness of the programmes and gave a fillip to operational acceptance of satellite-based broadcasting systems in the country.

The agencies responsible for education adopted satellite broadcasting in a big way. For *school education*, the Central Institute of Educational Technology (CIET) was set up, and several State Institutes of Educational Technology (SIETs) were created to provide the necessary infrastructure for ongoing operational use of satellite technology. For *higher education*, the University Grants Commission (UGC) approached ISRO for setting up the Consortium for Educational Communication (CEC) and several Educational Multimedia Research Centres (EMMRCs). For *open and distance learning*, the Indira Gandhi National Open University approached ISRO for the setting up of its broadcasting networks. All these networks were made operational on INSAT where bandwidth was provided to all educational agencies without any cost. In the meanwhile, ISRO experimented with the use of one-way video and two-way audio networks for training. These were found to be most effective for teacher training, and all the above educational agencies were oriented and trained to utilize these features (Tables 7.1 and 7.2).

In view of the great response by the education agencies to using satellite communications, ISRO proposed and launched the EDUSAT – a satellite totally dedicated to meet the needs of the education sector. Before the proposal was prepared, discussions were held with all the national educational agencies, and a presentation was made to the minister for HRD in the presence of heads of agencies such as the NCERT, UGC, IGNOU, AICTE and others. After agreement, the proposal was put up to the government, and funding approvals for the space system as well as the ground segment were received and implemented.

The experience of production of science programmes for SITE was transferred to Vigyan Prasar of the Department of Science and Technology (DST). This resulted in operational transmission of science programmes on several national and regional channels. The idea was to institutionalize a dedicated science channel, which did not materialize, but the quantum of science communication on national and regional television networks significantly increased.

7.2.3 Growth of Educational TV

So what started as an experiment in school broadcasting under SITE grew into a huge satellite-based educational operational system. It created a number of institutions for the utilization of technology (including space technology) for education. With further advancements in technology, all the above agencies are now using offline networks besides the online networks set up earlier. One example of an Internet-based offline use of educational technology is the National Programme on Technology Enhanced Learning (NPTEL) portal for engineering education.

7.2.4 Development Communications for Rural Audiences

One of the objectives of SITE was to reach out to rural adults with information on agriculture, health and family planning (population control) practices. The partner agency was the National Broadcaster, All India Radio to begin with and Doordarshan at a later stage. ISRO was responsible for overall evaluation and feedback, including the impact evaluation and audience feedback on the programmes presented. Additionally, ISRO took upon itself the task of producing and transmitting programmes for the rural audiences of the low-power transmitter installed at Pij in Kheda district of Gujarat. This transmission was carried out by ISRO from 1975 to 1989. This was a long-term intensive *experiment* in rural programme production jointly carried out with Doordarshan as a partner.

This experiment resulted in developing new approaches to rural communications. Normally rural instructional communications are top-down (information flows from expert/decision maker to audience), studio based and expert (subject matter) driven; among others the approach at Pij resulted in production of upwards or horizontal communication that was audience oriented, field oriented and participative. The inputs of the audience from feedback, research and direct audience participation became the driving force in communication design. Commenting on this approach, the December 1978 issue of *Seminar* said:

> the experience of Pij at the Space Applications Centre comes with a message to the Indian scene. The seriousness of approach, the humility, the capacity to asses both success and failure, the attempt to use TV in a vigilante role, guarding the rights of the totally helpless, giving strength where strength is needed has at last proved that it is possible for TV to fulfil

> its role in our conditions... If Doordarshan had thought fit to show the Pij programs through its TV channels this could have acted as catalyst for future programming... the experience must be disseminated more widely.

The Pij experiment received the UNESCO Prize for Development Communication in 1984.

Based on the Pij approach, training programmes were organized for Doordarshan personnel so that *technology transfer* could be achieved. A large number of training programmes were conducted.

Satellite technology led to a very rapid expansion of the national network reaching out to nook and cranny of the country through rebroadcast transmitters. It was hoped that the Pij approach of programming would be institutionalized by the user agency (Doordarshan), but it was too busy with expansion, growth and meeting the competition that this innovative, audience-oriented, research based-approach developed at Pij was lost. The priorities of the user agency were different. Rural audience was not its priority at that time. Therefore the great experiment and technology transfer efforts *did not result* in any kind of *institutionalization.* Development issues are local in nature, and if media has to play a meaningful role in development, the experience of Khed/Pij needs to be utilized and institutionalized by the operational agency. This however will not happen if development is not a priority of the broadcasters.

7.2.5 Telemedicine

ISRO promoted the application of satellite-based telemedicine in a big way. This was done more in collaboration with private software providers, private hospitals and government hospitals.

It was noticed that private IT software providers had been working on the development and use of telemedicine software. ISRO came forward to use this software over satellite networks to provide telemedicine facilities to remote hospitals. To begin with, an experimental network was established at a very well-known heart hospital – Narayana Hrudayalaya at Bangalore. This was connected to a tribal hospital of Karuna Trust, an NGO in Karnataka. Connectivity was also provided between the ICUs of Bangalore and Kolkata of Narayana Hrudayalaya. The feedback from doctors and patients was most encouraging, and the network was further expanded.

ISRO also attempted to provide connectivity to rural government hospitals and urban hospitals. Very remote locations such as the Andaman and Nicobar Islands and Leh in Ladakh were selected to connect to major hospitals such as AIIMS Delhi, PGI Chandigarh, etc. Terminals were also provided to several NGO hospitals such as Maa Amritamayi Hospital in Cochin and specialty hospitals including Sankara Nethralaya in Chennai. The remote hospitals found the facility most useful to be able to consult the specialty doctors from urban hospitals. Further mobile vans were provided to travel to rural areas and carry out eye camps and provide other telemedicine facilities. The network expanded with about 60 urban super-specialty hospitals connected to more than 350 rural hospitals.

The Telemedicine Society of India was involved in setting up CME networks to conduct educational programmes for students and doctors in remote areas. This became a regular feature for some medical colleges of Orissa. Similarly, the Federation of Obstetric and Gynaecological Societies of India (FOGSI) set up a network of receive terminals to conduct regular CME programmes.

The application of telemedicine was greatly boosted by the support of ISRO, which provided the connectivity and facilities for utilizing the networks. However despite some efforts, the Health Department of the Government of India did not get involved in a big way. Their priority was providing primary health care, whereas telemedicine was concerned with providing tertiary care. But urban and rural hospitals, and private hospitals, adapted the concept and continue to use it on a wide scale.

One of the reasons why telemedicine was not institutionalized in the government system could be that ISRO did not get involved in the development of telemedicine software. If it had done so, it could have developed relevant software, and the system could have been adopted by the government health department. ISRO completely relied on available software and therefore applications as required by government could not be promoted. Telemedicine was adapted by private hospitals, NGO hospitals and government hospitals in rural/remote areas. It did not become an integral part of the government health system.

7.2.6 Village Resource Centres

7.2.6.1 Community-Centred Convergent Application

Integrating the satellite communications capabilities with geospatial technologies, village resource centres were established in collaboration with NGOs working in the local area. Services such as telemedicine, tele-education, information on natural resources, advisories on agriculture, fisheries, livestock management, vocational training for skill improvement, alternative livelihood education, weather information, etc. were provided at the village level. These centres are established in collaboration with some well-known NGOs working in the local area. This delivers space-based services to the doorstep of the user. A network of about 400 such centres has been established and is serviced by Space.

7.2.6.2 Remote Sensing Applications

The operationalization of remote sensing applications too followed the same steps of joint experiments, technology development/transfer and institutionalization even more systematically with greater effectiveness. The following examples are illustrative of the process.

7.2.6.3 Crop Forecasting

The genesis of the remote sensing applications can be traced to a request made by a top agriculture scientist to the Chairman of ISRO to help in studying the spread of coconut wilt disease. This was long before any satellite was built or even the remote sensing (Earth observation) programme was conceived. This was followed by a request to try and study the area under paddy cultivation in a district in Andhra Pradesh and a district in Punjab. *Joint experiments* called Agricultural Resources Inventory and Survey Experiment (ARISE) and Crop Acreage and Production Estimation (CAPE) were undertaken. The procedures developed for CAPE were *continuously revised* and upgraded (*technology development and transfer)* to improve upon the accuracy and timeliness of crop estimates. The wide swath coverage and the ability to quantify different crops led to the national-level crop forecasting programme for a number of crops including sugar cane, potato, cotton, jute, mustard, sorghum, etc.

Realizing that remote sensing alone cannot provide a system for making multiple and reliable forecasts, a programme called Forecasting Agricultural Output Using Space, Agrometeorology and Land-Based Observations (FASAL) was *institutionalized.* This led to the establishment of the *Mahalanobis National Crop Forecast Centre* at the Indian Agriculture Research Institute (IARI), New Delhi.

7.2.7 National Natural Resources Management System

The concept of developing a National Natural Resources Management System (NNRMS) in which remote sensing technology would be integrated with conventional techniques to generate timely and accurate information for decision making was proposed by the ISRO to the Planning Commission of India. A national committee under the chairmanship of the Planning Commission was constituted. Fifty-nine well-defined experiments were conducted covering a vast variety of areas such as groundwater targeting, mineral exploration and fisheries. The results of these experiments were presented to senior officials, and the Planning Commission established nine committees to determine user/application needs for remote sensing; conceptualization and implementation of remote sensing space segments with necessary ground-based data reception, processing and interpretation facilities; and establishment of utilization systems for using remote sensing images and conventional data for various applications and resource management activities. The nine committees covered the following areas:

1. Agriculture and soils
2. Bio-Resources
3. Geology and mineral resources
4. Water resources
5. Ocean resources and meteorology

6. Cartography and mapping
7. Urban management
8. Rural development
9. Training and technology

Each standing committee is chaired by secretaries of the respective departments of the Government of India and consisting of experts from major user departments.

The Department of Space was identified as the nodal agency to establish the NNRMS. The Department of Space established several Regional Remote Sensing Service Centres and supported the development of state-level remote sensing centres. A centre – NESAC – was established at Shillong to meet the special needs of the North-East sector of the country. Thus, a well-spread out infrastructure was created to promote the use of remote sensing data for specific regional and local issues. A number of projects were taken up in a mission mode jointly with user agencies and state governments to effectively utilize the space imagery inputs.

Thus, under the NNRMS, a national organizational structure with user departments taking the lead in the applications of remote sensing imagery was created, and a national infrastructure of Regional Service Centres was established to meet the local needs of the states (Fig. 7.1).

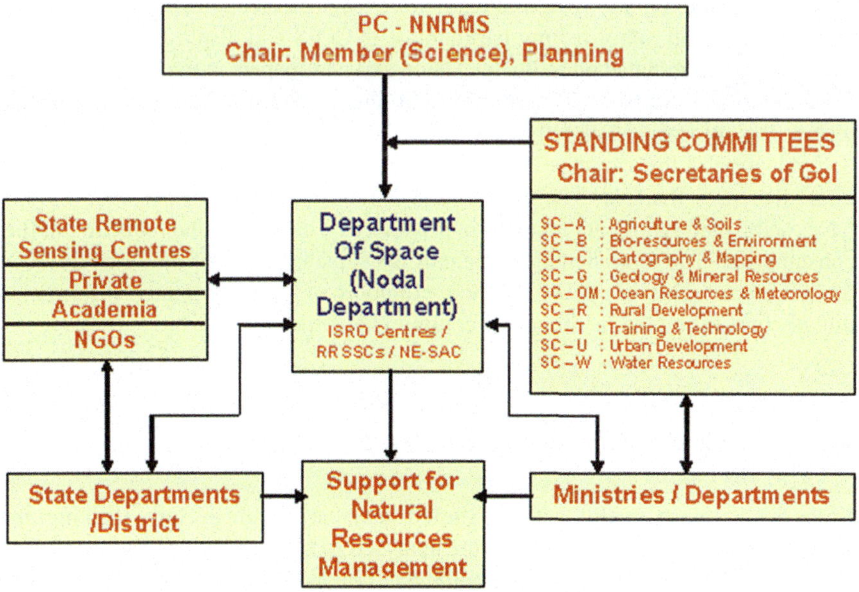

Fig. 7.1 NNRMS organizational structure

7.2.8 Integrated Mission for Sustainable Development

The drought in 1987–1988 prompted ISRO to examine the possibility of using remote sensing data to combat drought on a watershed basis. Watersheds in 21 drought-affected districts were characterized in terms of land cover, drainage, geomorphology, soil, etc., and action plans for siting water-harvesting structures, soil and water conservation, etc. were prepared. Remote sensing imagery was found to be ideally suited for the purpose.

The results of these watershed management studies were presented to a national-level committee, which defined the Integrated Mission for Sustainable Development to cover 126 districts and 92 blocks for action plan development.

7.2.9 Potential Fishing Zones

India has a 7500 Km-long coastline with a large potential for fishery development and a very significant population depending on fishing for their livelihood.

Jointly with the Earth System Science Organization (ESSO), ISRO developed the techniques for identifying potential fishing zones using ocean colour, sea surface temperature and surface wind vectors, surface currents and GIS along with several other variables. These are indications of food availability and favourable environmental factors for congregation of fish in a particular habitat.

Now the Indian National Centre for Ocean Information Services (INCOIS) of ESSO issues these 3-day forecasts regularly except on cloudy days. These advisories are sent to all major fishing harbours of INCOIS and displayed on electronic boards and disseminated on radio, television and newspapers in the local languages. It is also made available on websites, and about 25,000 fishermen subscribe to receive this information directly.

Evaluation of this exercise indicates that in 80% of the cases, the catch per unit effort increases fourfold. The search time is reduced by 30–70 per cent. Studies have established significant increase in productivity, catch size and reduction in fuel consumption. The total annual net benefit is estimated to be of the order 34,000–50,000 crores. This is one of the best examples of space-generated information directly benefitting the livelihood of common fishermen.

7.2.10 Disaster Monitoring and Support Systems

Space involvement in this activity started very early with cyclone tracking and warning systems. Satellite images were analysed to track the cyclone's genesis and its movement to predict the landfall point. Satellite reception systems were installed in coastal villages to give early warning of impending cyclones. This was a joint activity with the India Meteorological Department. The advance warning resulted in the saving of life and organization of better mitigation methods.

Extensive studies were further carried out on cyclone genesis, its track and landfall point and intensity estimation. The ground systems for weather forecasting were densified by adding more automatic weather stations (AWS) developed by ISRO, along with more Agromet Towers and Doppler radars. This resulted in the installation of a meteorological data processing system at IMD and SAC Ahmedabad, thus contributing to the improvement of weather forecasting and oceanic and atmospheric studies. ISRO has also developed a Meteorological and Oceanographic Satellite Data Archival Centre (MOSDAC) portal. This portal archives the meteorological and oceanographic data products from ISRO science missions and in situ observation network. The portal provides a variety of products and services on a wide spectrum of applications comprised of weather forecasting, cyclone prediction and other vital ocean and atmospheric parameters needed by national/international forecasting agencies, research organizations, educational institutions, individual researchers and students for advanced research.

7.2.11 Other Missions

As noted earlier, under the NNRMS, 59 experiments were initiated which grew to some 160 projects. This included a vast area of applications in, among others, the national drinking water mission, wetland inventory, geology and mineral exploration, coastal zone management, urban mapping, atmospheric and ocean studies, cyclone monitoring and the decision support centre for disaster management support programme. The list has only been growing.

7.2.12 Space Infrastructure to Support Development

To effectively support national development efforts in the country, the Space Department has established a fairly substantial organizational infrastructure. This includes chiefly:

- The Space Applications Centre (SAC) at Ahmedabad
- Development and Educational Communication Unit (DECU) at Ahmedabad
- The National Remote Sensing Centre (NRSC) including the Decision Support Centre (for Disaster Management) at Hyderabad
- The four Regional Remote Sensing Service Centres (at Jodhpur, Kolkata, Nagpur and Bangalore) and the North-Eastern SAC at Shillong to support the regional requirements of the states

SAC focuses on the design of space-borne instruments for ISRO missions and development and operationalization of applications of space technology for societal benefits. The applications cover communications, broadcasting, navigation, disaster monitoring, meteorology, oceanography, environment monitoring and natural resources survey.

SAC develops the ground transmit/receive systems (earth stations/ground terminals) and data/image processing systems. SAC has also been developing the algorithms and the software required for processing and product generation of data acquired from IRS satellites. It involves the design, development, operationalization and maintenance of software for remote sensing data processing related to earth, planetary and astronomical observations for Indian as well as the international user community. The software is transferred to NRSC, Hyderabad, for operational dissemination of data products to users.

DECU has focused on societal applications of communications. Beginning with Satellite Broadcasting, it conceptualized and implemented interactive networks for training, EDUSAT, tele-education and telemedicine and now has developed the large network of Village Resource Centres.

The National Remote Sensing Centre (NRSC) has the mandate of providing Earth observation (EO) data from space and aerial platforms, developing technologies and applications for natural resources management and providing support for monitoring and management of disaster capacity building for utilization of EO data.

The Decision Support Center (DSC) established at NRSC is a single-window delivery point for aerial- and space-enabled inputs together with other important data layers for use in disaster management of pre-, during and post-disaster phases.

The Regional Remote Sensing Centres at Kolkata, Jodhpur, Bangalore and Nagpur undertake studies specific to their states meeting user-specific needs, developing regional repositories and delivering localized actionable products and services.

Under NNRMS, a complete nationwide organization and infrastructure has been created to integrate the applications of space inputs in the development activities of the user agencies.

The Space Department has created a very large organizational infrastructure and has made substantial investments to support development applications. There is commitment at the policy level and in the government to provide support to development using space inputs. This should make space applications for development an ongoing reality in India.

7.2.13 Observations

1. In India, the Space Research Organisation has had a well-defined applications programme. It initiates applications in areas of national importance. This expedites the introduction and adoption of space inputs in user organizations. To carry out this task, it has established a Space Applications Centre and the NRSC with a large number of experts from areas of application.

2. The applications are jointly developed with the user agencies. The space organization also works towards development of necessary ground hardware/technology and software/techniques to utilize the space inputs. If this is not done, or if it is left to the user agency, it may take too long. If the task is subcontracted to a private organization, the cost is likely to escalate. This initiative by the space organization is essential for the timely and cost-effective utilization of space inputs.
3. Joint implementation builds confidence in the user agency, and it becomes self-reliant in the long run. Such self-reliance may not come in working with subcontracted private agencies.
4. Joint evaluation of the tools and techniques establishes the strengths and weaknesses of the space inputs. It gives feedback for continuous upgrading and improvement in the tools.
5. Joint working facilitates acceptance and integration of new inputs.
6. When the user agency and the research agency jointly present the results to decision makers, planners and funding agencies, the credibility is high, and support comes forth more easily.
7. Major user agencies such as agriculture could develop internal strengths over a period of time and may not need the support of ISRO after institutionalization. ISRO would only play a role of service provider for providing imagery, etc. However smaller user states and agencies may continue to need ISRO inputs.
8. It is essential that the application proposed is of direct interest and relevance to the user agency. If not, the user may not take interest or establish a long-term plan for its application.

7.2.14 Issues in Space for Development

1. Space technology is global in nature. It provides large coverage over large areas with large communication reach.
2. Development issues are local in nature. The action plans for sustainable development have to be prepared locally in the local context with sensitivity and understanding of local parameters preferably by the local actors.
3. There is need to adapt the global strengths of space to the local needs of the people. This process has to be made continuous (ongoing), viable, acceptable and self-sustaining.
4. This may not be a profitable business and not attractive for the private sector. The economics of the private sector are likely to make the process unviable and non-sustainable in the long run.
5. A suitable model needs to be evolved and adopted. It could employ government funding, international funding and collaboration with developing countries like India, and even some element of people's funding could be explored.

7.3 Concluding Remarks

In India, the Indian Space Research Organisation has taken the lead in creating national infrastructure for the spread and effective utilization of applications of space-based inputs towards the national development efforts of the country. It has worked jointly with the users and has demonstrated the successful operationalization and institutionalization of a large number of space-based development-oriented applications using satellite communications and Earth observation.

The programme is primarily government funded and implemented through government agencies. While this experience would be of great value to the international community, the organizational arrangements and processes would have to be tailored to the conditions of the adopting nations.

References

Bhatia BS (1999) Satellite communication for development and education – the Indian experience. Space Forum 5:125–146

Department of Space Indian Space Research Organisation. Indian Space Agency 27 Oct. 2017. https://www.isro.gov.in/

Navalgund R, Jayaraman V, Parth R (2007) Remote sensing applications: an overview. Curr Sci 93(12):1747–1766

Rao DP, Radhakrishnan K, Perumal A, Subramanian SK, Chenniah GCh, Murthy YVS, Hanumanta Rao G, Ramana Murthy J, Kameswara Rao SVC, Uday Bhaskar N. Integrated mission for sustainable development – a synergistic approach towards management of land and water resources. ICAR-National Institute of Agricultural Economics and Policy Research 27 Oct. 2017. http://www.ncap.res.in/upload_files/workshop/ws5_chapter9.pdf

Solanki HU, Dwivedi RM, Nayak S, Gulati DK, John ME, Somvanshi VS (2003) Potential fishing zones (PFZ) forecast using satellite data derived biological and physical processes. J Indian Soc Remote Sens 31(2):67–69

Chapter 8
Supporting Sustainable Development with Outer Space Activities

Simonetta Di Pippo

8.1 Introduction

As the Director of the United Nations Office for Outer Space Affairs (UNOOSA), I am appreciative of the opportunity to contribute to the ESPI Yearbook 2016 on the topic of space and sustainable development. The role that space activities and the international space community can play in support of sustainable development is invaluable. Space activities are responsible for many of the crucial technologies and capabilities that improve everyday life on Earth and society as a whole. Regardless of geographic region and level of development, every nation should be in the position to utilize space for its empowering capacities and ability to drive innovation and address global challenges. Advancements in space benefit all member states of the United Nations in one way or another, and thus these technologies support our common goals and efforts in addressing global challenges and implementing the 2030 Agenda for Sustainable Development.

8.2 The Role of UNOOSA

At UNOOSA, we work to bring the benefits of space to all of humankind. The Office promotes international cooperation in the peaceful uses of outer space and facilitates the use of space science and technology for sustainable economic and social development. To do this, the Office has three main roles: as a *global facilitator*, as a *capacity-builder* and as the *main gateway* to outer space in the United Nations system.

S. Di Pippo (✉)
United Nations Office for Outer Space Affairs (UNOOSA), Vienna, Austria
e-mail: simonetta.di.pippo@unoosa.org

C. Al-Ekabi, S. Ferretti (eds.), *Yearbook on Space Policy 2016*,
Yearbook on Space Policy, https://doi.org/10.1007/978-3-319-72465-2_8

As a global facilitator, UNOOSA plays a key role in the promotion of the peaceful uses of outer space and acts as a platform for international space diplomacy. While UNOOSA works closely with the 84 member states of the Committee on the Peaceful Uses of Outer Space (COPUOS), at the same time, it serves all 193 United Nations member states. Established in 1959, COPUOS supports efforts at the national, regional and global levels to maximize the benefits of space science, technology and applications and increase international cooperation in outer space activity. It provides a unique platform at a global level to monitor and discuss developments in the space agenda.

UNOOSA's role as a global facilitator can be seen in the many responsibilities UNOOSA has, particularly as secretariat to COPUOS and its subcommittees and working groups, executive secretariat to the International Committee on Global Navigation Satellite Systems (ICG) and permanent secretariat to the Space Mission Planning Advisory Group (SMPAG). The Office also facilitates cooperation between satellite data information providers and users of data for disaster recovery and risk reduction through the United Nations Platform for Space-based Information for Disaster Management and Emergency Response (UN-SPIDER).

Through UN-SPIDER, UNOOSA enables the sharing of best practices, especially for developing countries, on the use of space-based information for sustainable development, disaster risk reduction, changing climate adaptation and other environmental issues. UNOOSA conducts demonstration projects with partners to both evaluate and showcase the use of space-based information for sustainable development.

In our capacity-building role, UNOOSA helps to build the space capacity of non-space-faring countries, particularly developing countries, so that they too can access all the benefits of space. Through our Programme on Space Applications (PSA), the Office has established a long history of providing practical capacity-building and advisory services to developing countries. Those activities include workshops, training courses and pilot projects on topics such as remote sensing, satellite navigation, satellite meteorology, tele-education and basic space science. More than 18,000 participants have attended approximately 300 UNOOSA training activities in the last 45 years.

Through our UN-SPIDER programme, we also develop solutions to address the limited access developing countries have to specialized space-based technologies that can be essential in managing disasters and reducing disaster risks. UN-SPIDER aims to improve actions to reduce disaster risk or support disaster response operations through knowledge sharing and the strengthening of institutions in the use of space technologies, particularly through workshops, tailor-made technical advisory missions and the UN-SPIDER Knowledge Portal.

In 2010 UNOOSA launched our Human Space Technology Initiative (HSTI) to engage more countries in activities and international cooperation related to human space flight and space exploration. The aim of the HSTI is to increase the global benefit of space activities and make space exploration a truly international effort. HSTI provides a platform to exchange information, foster collaboration between partners from space-faring and non-space-faring countries and encourage emerging and developing countries to take part in, and benefit from, space research and applications. The HSTI has contributed to UNOOSA developing a more a holistic approach to capacity-building that is resulting in more tangible access to space for non-space-faring countries. Examples of such initiatives include the KiboCUBE programme with the Japan Aerospace Exploration Agency (JAXA), which offers educational and research institutions from developing countries the opportunity to deploy cube satellites from the Japanese Kibo module on board the International Space Station, and an agreement with the China Manned Space Agency (CMSA) to enable United Nations member states, particularly developing countries, to conduct space experiments and eventually fly astronauts and payload engineers on board China's future space station.

Lastly, as the gateway to outer space in the United Nations system, the Office leads UN-Space, a United Nations inter-agency mechanism to coordinate United Nations' agencies' activities for using space-related technologies for improving the human condition around the world. UNOOSA convenes annual sessions of the Inter-Agency Meeting on Outer Space Activities (UN-Space) to discuss current and future activities, emergency technologies of interest and other related matters among UN system entities.

In all the work of the Office, throughout all our activities and mandates, runs the theme of sustainable social and economic development and the contribution that space can bring to making this a reality.

8.3 Space and the 2030 Agenda for Sustainable Development

We have only one Earth and we are all dependent on its limited resources. It is therefore our duty to support and strengthen universal peace and to work together to address global challenges. On 25 September 2015, United Nations member states agreed to a 2030 Agenda for Sustainable Development and a set of 17 Sustainable Development Goals (SDGs) to end poverty, protect the planet and ensure prosperity for all, with no one left behind (Fig. 8.1). Within the framework of the Agenda and its goals, there is an emphasis on all nations and stakeholders acting in collaborative partnership to implement all goals and targets. And space science and technology are invaluable tools to help all countries achieve the Sustainable Development Goals.

Facilitating open access to space-derived data is an apt example of how countries can benefit from space activities for sustainable development. The nations of our world are faced with many global challenges, from the long-reaching impacts of climate change to the inequitable distribution of resources, which can only be collectively addressed by permitting fair and open access to substantive data. Data contributes to equal distribution of opportunities, broadens economic gain, fosters research and innovation and supports decision-making processes on the basis of accessible and transparent data. In this sense, access to space-based data can contribute directly to sustainable development. The sustainability impacts of space data applications include agriculture predictions, air and water quality tracking, ocean

Fig. 8.1 Sustainable Development Goals (Source: United Nations)

observations, public health mapping, telemedicine for remote areas and climate change monitoring. The list of earth-impacting space applications is nearly limitless, and many other valuable contributions are currently in development or being researched, with new findings and technological advancements being made on a seemingly daily basis.

8.4 UNOOSA as an Enabler of Space for Sustainable Development

UNOOSA has adapted its capacity-building role to incorporate the SDGs, acting as a facilitator and supporting countries and organizations in their need to access and use for cutting-edge space technology and data for the implementation of the SDGs. A key example of this new and innovative approach is the signing of an agreement with a private sector space actor, the Sierra Nevada Corporation (SNC), to provide countries, especially non-space-faring ones, affordable access to SNC's Dream Chaser spacecraft for space science experiments in low-Earth orbit that are related to the SDGs. Expected to be launched in 2021, this will be a unique opportunity for countries that do not have the infrastructure or financial resources for their own standalone space programme, enabling more equitable access to space and its benefits for sustainable development.

Another innovative step UNOOSA is taking is the establishment of a Space for Development Profile (SDP). The Profile will consist of a set of generic indicators, which can be expanded on a country basis, taking into account the priorities of a country and tailored to each country's development strategy, in collaboration with the country. It is expected that monitoring and evaluation processes will lead to greater learning, adjustment and decision-making. This continual process of feedback and adjustment will make capacity-building activities flexible and more responsive to new developments in the field of outer space activities and the environment within which they operate.

However, just measuring the indicators is not enough to improve them. UNOOSA is also developing a Space Solutions Compendium, a dedicated catalogue of solutions applicable at the national level that can be applied by the country in question. The Space Solutions Compendium will list providers that could contribute to the improvement of a given indicator. Solutions could include guidelines, training courses or other forms of assistance, bringing value to the countries that apply them. Thanks to the link between the Space for Development Profile and the Space Solutions Compendium, it will be possible to evaluate the impact of the solution, which will, on one hand, assist countries in the evaluation of the efficiency of the solution and on the other hand will help the provider of the solution to extract lessons learned.

With the support of UNOOSA as its secretariat, COPUOS also considers a range of sustainable development topics, including space and water, where the Committee recently noted how space-derived data is used extensively in water management, and space and climate change, where the Committee noted that adequate monitoring of and adaptation to climate change with the help of space technologies were crucial to tackling its adverse effects, thereby bringing global attention to these topics. The Committee's work is furthered by its two subcommittees, the Scientific and Technical Subcommittee and the Legal Subcommittee. The Scientific and Technical Subcommittee meets every year for 2 weeks to discuss questions related to the scientific and technical aspects of space activities. Topics for discussion include the use of space technology for socio-economic development and disaster management support, as well as issues that can impact the use of space for sustainable development, such as space weather and global navigation satellite systems. Similarly, the Legal Subcommittee meets every year for 2 weeks to discuss legal questions related to the exploration and use of outer space. Topics include the status and application of the five United Nations treaties on outer space, the definition and delimitation of outer space, national space legislation, legal mechanisms relating to space debris mitigation and international mechanisms for cooperation in the peaceful exploration and use of outer space.

The 2030 Agenda for Sustainable Development not only implies the use of space tools but also requires that space-related activities, as well as outer space environment itself, continue to be sustainable in the long term. The near-Earth environment is fragile, and the broadening of space applications, operations and increased strategic value of space has resulted in a growing need to enhance operational safety, asset security and long-term sustainability of outer space activities. COPUOS has proven instrumental in the creation of the five treaties and five principles of outer space and has thereby strengthened the international legal regime governing outer space. The Working Group on the Long-term Sustainability of Outer Space

Activities, established under the Scientific and Technical Subcommittee of COPUOS to identify areas of concern and propose measures for the long-term sustainability of outer space activities, has agreed on a first set of guidelines, with a second set under development. The sustainability of space itself at the forefront of UNOOSA and COPUOS' efforts as the international space community collectively looks to the future.

8.5 Next Steps: UNISPACE+50

The year 2018 will mark an important milestone for the Office and the international space community, as we recognize the 50th anniversary of the first United Nations Conference on the Exploration and Peaceful Uses of Outer Space (UNISPACE) held in Vienna in 1968 and acknowledge the contributions of subsequent UNISPACE conferences, convened in Vienna in 1982 and 1999. These global conferences examined the practical benefits of space science and technology and their applications, with a special focus on the needs of developing nations, global and regional development agendas and the benefits for society at large.

UNISPACE+50 will be an opportunity for the international community to examine where stronger space governance and supporting structures are required to protect the space environment and secure the long-term sustainability of outer space activities, as well as to set the future course of global space cooperation for the benefit of humankind. In this regard, the UNISPACE+50 process will be aligned with the 2030 Agenda for Sustainable Development, the United Nations Sendai Framework for Disaster Risk Reduction 2015–2030 and the Paris Climate Change Agreement. Through UNISPACE+50, UNOOSA and COPUOS seek to ensure that the benefits of modern collaborative space governance strongly support all nations in implementing these three agendas.

In order to guide preparations for UNISPACE+50, at its 2016 session COPUOS agreed on seven thematic priorities:

1. Global partnership in space exploration and innovation
2. Legal regime of outer space and global space governance: current and future perspectives
3. Enhanced information exchange on space objects and events
4. International framework for space weather services
5. Strengthened space cooperation for global health
6. International cooperation towards low-emission and resilient societies
7. Capacity-building for the twenty-first century

UNOOSA and member states have begun to work under these thematic priorities, with reports and recommendations to be fed into the UNISPACE+50 process and results.

The expected outcome of UNISPACE+50 is a new vision for international space cooperation and the global governance of outer space activities: Space 2030. Given the ever-increasing number of space-faring nations, continuing advances in space technologies, the evolving role of the commercial space industry and the focus on the 2030 Agenda for Sustainable Development, the international community must recognize and adapt to new global challenges. These unique conditions will be considered in Space 2030. We aim to ensure that space technology and its applications are effectively utilized to bring concrete benefits to all of humankind, paying special attention to the future space-faring nations and carefully considering the long-term sustainability of outer space activities for current and future generations.

Space 2030 will be based around four key thematic pillars: space economy, space society, space accessibility and space diplomacy.

Space Economy: Supporting the Development of Space-Derived Economic Benefits

Space economy is the full range of activities and use of resources that create and provide value and benefits to human beings in the course of exploring, understanding and utilizing space. This includes topics such as developing space technologies and infrastructure, increasing awareness of the benefits of a space economy for global sustainable development, addressing the economic rationale for space activities and discussing framework possibilities for the cooperation of private and public entities.

Space Society: Enabling the Evolution of Society and Societal Benefits Stemming from Space-Related Activities

Space society is focused on raising awareness of the benefits of space technologies and space-based services during decision-making processes for a sustainable society. It is important to shed light on how the integrated use of space applications and space technology can act as an essential driver for progression towards a sustainable environment on Earth.

Space Accessibility: Allowing All Communities to Use and Benefit from Space Technologies

Space accessibility addresses issues around coordination and communication among all relevant stakeholders in the promotion of the peaceful uses of space for humanity. Accessibility is geared towards the benefits of open-space data policies and practices to provide access to space. Capacity-building and education will support global efforts in the development of the space sector for the benefit of humanity, while equal and non-discriminatory access allows improvement of life on Earth, regardless of a country's scientific, technological and economic development.

Space Diplomacy: Building and Strengthening International Cooperation in Space Activities

The final Space 2030 pillar of space diplomacy aims to constructively engage international communities on the basis of equal footing and mutual respect, with the ultimate goal of addressing shared concerns and achieving shared objectives. Space diplomacy sets forth mechanisms for effective governance, the inclusion of emerging space countries, and methods to increase awareness of the benefits of space

Fig. 8.2 Space 2030 (Source: UNOOSA)

activities. The concept is crucial for effective decision-making among nations to address common challenges. Together, we must seek to build constructive, knowledge-based partnerships and use space diplomacy as a vehicle to foster an atmosphere of mutual trust.

Since none of these pillars should be elaborated in silos, it will be essential, as the global space community collectively moves forward, to see the various interlinkages and interdependencies as part of a holistic picture.

UNISPACE+50 and Space 2030 will help the international community achieve progress on a global scale. Our efforts are supplemented by space technology that supports our common goals to address global challenges and new, collectively-negotiated norms to ensure the benefits of space for future generations. We believe that, through this new perspective on space governance, we will further spotlight space as a force for providing improved well-being for all people and playing a valuable role in the attainment of the global development agendas (Fig. 8.2).

8.6 The Burgeoning International Space Community

As we prepare for UNISPACE+50 and Space 2030, the cooperation among the broader space community that this process will foster, including private actors, non-governmental organizations, civil society and more, is particularly exciting. The increasing private sector in particular is expected to extend its contribution and influence in the coming years. Thanks to this burgeoning international space community, increased access to space and the emergence of new socio-economic benefits derived from space will contribute to the implementation of the sustainable development agenda. To specifically involve the broader space community in UNISPACE+50 and Space 2030 preparations, the Office for Outer Space Affairs is organizing a series of high-level fora for a wide range of stakeholders to identify ways to harness space technology and applications for socio-economic development. The most recent forum was held in Dubai in November 2016 and resulted in the Dubai Declaration, which underscored the need for greater cooperation in outer space activities. The next forum will be held in November 2017, where the Office will engage interested parties in the broader space community in the UNISPACE+50 process to help us build Space 2030 and the future of international space cooperation.

8.7 Conclusion

In the 60 years since the launch of Sputnik 1, the world's dependence upon outer space activities and utilization of its benefits have steadily increased and will continue to do so. Space has changed our lives for the better. At the same time, we collectively face global challenges ranging from poverty, inequality and disease to climate change, pollution and deforestation. The 2030 Agenda for Sustainable Development and its 17 Sustainable Development Goals seek to address these challenges and ensure that no one is left behind. Space science, technology and applications are unique tools to help us understand and overcome global problems and achieve the Sustainable Development Goals. But to do this, we also need to maintain international cooperation in the peaceful uses and exploration of outer space. At the United Nations Office for Outer Space Affairs, we are working with the broader global space community to achieve these complementary goals. UNISPACE+50 and Space 2030 will solidify these efforts and protect outer space and its benefits for sustainable development, now and for generations to come. I am convinced that such cooperative endeavours will help us improve and preserve our home. We only have one Earth, so we need to look after it and each other. I hope you'll join us.

***

Chapter 9
Space Agencies' Perspective on Space for Sustainable Development

Josef Aschbacher, Clio Biondi Santi, and Wolfgang Rathgeber

9.1 Sustainable Development and UN Sustainable Development Goals (SDGs)

> *Sustainable development is development that meets the needs of the present, without compromising the ability of future generations to meet their own needs.*

The term sustainable development is interpreted in many ways, but its origins can be dated back to a development approach that aims to balance different, and often competing, needs against an awareness of the environmental, social and economic limitations that society faces.

Often the idea of 'sustainable development' is associated with the environment. But the focus of sustainable development is far broader than just the environment. It is also about ensuring a strong, healthy society. This means meeting the diverse needs of all people in existing and future communities; promoting personal well-being, social cohesion and inclusion; and creating equal opportunity.

How was it possible for world leaders to agree upon common objectives and plans for action regarding sustainable development? It all started in 1972 when governments met in Stockholm, Sweden, for the United Nations Conference on the Human Environment, to consider the rights of the human family to a healthy and productive environment. It was not until 1983 that the United Nations decided to create the World Commission on Environment and Development which defined sustainable development as 'meeting the needs of the present without compromising

J. Aschbacher (✉) · C. B. Santi · W. Rathgeber
European Space Agency, Frascati, Italy
e-mail: josef.aschbacher@esa.int

C. Al-Ekabi, S. Ferretti (eds.), *Yearbook on Space Policy 2016*,
Yearbook on Space Policy, https://doi.org/10.1007/978-3-319-72465-2_9

the ability of future generations to meet their own needs'. In 1992 the first United Nations Conference on Environment and Development was held in Rio de Janeiro, Brazil. It was here that the first agenda for environment and development was developed and adopted, also known as Agenda 21.

In the year 2000, the Millennium Development Goals (MDGs) were adopted by 189 states. They aimed to achieve a better world by 2015. In that year, the countries came together again to build on the many successes of the past 15 years and to go further. UN Resolution A/RES/70/1 of 25 September 2015, also known as 'Transforming Our World: The 2030 Agenda for Sustainable Development' (herein: Agenda 2030), is an intergovernmental agreement designed to guide global development efforts over the next 15 years, between 2015 and 2030. Its centrepiece, the Sustainable Development Goals (SDGs), replaces the MDGs. The new goals are designed to build on the MDGs and to 'complete what they did not achieve'.

While there were only eight MDGs, there are 17 SDGs, with 169 associated targets, and 304 proposed indicators. The SDGs are integrated and indivisible and balance the three dimensions of sustainable development: the economic, social and environmental. Spelled out, the 17 goals are:

1. *No Poverty* – End poverty in all its forms everywhere.
2. *Zero Hunger* – End hunger, achieve food security and improved nutrition and promote sustainable agriculture.
3. *Good Health and Well-being* – Ensure healthy lives and promote well-being for all at all ages.
4. *Quality Education* – Ensure inclusive and equitable quality education and promote lifelong learning opportunities for all.
5. *Gender Equality* – Achieve gender equality and empower all women and girls.
6. *Clean Water and Sanitation* – Ensure availability and sustainable management of water and sanitation for all.
7. *Affordable and Clean Energy* – Ensure access to affordable, reliable, sustainable and modern energy for all.
8. *Decent Work and Economic Growth* – Promote sustained, inclusive and sustainable economic growth, full and productive employment and decent work for all.
9. *Industry, Innovation and Infrastructure* – Build resilient infrastructure, promote inclusive and sustainable industrialization and foster innovation.
10. *Reduced Inequalities* – Reduce income inequality within and among countries.
11. *Sustainable Cities and Communities* – Make cities and human settlements inclusive, safe, resilient and sustainable.
12. *Responsible Consumption and Production* – Ensure sustainable consumption and production patterns.
13. *Climate Action* – Take urgent action to combat climate change and its impacts by regulating emissions and promoting developments in renewable energy.
14. *Life Below Water* – Conserve and sustainably use the oceans, seas and marine resources for sustainable development.

15. *Life on Land* – Protect, restore and promote sustainable use of terrestrial ecosystems, sustainably manage forests, combat desertification, halt and reverse land degradation and halt biodiversity loss.
16. *Peace, Justice and Strong Institutions* – Promote peaceful and inclusive societies for sustainable development, provide access to justice for all and build effective, accountable and inclusive institutions at all levels.
17. *Partnerships for the Goals* – Strengthen the means of implementation and revitalize the global partnership for sustainable development.

These 17 SDGs will stimulate action over the next 15 years. The interlinks and integrated nature of the SDGs are of crucial importance in ensuring that the purpose of the new Agenda is realized. The SDGs are associated with areas of critical importance for humanity and the planet:

- People: end poverty and hunger, in all their forms and dimensions, ensuring that all human beings can fulfil their potential in dignity and equality and in a healthy environment.
- Planet: protect the planet from degradation, including through sustainable consumption and production, sustainably managing its natural resources and taking urgent action on climate change, so that it can support the needs of the present and future generations.
- Prosperity: ensure that all human beings can enjoy prosperous and fulfilling lives and that economic, social and technological progress occurs in harmony with nature.
- Peace: foster peaceful, just and inclusive societies which are free from fear and violence. There can be no sustainable development without peace and no peace without sustainable development.
- Partnership: mobilize the means required to implement this Agenda through a revitalized Global Partnership for Sustainable Development, based on a spirit of strengthened global solidarity, focused in particular on the needs of the poorest and most vulnerable, and with the participation of all countries, all stakeholders and all people.

The 17 SDGs are new and different because of their universality, as they apply to every nation and every sector. Each and every goal is important in itself, but at the same time they are all connected. Governments decide how to incorporate SDGs into national policies: each country sets its own targets and indicators guided by the UN SDG framework but taking into account national development priorities.

Concretization of the SDGs can be achieved by three main steps: the first one is to indicate the goal (e.g. 'Zero Hunger'), then governments and organizations need to set certain targets to be able to analyse that goal and finally indicators show the measurable and actionable data collected, by monitoring progress, informing policy and ensuring accountability of all stakeholders. The main ten principles for global monitoring indicators are:

1. Being limited in number and globally harmonized.
2. Simple, single-variable indicators, with straightforward policy implications.

3. They allow for high-frequency monitoring.
4. Consensus based, in line with international standards and system-based info.
5. Constructed from well-established data sources.
6. Disaggregated.
7. Being universal
8. Mainly outcome-focused.
9. Science-based and forward-looking.
10. A proxy for broader issues or conditions.

9.2 How Space Can Contribute to Sustainable Development and the SDGs

Space assets and technology can help to implement the Sustainable Development Goals in many ways. As stated in Paragraph 76 of the 2030 Agenda, signatories agreed to 'promote transparent and accountable scaling-up of appropriate public-private cooperation to exploit the contribution to be made by a wide range of data, including Earth observation and geo-spatial information'. Implementing the Sustainable Development Goals is not just the work of the United Nations, nor UN Member States alone. It is a common task of a wide variety of actors, many of whom remain unaware of the many ways space technology can help them achieve their diverse projects.

Earth observation (EO), in particular, can provide unique benefits to projects supporting the SDGs by expanding monitoring capabilities. Earth observation satellites provide a unique source of information for anyone planning, carrying out or evaluating sustainable development projects. By showing a series of images over a period of time, they can help reconstruct a sequence of events. High-resolution images can be used for investigating highly targeted phenomena with a narrow field of vision during a specific time frame. Low-resolution imagery is better at depicting regional phenomena that may require more systematic and repetitive collection. EO reduces the cost of monitoring and makes reporting manageable, affordable and sustainable. Moreover, the data collected can also be used for other policy-making purposes.

Finally, observation and analysis can be done in a non-intrusive, objective and repeatable manner, providing for more fair and equitable decision-making. This last point should be highlighted – there are many technologies that will be helpful in achieving the SDGs, but most suffer from not being able to be used in an unbiased, across-the-board manner.

The European Space Agency (ESA) has developed a wide range of space programmes that provide useful contributions for sustainable development. Earth observation is a point in case. ESA's Earth observation programmes can be grouped into three categories, and all of them have the potential to contribute to the SDGs:

- The Earth Explorer missions are designed to address key scientific challenges identified by the science community while demonstrating breakthrough technology in observing techniques and featuring operational perspectives.
- In the domain of meteorological missions, ESA develops prototypes of satellites that are subsequently operated by EUMETSAT, on whose behalf ESA also procures recurrent satellite models.
- Moreover, ESA is the coordinator for the space component of the Copernicus programme. This EU-led space flagship programme features the dedicated Sentinel satellites, five of which have already been launched. Copernicus has kicked off a new era of operational Earth observation, leaving the initial 'pure science and technology' phase behind, and meeting with huge user interest. Around 28 PB of data have been downloaded so far, and Copernicus information is used in a wide variety of domains, including sustainable development.

There are ten main SDGs that can be related to space:

1. *SDG 2* – calls for zero hunger. Managing the health of livestock is one path to that goal. ESA co-founded the VGTropics project, an information system to manage animal health data in data-sparse environments, such as developing countries in Africa. Livestock survey planning, livestock distribution, data analysis and syndromic surveillance are all supported and facilitated by a satellite network, including satellite navigation, GPS units, satellite-based telecommunication services and satellite-Earth observation. Thus, VGTropics works to offset weak capacity in some African countries to conduct diagnostics and gather coordinated information. The commercialization of VGTropics started at the end of 2015.
2. *SDG 3* – promotes good health and well-being. Part of this is fighting the health epidemic unleashed by Ebola. During the Ebola crisis, ESA supported laboratories by providing them with satellite data thanks to an inflatable satellite antenna. This technology facilitated rapid and reliable diagnosis. The so-called B-Life system, developed within ESA's ARTES Integrated Applications Promotions programme, was used to support the Ebola treatment centre in N'Zerekore, a remote area of Guinea. B-Life enabled collaboration in real time between the on-the-ground emergency team and St Luc's Hospital in Belgium, enabling treatment plans to be modified as patient blood samples were analysed. In December 2014, the B-Life service was registered as part of the European Emergency Capacity Response within the European Mechanism for Civil Protection managed by the European Commission.
3. *SDG 4* – focuses on quality education. One way ESA promotes this goal is through satellite-enabled links to rural schools in South Africa and Italy. ESA supports a project that will change positively the situation of remote rural schools. In partnership with Luxembourg's satellite broadband operator SES Techcom Services, ESA's Advanced Research in Telecommunication Systems Programme and Openet Technologies unveiled Satellite Way for Education (Sway4edu2) to guarantee Internet access for rural schools in South Africa and Italy that have been equipped with satellite terminals and other required

materials. Such Internet-based eLearning for teachers and students fosters cultural integration among schoolchildren, boosts language learning and raises environmental awareness thanks to specific courses on local wildlife heritage and sustainable approaches to daily living.

4. *SDG 6* – supports clean water and sanitation. In 2002, ESA worked with UNESCO to launch the TIGER initiative to use Earth observation technology for improved integrated water resources management in Africa. Exploiting EO technology fills existing information gaps for effective and sustainable water resources management at national to regional scales. Guided by its own international Steering Committee, TIGER received the endorsement of the African Ministers' Council on Water. Now, the TIGER initiative aims particularly at supporting capacity-building activities and development projects in some 42 African countries. Delegates from 19 African and 10 European countries participated in TIGER's 2016 workshop held in Addis Ababa.
5. *SDG 7* – research for affordable and clean energy. As part of the Horizon 2020 Space Work Programme 2014, the European Commission funded a Programme Support Activity (PSA) for the implementation of a Strategic Research Cluster (SRC) on 'In-Space electrical propulsion and station keeping'. Electric propulsion has been identified by European actors as a strategic technology for improving European competitiveness in different space areas such as in-space operations and transportation.
6. *SDG 11* – supports sustainable cities and communities. With air pollution linked to millions of deaths around the world, it has never been more important to monitor the air we breathe. The Sentinel-5 instrument will be very important to continue the monitoring of our atmosphere in an operational system. Delivering important data on the composition of the atmosphere, Sentinel-5 is set to make a step change in monitoring and forecasting global air quality. This state-of-the-art instrument will be installed on the polar-orbiting MetOp Second Generation satellite. It will monitor the composition of Earth's atmosphere globally on a daily basis by measuring trace gases – such as ozone, sulphur dioxide, methane and carbon monoxide – and aerosols that affect air quality and climate.
7. *SDG 12* – supports responsible consumption and production. ESA and the European science community and research institutions are pushing the boundaries of science and technology, putting Europe at the forefront of scientific development and giving it a key role in industrial applications. ELIPS (European Programme for Life and Physical Sciences in Space) is ESA's research programme for science and applications in microgravity, helping to improve our life on Earth and enable humankind's long-term presence in space. The ISS offers a fantastic vantage point for studying Earth and its climate, the Sun and its varying characteristics and all the particles and radiation coming from the whole Universe.
8. *SDG 13* – monitors climate change. Combating desertification is essential to ensuring the long-term productivity of inhabited drylands. ESA initiated the DesertWatch project with the objective of developing a geoinformation system

for assessing and monitoring land degradation using EO technologies and for supporting national authorities in reporting to the UNCCD. While the DesertWatch project originally focussed on the Northern Mediterranean countries (Annex IV countries), this extension will study and demonstrate how the DesertWatch approach can be adapted to respond to the needs of non-Annex IV countries and, in particular, of those emerging from developing countries where access to field measurements is scarce and highly dispersed in quantity and quality.

9. *SDG 14* – preserves life below water. Satellite monitoring of wildlife has become a critical tool for environmental research and nature preservation. In acknowledgement of this, ESA has been supporting the development and testing of a new low-cost wildlife tag. Currently being developed in an ongoing ARTES 5.1 activity, the new system works in conjunction with the Argos satellite monitoring system, a non-profit initiative established in 1978 that is dedicated to helping the research community monitor and better understand the environment. The new tag developed in this ARTES 5.1 activity has a tiny transmitter powered by a battery or a solar panel that sends messages of short duration (less than 1 s) to the Argos constellation, whose six satellites circle at 850 km in polar sun-synchronous orbit, meaning they pass over a given location on Earth at approximately the same time each day.
10. *SDG 15* – preserves life on land. Implementation of SAR processing capabilities is done at the Observatoire Satellitale des Forêts d'Afrique Centrale (OSFAC) to demonstrate innovative forest degradation and deforestation monitoring based on SAR and optical satellite imagery for REDD+ purpose. The overall goal of 'SAR for REDD' is to provide the user organization *Observatoire Satellitale des Forêts d'Afrique Centrale* (OSFAC) with an automatic and operational SAR preprocessing tool, to enhance their remote sensing and analysis ability beyond the commonly used optical data sets. It addresses the GFOI research priorities of sensor interoperability and forest degradation over a site in the Democratic Republic of Congo.

It should be added that there are less obvious applications of Earth observation data, for example, to SDG 10 on reducing inequalities and to SDG 16 on peace, justice and strong institutions. To be successful, sustainable development efforts to improve living conditions must include a focus on promoting effective governance and accountable institutions. Support for free and fair elections is a key component of that strategy. Earth observation data could be incorporated at every stage of these types of projects. For instance, satellite imagery and maps could be used pre-election to improve the logistics and planning of elections in developing countries. It could be used to conduct an informal census in rural areas to confirm official numbers and inform better distribution outreach efforts. It could assist in setting election district boundaries and polling station locations to maximize potential voter turnout. Weather and topography information can help in figuring out when to hold an election in countries or regions that experience extreme weather or have difficult terrain. Simply getting election materials to remote areas can be difficult, and satellite data could be used to improve those efforts as well.

9.3 Conclusion

Promoting how space can support and contribute to sustainable development requires working alongside those who are implementing measures and assessing progress towards the SDGs. Sustainable development concerns all of us. As a responsible international organization, ESA uses its technology for humankind's development. Tackling societal challenges is one of ESA's priorities. ESA and its partners have achieved a lot in this regard, and they will carry on in their efforts. However, there is a need to increase awareness of how ESA's programmes, be they related to Earth observation, satellite telecommunications, navigation, human spaceflight or technology, can help accomplish the SDGs.

Chapter 10
Earth Observation for Humanitarian Operations

Stefan Lang, Petra Füreder and Edith Rogenhofer

Abstract Large-scale population displacements have ever increased the need for more effective humanitarian assistance. Humanitarian organizations require up-to-date reliable information about the situation on-site. Field-based surveys are often limited in crisis situations due to time or accessibility constraints. Geospatial and Earth observation (EO) technologies have increasingly become popular in the humanitarian community. An EO-based information service was set up for Médecins Sans Frontières (MSF) which provides dedicated geospatial information products in support of their operations. The core of the service portfolio is population monitoring using dwelling extraction from (multi-temporal) very high-resolution (VHR) satellite imagery. The service is mainly requested for refugee and IDP (internally displaced people) camps. Additional services on environmental resources, including groundwater, are provided on demand. As of mid-2017, over 350 maps at 60 locations in over 20 countries have been produced.

10.1 Introduction

Large-scale population displacements within complex emergency situations caused by armed conflicts and protracted crises over recent years, in particular in the Middle East and Eastern Africa, have ever increased the need for more effective humanitarian assistance. According to official UNHCR figures, by the end of 2016, more than 65 million people had been forced to flee their homes at a rate of up to 20 people every minute.[1] When reaching out to refugees or IDPs (internally displaced persons, not crossing international boundaries), humanitarian organisations require up-to-

[1] 'Figures at a Glance'. 29 June 2017. UNHCR 1 Oct. 2017 http://www.unhcr.org/figures-at-a-glance.html

S. Lang (✉) · P. Füreder
Department of Geoinformatics – Z_GIS, University of Salzburg, Salzburg, Austria
e-mail: Stefan.Lang@sbg.ac.at; petra.fuereder@sbg.ac.at

E. Rogenhofer
Médecins Sans Frontières, Vienna, Austria
e-mail: Edith.ROGENHOFER@vienna.msf.org

C. Al-Ekabi, S. Ferretti (eds.), *Yearbook on Space Policy 2016*,
Yearbook on Space Policy, https://doi.org/10.1007/978-3-319-72465-2_10

date reliable information about the situation on-site and the local/regional context.[2] The need for effective information permeates all stages of humanitarian assistance, from mission planning in the immediate emergency phase to providing care and maintenance in the longer run, including the availability of natural resources such as ground and surface water and determining the nutrient capacity of the immediate environment.

Obtaining information on-site by gathering evidence on the ground is often limited in crisis situations, with time and accessibility being the most limiting factors.[3] Situational awareness in complex humanitarian crises therefor relies on satellite image mapping[4] and other geospatial technologies. The optimisation of information flows in crisis situations is a critical asset in the humanitarian domain, and technological innovation is key as in many other areas, while the usefulness of any new 'gadget' will be strictly tested according to whether it ultimately helps to save lives or not. Geospatial and Earth observation (EO) technologies have become increasingly popular in the humanitarian community.[5] But turning these data into relevant geospatial information products[6] for humanitarian actors remains a challenge.[7] Humanitarian aid organisations play a key user role in this respect; they are the ones to adopt, test, improve and further develop any new technology, in close collaboration with those providing it. Over recent years, projects and initiatives have been launched in which research institutions and humanitarian actors have shared both technological and practical experience in mutual exchange. GIS/EO has become one of the key enabling technologies in humanitarian action. A large variety of different optical and radar sensor types, as well as remotely piloted aircraft systems ('drones'), have begun to provide many kinds of spatial, temporal and spectral resolutions.

---

[2] Lang, S., Schoepfer, E., Zeil, P. and Riedler, B. (2017) Earth observation for humanitarian assistance. GI Forum – Journal for Geographic Information Science, 1/2017, pp. 157–165.

[3] Lang, S., Füreder, P., Kranz, O., Card, B., Roberts, S. and Papp, A. (2015) Humanitarian emergencies: causes, traits and impacts as observed by remote sensing. in Thenkabail, P., (ed.) Remote Sensing Handbook, New York: Taylor and Francis. pp. 483–512.

[4] Voigt, S., Schoepfer, E., Fourie, C. and Mager, A. (2014) Towards semi-automated satellite mapping for humanitarian situational awareness. in Global Humanitarian Technology Conference (GHTC): IEEE. pp. 412–416.

[5] Bjorgo, E. (2001) Supporting humanitarian relief operations. in Baker, J. C., O'Connell, K. M. and Williamson, R. A., (eds.) Commercial observation satellites: at the leading edge of global transparency, Santa Monica, CA.: ASPRS RAND. pp. 403–427; Lang, S., Tiede, D., Hölbling, D., Füreder, P. and Zeil, P. (2010) EO-based ex-post assessment of IDP camp evolution and population dynamics in Zam Zam, Darfur. International Journal of Remote Sensing, 31(21), pp. 5709–5731; Voigt, S., Kemper, T., Riedlinger, T., Kiefl, R., Scholte, K. and Mehl, H. (2007) Satellite Image Analysis for Disaster and Crisis-Management Support. IEEE Transactions on Geoscience and Remote Sensing, 45(6), pp. 1520–1528.

[6] Füreder, P., Lang, S., Hagenlocher, M., Tiede, D., Wendt, L. and Rogenhofer, E. (2015) Earth observation and GIS to support humanitarian operations in refugee/IDP camps. in Palen, Büscher, Comes and Hughes, (eds.) Geospatial Data and Geographical Information Science – Proceedings of the ISCRAM 2015 Conference, Kristiansand.

[7] Lang et al. (2017).

In all mandates and missions, humanitarian aid is obliged to ensure that humanitarian action is targeted to the right people at the right place at the right time. New tools and techniques may help, if used in the right way. Non-governmental organisations (NGOs) such as *Médecins Sans Frontières* (MSF) rely vitally on reliable (i.e. not misleading), up-to-date (often 'real-time') information to plan, direct and, at times, modify their operations and logistics on the ground.[8]

10.2 Earth Observation in Support of Humanitarian Action

The term Earth observation (EO) is a broad concept that encompasses all common observational instruments as well as prediction and forecasting models to integrate and utilise these measurements. The Group on Earth Observation (GEO), formed by governments and international organisations, strives to pool the variety of observation systems in a system of systems (GEOSS) for monitoring the Earth in its entirety to address global key challenges and the most relevant societal priority areas including *natural and human-induced disasters.*[9] EO from space using satellite remote sensing[10] is one of the major technological contributors to GEOSS, comprising the space infrastructure (satellite sensors) and data provisioning and archiving subsystem, and the image analysis and information extraction subsystem, including information delivery and interoperable usage.

Humanitarian response benefits from the general assets of remotely sensed data as compared to conventional field mapping.[11] Table 10.1 summarises those characteristics of remote sensing data with a focus on humanitarian action. In combination with advanced information extraction techniques,[12] EO-based technology has evolved to become a decision-supporting tool for humanitarian professionals. Due to its observational power, ubiquitous usage and availability, remotely sensed data and derived information products complement time and effort-intensive field-based surveys and thus enrich the pool of spatially aware technologies for humanitarian relief support.

[8] Füreder et al. (2015).

[9] 'Building a Global System of Systems.' GEO 1 Oct. 2017 http://www.earthobservations.org/documents/geo_brochure.pdf

[10] Li, D. (2009) An overview of Earth observation and geospatial information service. in Li, D., Shan, J. and Gong, J., (eds.) Geospatial technologies for Earth observation, New York: Springer. pp. 556.

[11] Lang et al. (2015).

[12] Lang et al. (2010); Spröhnle, K., Tiede, D., Schoepfer, E., Füreder, P., Svanberg, A. and Rost, T. (2014) Earth observation-based dwelling detection approaches in a highly complex refugee camp environment — a comparative study. Remote Sensing, 6(10), pp. 9277–9297; Tiede, D., Krafft, P., Füreder, P. and Lang, S. (2017) Stratified Template Matching to Support Refugee Camp Analysis in OBIA Workflows. Remote Sensing, 9(326).

Table 10.1 General assets in using remote sensing data and their potential as translated to humanitarian action

Asset	Potential for humanitarian action
'From a distance'	No direct access required
	Objective source of information
	Cost-effective compared to field data collection
'Area-wide coverage'	Same conditions over large areas
	Time-efficient data collection
'Level of detail'	Control of level of detail
	Overview vs. detailed information
'Global availability'	Comparable data
	Fast and easy access to data
'Time series, monitoring' (including retrospectively)	Objective source of documentation on how a situation looked before a critical situation emerged
	Validate other sources of information (e.g. from field data)
	Ex post evidence on past situations, events, changes

Source: Lang et al. (2015)

Remote sensing as an objective imaging technology captures data over large areas under similar conditions, taken from a constant orbital view. Thus, remote sensing entails a democratising component, designed to reveal the situation *as is*, non-distorted, non-manipulated and potentially accessible to everyone. In addition to an on-demand assessment of the situation on the ground, the regular acquisition of remote sensing data also enables monitoring constant or changing ground conditions using standardised techniques. This can happen across various scales and time periods over the same region, as was demonstrated in the case of the IDP camp Zam Zam in the Darfur region in Sudan.[13] The archiving routines of remote sensing data enable retrospective views, where specific developments can be historically assessed, not just ex post,[14] e.g. the growth or dismantlement of a camp (see Fig. 10.1). New systems such as the Sentinel fleet, installed by the European Copernicus Programme,[15] deliver high-frequency regular overviews on a global scale. Its general potential to monitor camps and camp-like structures has been assessed by Wendt et al.[16] While we here focus on optical EO satellite data and

[13] Hagenlocher, M., Lang, S. and Tiede, D. (2012) Integrated assessment of the environmental impact of an IDP camp in Sudan based on very high-resolution multi-temporal satellite imagery. Remote Sensing of Environment, 126, pp. 27–38; Kranz, O., Sachs, A. and Lang, S. (2015) Assessment of environmental changes induced by internally displaced person (IDP) camps in the Darfur region, Sudan, based on multi-temporal MODIS data. International Journal of Remote Sensing, 36(1), pp. 190–210.

[14] Lang et al. (2010).

[15] Regulation (EU) No 377/2014 of the European Parliament and of the Council of the 3 April 2014 establishing the Copernicus Programme and repealing, 911/2010 (2014).

[16] Wendt, L., Lang, S. and Rogenhofer, E. (2017) Monitoring of refugee and IDP camps with Sentinel 2 imagery – a feasibility study. GI Forum – Journal for Geographic Information Science, 1/2017, pp. 172–182.

Fig. 10.1 Evolution of the IDP camp Zam Zam, Darfur, Sudan, as observed by VHR imagery (Quickbird) from 2002, 2008 and 2013 (Source: Lang et al. in Taylor and Francis (2015). This image was originally published in Lang et al. (2015))

related techniques of passive remote sensing, in general also active, and particularly radar, remote sensing has high prospective potential in the field of humanitarian assistance due to its weather and daylight independence.[17]

Remote sensing products need to undergo certain quality checks. In terms of the accuracy and reliability of the information content, they are usually evaluated against ground reference data.[18] However, in humanitarian crisis situations, ground records are usually scarce and difficult to obtain and rarely have in-field assessments been carried out in parallel or in a similar time window.[19] In this respect, independent re-evaluations of image interpretation or plausibility checks can be conducted to compensate for the lack of ground reference data. In fact, this poses challenges to the use of remote sensing from a scientific point of view, but otherwise

[17] Braun, A., Lang, S. and Hochschild, V. (2016) Impact of refugee camps on their environment – a case study using multi-temporal SAR data Journal of Geography, Environment and Earth Science International, 4(2), pp. 1–17.

[18] Congalton, R. G. and Green, K. (1998) Assessing the accuracy of remotely sensed data: principles and practices, Boca Raton: Lewis Publishers.

[19] Checchi, F., Stewart, B. T., Palmer, J. J. and Grundy, C. (2013) Validity and feasibility of a satellite imagery-based method for rapid estimation of displaced populations. International Journal of Health Geographics, 12(4), pp. 1–12.

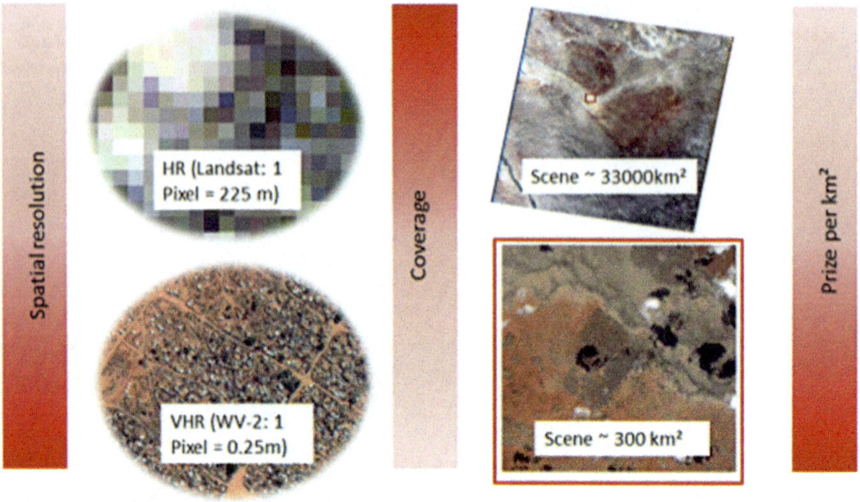

Fig. 10.2 Trade-off between spatial resolution, area being covered and costs of satellite remote sensing data (Source: Lang et al. in Taylor and Francis (2015). This image was originally published in Lang et al. (2015))

it underlines the strong need for such data, precisely because there is little or no independent information available. This paradoxical situation may be considered a key aspect in the determination of the usefulness of remote sensing technology in the humanitarian domain as well as in similar fields of applications. Next to the correctness of a mapping product, the matching of the actual content with specific user needs and the way this is presented are critical elements of product usability. In addition, products to be delivered need to fit into the standard workflows of the receiving organisations. So, it is not just a matter of 'doing the right thing' but 'doing the right thing *right*'.[20]

Next to its usability, the practical use of remote sensing data requires a demand-driven trade-off between technical merits and costs[21]; see Fig. 10.2. Observing large areas with very high resolution (VHR), i.e. submeter resolution, is not feasible, neither technically nor cost-wise. On the contrary, we cannot expect great spatial detail from a high-resolution (HR) scene covering thousands of square kilometres, as, e.g. Landsat-8 or Sentinel-2.

With respect to information products, there are countless options, but not all of them prove to be useful in the context of humanitarian operations. The tailoring of information products according to user needs is an interactive and iterative process, requiring both technical and organisational understanding from either side. Figure 10.3 shows a generalised view of themes being addressed and translated in a

[20] Zeil, P. and Lang, S. (2009) Do have clients a role in validation? . in Broglia, M., (ed.) Validation of geo-information products for crisis management, JRC, Ispra.

[21] Füreder et al. (2015).

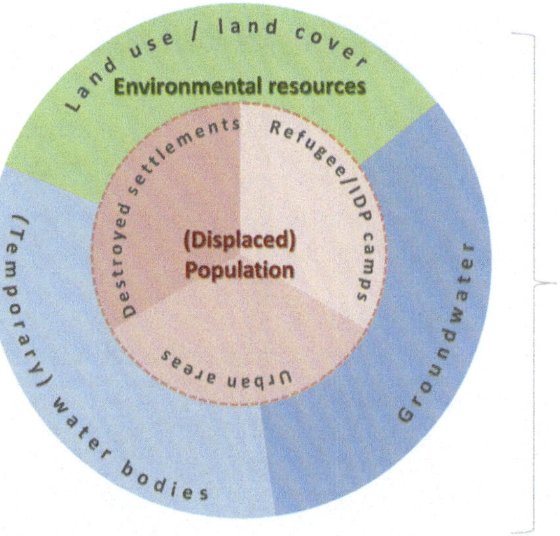

Product type name	Product type
Situation map	DAT-SIM
Camp structure	POP-CST
Camp extent	POP-CXT
Dwelling extraction	POP-DWX
Dwelling density	POP-DWD
Dwelling change detection	POP-DWC
Camp evolution	POP-CEV
Population estimation	POP-PES
Built-up structure	POP-BUS
Built-up change detection	POP-BUC
Graveyard detection	POP-GVD
Destroyed dwellings	POP-DSD
Hydrogeological reconnaissance map	ENV-HYR
Riverbed analysis (jetting)	ENV-RAJ
Land use land cover change	ENV-LUC
Land use land cover	ENV-LUL
Habitat modelling	ENV-HMO

Fig. 10.3 Abstract view of information product portfolio in a humanitarian assistance context (left), translated into a list of EO-based information products (See Schörghofer et al. (2017) for details)

set of information products. A selection of those is shown in the right part of the figure and illustrated in Fig. 10.4. The Austrian Space Programme-funded project EO4HumEn (2013–2016) set up a threefold EO-based information service to support the humanitarian operations of *Médecins Sans Frontières* (MSF). It comprised services on (1) population monitoring, (2) groundwater exploration and (3) environmental impact assessment in refugee/IDP camps. The follow-up project EO4HumEn+ (2016–2018) was designed to extend both the service portfolio as well as the board of user organisations. While information on (displaced) populations remains the core of the service portfolio, the previous focus on refugee/IDP camps has been now extended to urban settings. The outer circle denotes the secondary information needs on environmental resources, hydrological conditions, land use changes and land degradation, among others.

EO-based information on the number and composition of displaced populations[22] is critical in each phase of humanitarian crisis response.[23] In the emergency phase, satellite images provide fast information on the situation on the ground, whether the people in need are gathering in distinct camps or more complex urban settings. This can save important time and resources (staff, logistics and costs) as it

[22] Ehrlich, D., Lang, S., Laneve, G., Mubareka, S., Schneiderbauer, S. and Tiede, D. (2009) Can Earth observation help to improve information on population? Indirect population estimations from EO derived geospatial data. in Jasani, B., Pesaresi, M., Schneiderbauer, S. and Zeug, G., (eds.) Remote Sensing from Space Supporting International Peace and Security, Berlin: Springer. pp. 211–237.

[23] Füreder et al. (2015).

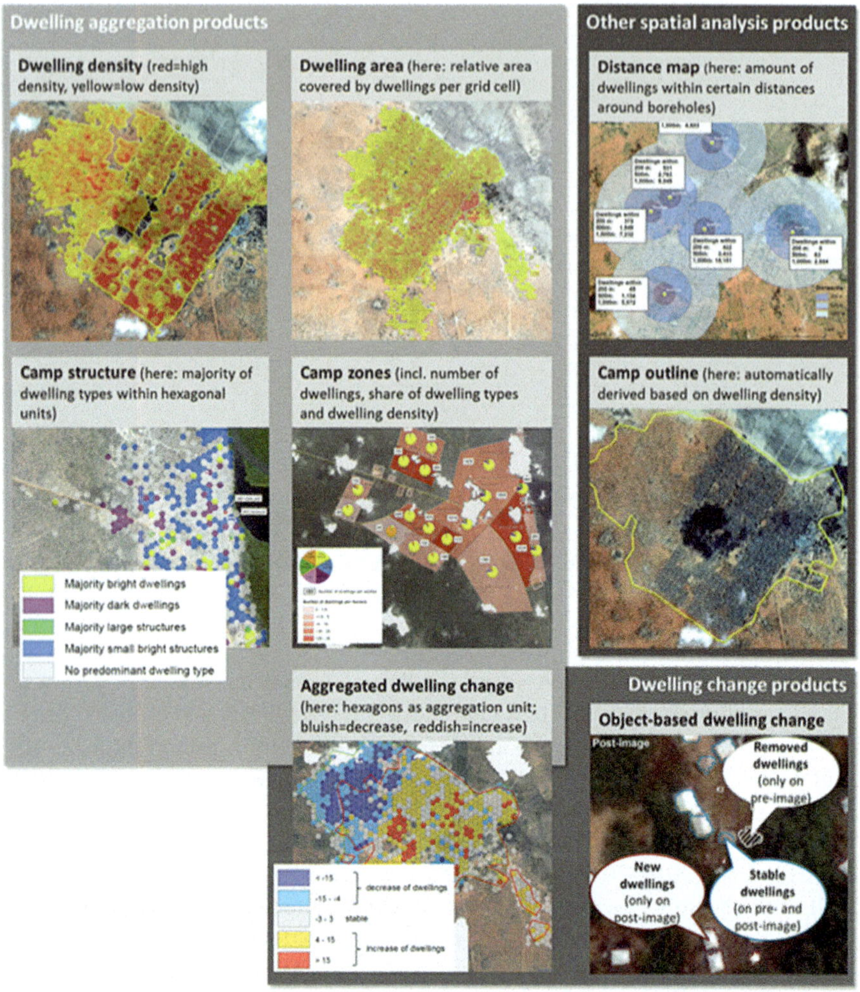

Fig. 10.4 Added-value products based on the extraction of dwellings (Source: Füreder et al. at ISCRAM (2015). This image was originally published in Füreder et al. (2015))

also provides means to verify indications on human presence, to direct humanitarian assistance. For camps in the setup and construction phase with highly dynamic and self-settling activities, satellite images can be used to monitor camp evolution, providing indicators of population estimates and assisting in planning logistical infrastructure and services such as healthcare or vaccination campaigns. This information can also be valuable for semi-permanent camps in the care and maintenance phase. The information obtained by extracting and characterising single dwellings can be enriched by spatial analysis. Density maps show areas where the distribution of dwellings can be optimised. Distance maps (dwellings to boreholes, latrines, hospitals, etc.) can be calculated to support camp planning. See Fig. 10.4 for further examples.

One of the most critical tasks for humanitarian organisations is the supply of sufficient potable water. While the immediate demand can be met by purification of surface water or transport from other areas ('water trucking'), the exploitation of local groundwater resources is usually the best option in the longer term. This requires an assessment of the hydrogeological situation within a short time frame, despite missing or unavailable detailed geological or hydrogeological maps. Workflows and procedures to rapidly provide hydrogeological reconnaissance maps based on freely available remote sensing and geological data are discussed by Wendt et al.[24]

Land use and land cover (LULC) are important information to better understand the geographical context of a camp, in terms of the ecosystem conditions and the provision of natural resources for food, fuel wood, building material, etc. Multiple sources of remotely sensed data can be used to accomplish this in a series of scale levels. LULC change analysis can then be utilised to assess the carrying capacity of a hosting space and the potential depletion of natural resources due to the presence of larger or more pertinent camps using multi-temporal SAR data[25] or VHR optical data.[26]

10.3 EO-Based Information Services for Humanitarian Organisations: A Dedicated Information Service for *Médecins Sans Frontières* (MSF)

The previous chapters have outlined that mission-critical information on refugee and IDP camps can support planning of emergency response and relief using multi-temporal and multi-scale information from satellite imagery and GIS data. Drawing from this, an EO-based information service[27] provides dedicated geospatial information products in support of humanitarian operations. The developed products are delivered as tailor-made maps, online Web services and advanced cartographic visualisations, utilised and validated by teams on the ground. The information service was set up for, and collaboratively designed with, *Médecins Sans Frontières* (MSF), an independent and international humanitarian organisation operating around the world. The service has been operational since 2012 under constant refinement of the product and information request level. This may be considered a typical case for a 'single-user' service, in particular in its original conceptualisation.[28]

---

[24] Wendt, L., Hilberg, S., Robl, J., Hochschild, V., Rogenhofer, E., Füreder, P., Lang, S. and Zeil, P. (2014) Assisting the exploration of groundwater near refugee/IDP camps using remote sensing and GIS. in Gemeinsame Tagung 2014 der DGfK, der DGPF, der GfGI und des GiN Hamburg.

[25] Braun et al. (2016).

[26] Hagenlocher et al. (2012).

[27] 'Special Session @ GI Forum Salzburg, 4–7 July 2017.' Earth4HumEn+ 1 Oct. 2017 www.zgis.at/humanitarian-services

[28] Lang et al. (2017).

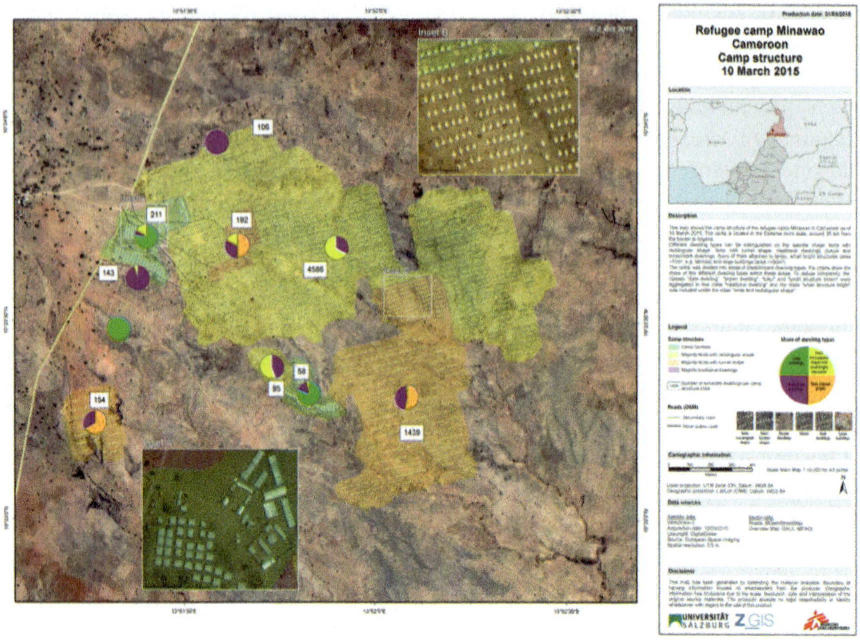

Fig. 10.5 Zoning map of Minawao refugee camp in Cameroon, showing amount, distribution and share of different dwelling types (© Credits Z_GIS, MSF, Source: Lang et al. (2017). This image was originally published in Lang et al. (2017))

The population monitoring service offers estimates of actual camp populations using dwelling extraction from (multi-temporal) VHR satellite imagery. Based on such information, further added-value products are provided to analyse internal camp structure or camp evolution (see Fig. 10.4). Additional services on environmental resources, including groundwater, are provided on demand. As of mid-2017, Z_GIS had produced over 350 maps detailing the situation on the ground at 60 locations in over 20 countries, predominantly in Eastern Africa, with more than 110 maps in South Sudan alone. Camp monitoring has been done up to nine times over the same location, for camps in Cameroon (see Fig. 10.5), Kenya and South Sudan. Products are provided in a variety of formats: static PDF maps, Google Earth overlays, dynamic online maps or image files (jpg, tif, png). These are used to organise vaccination campaigns (against malaria, cholera, yellow fever, etc.); for medical, food and shelter logistics; for monitoring the habitat of resistant vector-borne diseases; for drilling water wells; for estimating casualties after camp attacks; and for assessing damaged areas after floods or earthquakes. Z_GIS combines Earth observation data with advanced image analysis techniques, such as object-based image analysis (OBIA),[29] thus has tailored several information products for MSF, such as population estimation, hydrogeological reconnaissance, land cover (change) and others.

The population estimation service utilises (multi-temporal) very high-resolution (VHR) satellite imagery to derive information on the amount and spatial distribution of dwellings, dwelling density, camp extent, internal camp structure and general growth of the camp.[30] Aggregated information of single dwellings provides a better overview and reveals certain patterns within a camp (see Fig. 10.4, left). Dwelling density calculations (see Fig. 10.4, top) show where the population is concentrated in a refugee camp. The aggregation of dwellings to camp zones (Fig. 10.4, middle) supports camp planning and management. Camp structure analyses (Fig. 10.4, middle) show predominant dwelling types within definable regular reporting units (e.g. regular grids, hexagons). Multi-temporal analyses and change assessments assist camp monitoring (Fig. 10.4, bottom).

Figure 10.5 shows a zoning map of the Minawao refugee camp in Cameroon based on the distribution and share of different dwelling types. Since 2011, violent attacks by Boko Haram and the counterinsurgency operations of the Nigerian army have forced hundreds of thousands of people from northeast Nigeria to seek refuge in Cameroon, Chad and Niger. During the past 2 years, violence has increasingly spread from Nigeria into the three neighbouring countries, causing further displacement. By the end of 2016, there were around 86,000 refugees and 198,000 internally displaced people in Cameroon. In response, MSF scaled up its activities in several locations in the north of the country, providing healthcare, including maternal services and nutritional support. In the Minawao camp, which is administered by the UN refugee agency UNHCR, MSF staff carried out 58,147 consultations during the year. They also improved water and sanitation, trucking 3000 cubic metres of water per week, and assisted with the construction of 32 kilometres of pipes to find a permanent solution to the scarcity of water in the camp.

Beyond the regular standard product portfolio, ad hoc information derived from EO data is requested by MSF from time to time. Füreder et al.[31] point out some examples, such as land cover information for human African trypanosomiasis/sleeping sickness (HAT) and malaria vector control program planning to display the natural habitats of the tsetse fly and anopheles mosquito, support for cholera intervention and information on regularly flooded areas for planning suitable sites for hospitals.

[29] Blaschke, T., Hay, G. J., Kelly, M., Lang, S., Hofmann, P., Addink, E., Feitosa, R. Q., Van der Meer, F., Van der Werff, H., Van Coillie, F. and Tiede, D. (2014) Geographic Object-based Image Analysis: a new paradigm in Remote Sensing and Geographic Information Science. International Journal of Photogrammetry and Remote Sensing, 87(1), pp. 180–191.

[30] Füreder et al. (2015).

[31] Füreder et al. (2015).

10.4 Outlook

This chapter has highlighted some of the general assets of remote sensing technology and its usefulness for humanitarian operations, illustrated by an operational information service provided to MSF. Currently, the service portfolio is being extended and adapted to specific information needs by other global humanitarian actors such as the International Committee of the Red Cross (ICRC), the (Austrian) Red Cross, SOS Children's Villages and Groundwater Relief, according to their primary mandates and fields of action. Challenges faced by different actors in running such services operationally are often generalizable; they apply to data acquisition, data integration, interoperability and data exchange routines. Working in trans-organisational settings, where several humanitarian actors use, produce and share data, poses technical challenges. The first experiences of a collaborative online mapping platform called CMaP[32] show that the technical means to support this endeavour exist. As pointed out by Lang et al.,[33] the usage of EO data for humanitarian crisis mapping has increased tremendously during the past few years, a trend that can be expected to continue due to the rising number of remote sensing systems that have increased awareness within the humanitarian community. In addition, the emergence of new technologies, such as portable ground sensors, including drones as well as 'human sensors', based on signals or social media use, will enable new ways of humanitarian support but at the same time will raise new issues of ethics, privacy and potential misuse. Thus, next to the technical prerequisites, mutual trust building is a key requirement for such services to be taken up sustainably.

Acknowledgements Research leading to this article has been co-funded by the Austrian Research Promotion Agency (FFG) under the Austrian Space Application Programme (ASAP) within the projects EO4HumEn (*EO-based services to support humanitarian operations: monitoring population and natural resources in refugee/IDP camps*, contract no. 840081) and EO4HumEn+ (*extended EO-based services for dynamic information needs in humanitarian action*, contract no. 854041). Direct funding has been received from Médecins Sans Frontières (MSF) Austria and the Karl Kahane Foundation.

References

Bjorgo E (2001) Supporting humanitarian relief operations. In: Baker JC, O'Connell KM, Williamson RA (eds) Commercial observation satellites: at the leading edge of global transparency. ASPRS RAND, Santa Monica, pp 403–427

Blaschke T, Hay GJ, Kelly M, Lang S, Hofmann P, Addink E, Feitosa RQ, Van der Meer F, Van der Werff H, Van Coillie F, Tiede D (2014) Geographic object-based image analysis: a new

[32] Schörghofer, R., Lang, S., Wendt, L. and Riedler, B. (2017) CMaP – collaborative mapping platform for humanitarian organisations. GI Forum – Journal for Geographic Information Science, 1/2017, pp. 207–216.

[33] Lang et al. (2017).

paradigm in remote sensing and geographic information science. Int J Photogramm Remote Sens 87(1):180–191
Braun A, Lang S, Hochschild V (2016) Impact of refugee camps on their environment – a case study using multi-temporal SAR data. J Geogr Environ Earth Sci Int 4(2):1–17
Checchi F, Stewart BT, Palmer JJ, Grundy C (2013) Validity and feasibility of a satellite imagery-based method for rapid estimation of displaced populations. Int J Health Geogr 12(4):1–12
Congalton RG, Green K (1998) Assessing the accuracy of remotely sensed data: principles and practices. Lewis Publishers, Boca Raton
Ehrlich D, Lang S, Laneve G, Mubareka S, Schneiderbauer S, Tiede D (2009) Can Earth observation help to improve information on population? Indirect population estimations from EO derived geospatial data. In: Jasani B, Pesaresi M, Schneiderbauer S, Zeug G (eds) Remote sensing from space supporting international peace and security. Springer, Berlin, pp 211–237
Füreder P, Lang S, Hagenlocher M, Tiede D, Wendt L, Rogenhofer E (2015) Earth observation and GIS to support humanitarian operations in refugee/IDP camps. In: Palen, Büscher, Comes, Hughes, (eds) Geospatial data and geographical information science – proceedings of the ISCRAM 2015 conference, Kristiansand
Hagenlocher M, Lang S, Tiede D (2012) Integrated assessment of the environmental impact of an IDP camp in Sudan based on very high resolution multi-temporal satellite imagery. Remote Sens Environ 126:27–38
Kranz O, Sachs A, Lang S (2015) Assessment of environmental changes induced by internally displaced person (IDP) camps in the Darfur region, Sudan, based on multi-temporal MODIS data. Int J Remote Sens 36(1):190–210
Lang S, Tiede D, Hölbling D, Füreder P, Zeil P (2010) EO-based ex-post assessment of IDP camp evolution and population dynamics in Zam Zam, Darfur. Int J Remote Sens 31(21):5709–5731
Lang S, Füreder P, Kranz O, Card B, Roberts S, Papp A (2015) Humanitarian emergencies: causes, traits and impacts as observed by remote sensing. In: Thenkabail P (ed) Remote sensing handbook. Taylor and Francis, New York, pp 483–512
Lang S, Schoepfer E, Zeil P, Riedler B (2017) Earth observation for humanitarian assistance. GI Forum – J Geogr Inf Sci 1:157–165
Li D (2009) An overview of earth observation and geospatial information service. In: Li D, Shan J, Gong J (eds) Geospatial technologies for Earth observation. Springer, New York, p 556
Schörghofer R, Lang S, Wendt L, Riedler B (2017) CMaP – collaborative mapping platform for humanitarian organisations. GI Forum – J Geogr Inf Sci 1:207–216
Spröhnle K, Tiede D, Schoepfer E, Füreder P, Svanberg A, Rost T (2014) Earth observation-based dwelling detection approaches in a highly complex refugee camp environment—a comparative study. Remote Sens 6(10):9277–9297
Tiede D, Krafft P, Füreder P, Lang S (2017) Stratified template matching to support refugee camp analysis in OBIA workflows. Remote Sens 9:326
Voigt S, Kemper T, Riedlinger T, Kiefl R, Scholte K, Mehl H (2007) Satellite image analysis for disaster and crisis-management support. IEEE Trans Geosci Remote Sens 45(6):1520–1528
Voigt S, Schoepfer E, Fourie C, Mager A (2014) Towards semi-automated satellite mapping for humanitarian situational awareness. In: Global Humanitarian Technology Conference (GHTC): IEEE. San Jose, pp 412–416
Wendt L, Hilberg S, Robl J, Hochschild V, Rogenhofer E, Füreder P, Lang S, Zeil P (2014) Assisting the exploration of groundwater near refugee/IDP camps using remote sensing and GIS. In: Gemeinsame Tagung 2014 der DGfK, der DGPF, der GfGI und des GiN Hamburg
Wendt L, Lang S, Rogenhofer E (2017) Monitoring of refugee and IDP camps with Sentinel 2 imagery – a feasibility study. GI Forum – J Geogr Inf Sci 1:172–182
Zeil P, Lang S (2009) Do have clients a role in validation? In: Broglia M (ed) Validation of geo-information products for crisis management. JRC, Ispra

Chapter 11
The Field, Its Needs and New Technologies

Andreas Papp and Leonora Barclay

There are currently over 65 million forcibly displaced persons worldwide. Over half of refugees are under the age of 18.[1] In crisis situations, children and young people are the most vulnerable to abuse and exploitation. To strengthen our emergency preparedness and response, SOS Children's Villages is participating in two initiatives relating to new technologies. First, SOS Children's Villages is partnering with Allianz and the German Aerospace Center (DLR) for a study on how satellite technology and other space infrastructure can support our work. In addition, SOS Children's Villages is partnering with the Department of Geoinformatics at the University of Salzburg to pilot vulnerability studies, which, in combination with local knowledge of an area, could support SOS Children's Villages in preparing for crisis situations as well as improving reactiveness. While these partnerships take different approaches, both can be used as a basis for strategies in building NGO programmes.

11.1 Introduction

SOS Children's Villages was founded in response to an emergency – the many children without parental care on the streets of Austria. Today, the organisation is active in 134 countries and territories, providing a range of services to children and families, including alternative care in SOS Children's Villages, youth programmes, family strengthening programmes and emergency response.

In crisis situations, children and young people are the most vulnerable to abuse and exploitation, as well as missing out on education. SOS Children's Villages

[1] "Figures at a Glance." 2015. UNHCR 17 Jun. 2016 http://www.unhcr.org/figures-at-a-glance.html

A. Papp (✉) · L. Barclay
SOS Children's Villages International, Vienna, Austria
e-mail: Andreas.Papp@sos-kd.org

C. Al-Ekabi, S. Ferretti (eds.), *Yearbook on Space Policy 2016*,
Yearbook on Space Policy, https://doi.org/10.1007/978-3-319-72465-2_11

responds to these needs in a number of ways: implementing Child Friendly Spaces where children and families are provided with a safe space and, depending on the context, receive food, educational, medical and psychosocial support; providing interim-care for unaccompanied children and, if needed, long-term care such as foster families for unaccompanied minors, where family reunification is not possible; and also by sharing resources such as food and water, as well as giving families a roof over their heads in order to prevent family separation. Child protection is another core part of SOS Children's Villages' work, keeping children and young people safe during and after crisis situations. In June 2016, SOS Children's Villages was involved in 25 emergency programmes around the world.

The recently agreed Sustainable Development Goals (SDGs) call on all civil society to participate in reaching these goals. Five of the SDGs relate directly to SOS Children's Villages and will form a central part of our work over the coming decades: no poverty, quality education, decent work and economic growth, reduced inequalities and peace and justice. In response to this call, SOS Children's Villages has aligned our internal strategy with the SDGs; the strategy sets out our journey as a federation until 2030 and supports the organisation in focusing on the five SDGs of particular relevance. This includes one of the strategic initiatives of the SOS Children's Villages Strategy 2030, which aims to strengthen families by increasing our focus on preventing family separation and providing emergency response.

One of the current focuses of SOS Children's Villages regarding emergencies and disasters is to strengthen our preparedness, particularly in high-risk countries. A key part of this is harnessing new technologies to improve both the speed and efficiency of emergency response activities. In this chapter, we will highlight two initiatives in this area: the partnership with Allianz and DLR[2] and the partnership with Z_GIS, the Department of Geoinformatics at the University of Salzburg. These two initiatives demonstrate both the advantages of partnerships with others and the different benefits of new technologies.

11.2 Utilising Experiences from Other Sectors

The aim of the partnership with Allianz and DLR is to improve the safety of SOS programme beneficiaries and co-workers across the globe as well as to minimise the damage incurred by disasters. One focus area for the SOS Children's Villages emergency response team is to have an early warning system, sending a warning and additional information about the threat when a disaster approaches or occurs. This would allow us ahead of the disaster to prepare children and families, providing food and other resources, which would minimise the impact of the crisis.

SOS Children's Villages is positioned in 134 countries, which gives us the opportunity to reach out as first responder to affected communities in the local area. For example, in Nepal, SOS Children's Villages together with local partners provided services

[2] DLR: currently established via a project collaboration.

to around 30,000 people in the first 6 weeks after the earthquakes.[3] There is a clear demand across the NGO sector for partnerships that would support the work of NGOs.

This aim made the Earth Observation Center (EOC) of DLR a strong partner of choice, as DLR focuses on supporting the entire disaster management cycle. It operates the "Center for Satellite Based Crisis Information" (ZKI), which provides a 24/7 service for the rapid provision, processing and analysis of satellite imagery during natural disasters, for humanitarian relief activities and civil security issues. Satellite imagery is also used to extract key information in emergency situations, which can be used to ensure an effective response in the immediate aftermath of an emergency. DLR also supports the development of early warning systems in the domain of natural hazard prevention that could provide crucial support in light of the emergency preparedness work of SOS Children's Villages.

It is only once we have addressed the needs of our beneficiaries that we can also reach out to the most vulnerable children and families in the affected area; thus this would enable us to expand our support as well as ensure it is targeted.

The initial study conducted by DLR aims to examine how satellite imagery and other new technologies can support SOS Children's Villages during and after crisis situations and in which contexts they can be used to save lives. The study centres on three areas where technology could support the work of SOS Children's Villages in emergency prevention and response. These are:

- Preparedness and early warning: analysing how a preparedness and early warning system could look, including content-related design and technical specifications. This would support us in accessing and supporting disaster areas, such as providing information on the accessibility of roads and, when flooding occurs, the extent of the flooding. A specific early warning system for SOS Children's Villages could assess the potential risks of natural disasters, such as earthquakes or floods as well as local outbreaks of violence, enabling SOS Children's Villages to better prepare for these situations.
- Emergency mapping: examining additional uses for new technologies. This includes exploring how satellite-based emergency mapping can provide information about an emergency or crisis situation and how this can best support SOS Children's Villages.
- Humanitarian technologies: examining additional technical topics and options for SOS Children's Villages. This could include humanitarian telemedicine, teleteaching, energy supply, safety techniques, logistics, infrastructure management, refugee movement monitoring, agricultural land use, food security and water supply.

The study was presented in July; based on its results, SOS Children's Villages will decide upon which tools have the largest capacity and how best these can be utilised in our work. There are big challenges ahead due to climate change and its

[3] "Emergency Response Programme Weekly Situation Report" 26 Apr.–24 May 2015, SOS Children's Villages International 27 June 2016 https://www.sos-childrensvillages.org/publications/news/humanitarian-emergencies/nepal-earthquake

impact. NGOs need to respond by building partnerships where each sector brings knowledge and experience, so we can work together to find solutions to these challenges. There is much to be gained on both sides from sharing knowledge and resources; this approach would particularly strengthen the work of NGOs.

11.3 Harnessing Space Infrastructure with the University of Salzburg

As outlined above, SOS Children's Villages provides a range of services in the area of emergency preparedness and response. There are, therefore, clear benefits of access to geoinformation, which could provide key information that would otherwise not be available, such as the accessibility and status of roads and water services. SOS Children's Villages International will partner with the Z_GIS department at the University of Salzburg as part of their EO4HumEn+ project, which supports humanitarian organisations by providing Earth observation based on satellite images and geographic information.

There are SOS Children's Villages all over the world, which can be used as a local basis in order to provide support and shelter after a disaster. As an established member of the community, in addition to the quality of the facilities in the SOS Children's Villages, the organisation is regularly asked to act as a hub for partner organisations during emergency situations.

The partnership with Z_GIS is an opportunity to prepare for such situations by, together with the SOS Children's Villages local member association of the relevant country, collecting information before an emergency to be able to roughly predict an emergency as well as the future needs of the SOS Children's Villages and surrounding communities. For example, in an area vulnerable to drought, where are the nearest water sources? How many people live in the area and would they be likely to go in the direction of an SOS Children's Village in case of emergency? If the SOS Children's Village has this information, it can anticipate the needs of the local population in emergency situations by having extra supplies and resources available.

This partnership with Z_GIS means SOS Children's Villages can support local authorities and organisations in responding to a crisis and bringing in resources, such as fire brigades and relief authorities. By keeping the tool up-to-date, SOS Children's Villages could ensure that the local authorities are kept up-to-date and aware of where their resources are most needed. As new responders, agencies and NGOs arrive, they will be able to have an overview of the services and activities in each area, so relief activities can be coordinated and managed.

This is a good opportunity for SOS Children's Villages to strengthen our role as a centre for emergency support in the immediate aftermath of a crisis and during the period before other organisations arrive on the scene and establish their own facilities.

Planning in this regard would also be based on vulnerability studies in high-risk countries, allowing SOS Children's Villages to effectively address the needs of the most vulnerable children and families facing humanitarian crisis and disasters around the world. In combination with the information provided by the DLR early warning system outlined above, the information gained from the vulnerability studies will support us in tailoring and rolling out contingency plans and programmes in these countries.

Thus, SOS Children's Villages can pre-position the necessary materials in these areas, such as in anticipation of flooding or other disasters. In addition, these vulnerability studies will focus on sudden as well as slow-onset emergencies. For example, if there is a malaria outbreak every year, the vulnerability studies could highlight which areas are affected the most, highlighting where the pre-positioned material should be sent.

There is a particular need for this approach in countries without strong civil services. By predicting possible risks, SOS Children's Villages in countries in the Global South can have an emergency pack, with both general supplies, such as food and water, as well as resources specific to the local area, such as malaria tablets and mosquito nets in areas where malaria is a risk. The support given by SOS Children's Villages to our beneficiaries as well as to other affected children and families could thus be tailored to local needs, making it more effective, as the necessary supplies and resources would already be there when a disaster or crisis situation occurs.

Vulnerability studies are, however, only part of the picture. The advantage of working in partnership with Z_GIS is that, as SOS Children's Villages are established members of the community, each local member association is an expert on their own local area. Thus, the technology would support rather than guide them as they make decisions based on up-to-date information as well as their own local knowledge in terms of coping mechanisms and capacities.

Each area has a complex mix of factors that could affect a crisis situation, so by combining vulnerability studies with local knowledge, SOS Children's Villages can increase our resources and support where needed.

As the impact of climate change will create more crises, including extreme weather hazards, scarcity of goods and natural resources as well as conflicts, there will be an increased demand for our emergency preparedness and response services. SOS Children's Villages must respond to this anticipated increased demand; this means we must reach out to partners to establish how best these demands can be met.

11.4 Conclusions and Recommendations

New technologies such as geoinformation give a deeper understanding of a local situation and local developments, including local pressures such as a lack of water, the geographical terrain and natural disasters. It is vital that NGOs have access to this information and so can utilise it to reach the most vulnerable individuals.

As a consequence, it would be beneficial for NGOs to embrace these new technologies in order to strengthen their own services. By engaging in partnerships, SOS Children's Villages is actively taking steps to improve and expand our emergency preparedness and response, pooling resources with partners in areas that we could not achieve alone.

The partnerships with DLR and Z_GIS pool the knowledge of both the NGO and research institutions and bring benefits to both partners. Often, a factor that limits NGOs in their use of space technology is a lack of resources; by not building the resources internally, NGOs must reach out to partners in order to fully use space/ satellite technology. This is also true in the other direction; the space industry and research institutes rarely give presentations at NGO conferences and workshops.

On the other hand, corporations are taking up their social responsibility and working in partnerships with the space community, meaning they can now be the partners of choice for NGOs working in emergency preparedness and response.

There is a demand for these partnerships and the experience of SOS Children's Villages shows that this can work effectively. We have outlined the range of benefits that space technology can bring to NGOs in the field of emergency preparedness and response. These advantages only come from cooperation with partners.

In our experience, NGOs frequently have a lack of presence in workshops and congresses on new technologies. This must be addressed. With advantages for both corporate partners and NGOs, there is a clear need for more dialogue between the sectors.

Time and again, change in NGOs comes from bottom-up demands; issues or requests are first highlighted in the field and then passed on to the central office. We reach out to colleagues working in this area and ask them to consider the strengths and advantages of such partnerships.

An additional advantage of partnerships is sharing experiences and knowledge and the benefits this reaps. A core part of the partnership with both Allianz and DLR as well as with Z_GIS is that geoinformation by itself is not enough – to gain the full benefits of such partnerships, the NGO must also share its own experiences in the local area. Thus, a partnership must be based on dialogue and pooling resources, rather than seen as a transaction.

NGOs are often hesitant to invest donor funds in research and development, because innovation always has the risk of failure. Instead, they choose to invest donor funds in areas with clear impact and little risk. However, with partnerships and cooperation such as those outlined above, we in the NGO sector have access to knowledge and resources to support us in learning which tools we can use and how.

As NGOs cannot invest huge amounts of funds in research and development, it is to our benefit when research institutes and corporate partners help us build our resources. For example, exchange programmes and partnerships would enable the sectors to understand each other better, leading to an increased awareness of the needs of each sector.

Chapter 12
Financial and Nonfinancial Aspects of Sustainable Development

Alfredo Roma and Alessandra Vernile

Abstract The Addis Ababa Action Agenda, the adoption of the 2030 Agenda, the Sustainable Development Goals (SDGs) and the adoption of the Sendai Framework for disaster risk have been important achievements for the global community. The COP 21, held in Paris in 2015, represents the first major step towards the achievement of the targets set by the Sustainable Development Goals agreed by the international community in September 2015. Such goals may be achieved through the involvement of public and private actors, especially investing in school and culture. Space can play an important role through its infrastructure like Earth observation, navigation and connectivity. The SDGs are also in line with the two goals of the World Bank: ending extreme poverty and boosting shared prosperity. The contribution of the World Bank to the Sustainable Development Goals is related to the data revolution proposed by the international leaders during the High-Level Panel of Eminent Persons on the Post-2015 Development Agenda. Even drones can help the reaching of the SDGs by collecting imagery for disaster risk reduction. The activities of private actors in the space sector increased exponentially in the last years thanks to the proliferation of start-ups like Planet, which provides a complete imagery of the Earth. Finally, it would be useful to set up a public-private partnership (PPP) scheme that will help to move to more sustainable business models.

The new approach to development finance through the Addis Ababa Action Agenda, the adoption of the 2030 Agenda and the Sustainable Development Goals, the adoption of the Sendai Framework for Disaster Risk Reduction and finally the COP21 with the Paris Climate Agreement were all important achievements for the global community in 2015. All of them showed that the global community could work together.

The COP21, held in Paris at the end of 2015, was the first major step towards achieving the targets set by the Sustainable Development Goals, agreed by the international community in September 2015, a few months before the event in Paris. On the occasion of the G20 meeting, China and the USA ratified the Paris Agreement

A. Roma (✉) • A. Vernile
European Space Policy Institute (ESPI), Vienna, Austria
e-mail: alfredo.roma@astlegal.com; Alessandra.Vernile@espi.or.at

C. Al-Ekabi, S. Ferretti (eds.), *Yearbook on Space Policy 2016*,
Yearbook on Space Policy, https://doi.org/10.1007/978-3-319-72465-2_12

marking a decisive step forward for the whole international community. During the meeting held in Antalya, Turkey, it was shown that it is necessary to act fast and decisively on global matters, especially regarding financial stability.[1]

The day before the G20, the consortium Climate Transparency published the report "Brown to Green: Assessing the G20 transition to a low-carbon economy". The report describes the Paris Agreement's mission and what to do to support it. The report, which contained statistics about decarbonisation, showed how it was progressing, even if very slowly.[2]

Decarbonisation is important since it is one of the key issues towards sustainable development. In this context, it is important to understand the role of space and how it is entwined with sustainable development.

In this frame, the Millennium Development Goals in 2000 before, and the Sustainable Development Goals (SDGs) now, can help to understand what sustainable development is and to add tiles to the mosaic of sustainability. Through the SDGs today, it can be seen how space can really support the achievement of each goal to end poverty, protect the planet and guarantee prosperity for all.

Each goal can be achieved through the involvement of public and private actors and civil society. Investing in schools and culture would bring an enormous return in terms of health improvement, hunger reduction, inequality reduction, development of economic activities in situ, development of clean energy and the reduction of terroristic attacks.

The role of space can substantially contribute to these goals through its infrastructures. Earth observation (EO) is an important tool because of its contribution to the optimisation of agriculture and pollution monitoring. Today, also connectivity has an important role not fully recognised in the list of SDGs. Connectivity is important for telemedicine, e-learning and access to information. Also, drones guided through space infrastructures can play a useful role.

The actors in this process are the United Nations, national governments and national space agencies, the European Union, the European Space Agency, Eumetsat and the manufacturing industry and operators. The involvement of diverse international actors should stimulate international cooperation, in accordance with the basic principle of the Outer Space Treaty (OST) and the other space treaties establishing that space belongs to all human beings who should be the beneficiaries of all the advantages coming from space exploitation and exploration.

---

[1]The Antalya G20 took place on 15–16 November 2015. The G20 members are Argentina, Australia, Brazil, Canada, China, France, Germany, Italy, India, Indonesia, Japan, Mexico, Republic of Korea, Russia, Saudi Arabia, South Africa, Turkey, the United Kingdom, the USA and the European Union. The G20 members have invited Spain as a permanent invitee. Additionally, Zimbabwe was invited, as they are the 2015 Chair of the African Union, Malaysia was invited as they are the 2015 Chair of the Association of Southeast Asian Nations (ASEAN), Senegal was representing the New Partnership for Africa's Development, and Azerbaijan and Singapore were also invited; see further, "2015 Turkey G20" G20.org.tr 23 Oct. 2017 http://g20.org.tr/

[2]"Brown to Green: Assessing the G20 transition to a low-carbon economy" Sep. 2016, Climate-Transparency.org 23 Oct. 2017 http://www.climate-transparency.org/wp-content/uploads/2016/08/Brown-to-Green-Assessing-the-G20-transition-to-a-low-carbon-economy.pdf

12.1 The Involvement of Public Institutions: The Case of the World Bank[3]

The World Bank has a strong relationship with the United Nations that goes back to 1947 when the institution was designated as an independent specialised agency of the UN system.[4] Through this capacity, the World Bank has collaborated with the UN in every region and sector over the years. This engagement has deepened because of the Millennium Development Goals (MDGs) and, today, the Sustainable Development Goals (SDGs).

The SDGs are in line with the two goals of the World Bank: ending extreme poverty and boosting shared prosperity.[5] The World Bank has made a commitment to countries all around the world to deliver on the 2030 Agenda through three critical areas:

- Finance
- Data
- Implementation and support of country-led and country-owned policies[6]

Global poverty has fallen steadily since the 1980s, except in Africa where poverty has been halved more recently, thanks to intensified efforts to boost incomes and goad the resilient behaviour of individuals who still live in extreme poverty, especially in sub-Saharan Africa.

In 2013, 767 million people lived below the international poverty line on less than $1.90 per day, showing that global poverty had decreased from about 28% in 1999 to 11% in 2013. In 2016, just under 10% of the world's workers were living with their families on less than $1.90 per person per day, down from 28% in 2000. In less developed countries, nearly 38% of workers were living below the poverty line.[7]

---

[3] Anderson, Edward Charles. "International Financial Institutions: Space in development". Presentation at the 10th ESPI Autumn Conference "Space for Sustainable Development". Sept. 15–16 2016, ESPI, Vienna. http://www.espi.or.at/images/10th_autumn_conf/Files/Presentations/Panel_3/9.-ESPI-Space-for-Development---Anderson---2016.pdf

[4] "The United Nations-World Bank Partnership." United Nations Peacebuilding Support Office 23 Oct. 2017 http://www.un.org/en/peacebuilding/pbso/unwb.shtml

[5] World Bank Group. "*A Measured Approach to Ending Poverty and Boosting Shared Prosperity: Concepts, Data, and the Twin Goals*." Policy Research Report; World Bank. Washington DC, 2015. 23 Oct. 2017 https://openknowledge.worldbank.org/handle/10986/20384; See also: Jim Yong Kim. "World Bank President: Ending Extreme Poverty Possible by 2030". Georgetown University, Washington DC, 2 Apr. 2013 https://www.georgetown.edu/news/president-of-the-world-bank.html

[6] "Sustainable Development Goals (SDGs) and the 2030 Agenda." The World Bank 23 Oct. 2017 http://www.worldbank.org/en/programs/sdgs-2030-agenda

[7] "Sustainable Development Goal 1." Sustainable Development Knowledge Platform 23 Oct. 2017 https://sustainabledevelopment.un.org/sdg1

The World Bank is highly committed to finance development. Funds for development are composed of:

- Domestic resources, which are the largest available source of funding for countries' development plans. To unlock these resources, countries must build effective tax regimes and government institutions, to improve public spending.
- Official development assistance, which is a particularly important source of funding for the poorest countries. These funds must catalyse and leverage new development resources. Each year the global community provides $135 billion.
- Private sector finance is the largest potential source of funding. To unlock these resources, each country must improve the business climate, develop local capital markets and mitigate risk for investment assets.[8]

The World Bank is committed to financing for development through loans, grants, equity investments and guarantees to partner countries and private businesses, with a flood of investments of $60 billion. The major focus is on Southern Africa, where $15 billion has been allocated in the form of grants. The other investments of the World Bank – in Latin America and the Caribbean ($10 billion), Middle East and North Africa ($5 billion), Europe and Central Asia ($10 billion), East Asia and the Pacific ($9 billion) and South Asia ($11 billion) – are a combination of loans, grants, equity investments and guarantees to partner countries and private businesses.[9] A relevant aspect is the technical assistance offered to recipient countries. Through this, the World Bank tries to assist existent local technical structures to help the countries build a favourable tax regime, practise good governance and establish relevant institutions to stimulate private investments.

The contribution of the World Bank to sustainable development is related to the data revolution requested by international leaders during the High-Level Panel of Eminent Persons on the Post-2015 Development Agenda.[10] The data revolution for sustainable development takes advantage of new technologies, crowd sourcing and improved connectivity. In particular, mass participation and open government initiatives were stressed as relevant topics that would help the quality of statistics and information to all citizens to guarantee the empowerment of people.

Embracing an open data policy would help to achieve the objective of open development. Open development is understood as the emerging set of possibilities to catalyse positive changes through open information-networked activities in international development. Positive development can emerge through new models of engagement and innovation that involve the society more through more collaborative tools and is driven by beneficiaries.[11]

---

[8] Anderson, Edward Charles (2016). *In* ibid.

[9] Ibid.

[10] "The Report." High-Level Panel of Eminent Persons on the Post-2015 Development Agenda 23 Oct. 2017 http://www.post2015hlp.org/the-report/

[11] Smith, M. L., and Katherine M.A. Reilly. "Open Development. Networked Innovation in International Development". (2014) The MIT Press, Massachusetts, Boston.

In this framework, the role of the government needs to be sustained by private actors, in particular locally established start-ups. The involvement of private actors in an open sustainable development process has been shown to be highly useful due to the elements that characterise a start-up:

- Private financing
- Low-cost services
- High rate of returns
- Major flexibility
- More independence from public institutions

A major contribution to sustainable development is made by space applications in the field of communications, Earth observation, remote sensing and navigation. When looking at the initiatives undertaken by the World Bank in Africa, it can be seen that the majority of them are in the fields of navigation and communications. Some examples follow.

The OpenStreetMap project in Africa is an open free map data project that has tremendous benefit for humanitarian and economic development. The principal objectives are to be the connecting point between humanitarian actors and open mapping communities, to provide remote data creation during crises, to collect and organise existing data sources, to support deployments to the field, to be a distribution point for free data and to develop open knowledge and tools.[12] Since its first usage from August 2005 to January 2016, the number of users in Africa increased exponentially.

An example that shows the importance of having access to open data and SDGs is the involvement of the World Bank Group in initiatives in Tanzania.[13] Among these, there is "Dar Ramani Huria".[14] The project is a community-based mapping project based in Dar es Salaam. Dar Ramani Huria has the objective of training local university students and community members in the city, to use OpenStreetMap to create sophisticated and accurate maps of Dar es Salaam. The city suffers from devastating floods that cause millions of dollars of damages. The damages caused by floods could be prevented with adequate planning, through sophisticated and updated maps. These maps help communities to draw residential areas, roads, streams, floodplains and other features, relevant for disaster prevention and responses to areas not signalled on the map. The maps developed are then combined with other data with InaSAFE,[15] to obtain a complete image of the areas to guarantee a disaster response.

---

[12] "About." OpenStreetMap 23 Oct. 2017 https://blog.openstreetmap.org/about/

[13] To know more about Tanzania: "Overview." The World Bank 23 Oct. 2017 http://www.worldbank.org/en/country/tanzania/overview

[14] Swahili expression for "Dar Open Map".

[15] InaSAFE has been jointly developed by the Indonesian government-BNPB, the Australian government, the World Bank-GFDRR and independent contributors. These agencies and the individual software developers of InaSAFE take no responsibility for the correctness of outputs from InaSAFE or decisions derived as a consequence. "InaSAFE." Inasafe.org 23 Oct. 2017 http://inasafe.org/

The World Bank's Global Facility for Disaster Reduction and Recovery (GFDRR)[16] supports the Dar Ramani Huria project. Since the launch of the project in 2015, two million citizens have benefitted from the renewal of local maps that were 20 years old. Twenty-nine wards, 1254 km waterways and 3396 km of roads have been mapped. This has involved the training of 460 experts in mapping and the establishment of ten disaster prevention teams. Thanks to these innovative approaches in Tanzania, the risks from urban flooding and the rapid urbanisation have been reduced.[17]

Drones are a consistent support for mapping; they were initially used in Tanzania as an experiment in the frame of the "Dar Ramani Huria" project. In February 2015, the World Bank involved the Swiss non-profit organisation Drone Adventures. The objective was to test the idea of collecting and using drone imagery for disaster risk reduction. Initially, the drones were not considered as a tool for collecting data, but the possibility of having high-quality images at a good price was considered a relevant incentive for using drones and letting them contribute to the mapping exercise.[18]

The use of drones has been evaluated both by the Tanzanian government and the World Bank: both institutions consider mapping as the baseline to understand the development of cities and help the improvement of existing infrastructures.

In this regard, the World Bank intends to continue to use drones, only if backed by a favourable regulatory environment. The set-up of the right context to continue to use drones is possible only if the Tanzanian government is still interested in collaborating with the World Bank to set up a regulation on the use of drones in place.[19]

12.2 The Involvement of Private Actors: Planet[20]

To achieve the SDGs, the involvement of private actors is important to reduce the gap when the only economic assistance comes from public institutions. International organisations and governments have already defined their efforts in global

[16] Established in 2006, the Global Facility for Disaster Reduction and Recovery (GFDRR) is a partnership of 39 countries and 8 international organisations committed to helping developing countries reduce their vulnerability to natural hazards and adapt to climate change. The partnership's mission is to mainstream disaster risk reduction (DRR) and climate change adaptation (CCA) in country development strategies by supporting a country-led and managed implementation of the Hyogo Framework for Action (HFA). GFDRR's Partnership Charter, revised in April 2010, sets its original mission, rationale and governance structure. "Funding Structure & Partnerships." GFDRR.org 23 Oct. 2017 https://www.gfdrr.org/funding-structure-partnerships

[17] Dar Ramani Huria, "What We Do." 23 Oct. 2017 http://ramanihuria.org/

[18] "World Bank Using UAVs for Disaster Risk Reduction in Tanzania." 19 Aug. 2015. OPEN DRI 23 Oct. 2017 https://opendri.org/world-bank-using-uavs-for-disaster-risk-reduction-in-tanzania/

[19] "Flood Mapping for Disaster Risk Reduction: Obtaining High-Resolution Imagery to Map and Model Flood Risks in Dar es Salaam. Open Data for Resilience Initiatives." Sept. 2016 OPEN DRI 23 Nov. 2017 https://opendri.org/wp-content/uploads/2016/09/1case-studies-dar-es-salaam-final2.pdf

[20] Zolli, Andrew. "Financial and Non-financial aspects of Sustainable Development-Planet Labs". Presentation at the 10th ESPI Autumn Conference "Space for Sustainable Development". Sept. 15–16 2016, ESPI, Vienna. http://www.espi.or.at/10th-autumn-conf

development and sustainable development.[21] The contribution of private entities has a double advantage: on the one hand, it would offer further economic support to public institutions, and, on the other, such involvement impacts on the socio-economic assets both of the countries who offer their support to development initiatives and of the countries who welcome these initiatives, through economic growth, job contribution and tax benefits.[22]

The private sector has also a key role in providing specific skills and innovative solutions in coordination with implementation models developed in universities or research centres. The involvement of private entities in the sustainable development process also includes the involvement of civil society, which plays a relevant supporting role by pointing out issues that are important for citizens, helping institutions in elaborating credible initiatives that will benefit the society itself. In this framework, governments have to provide the policy context and risk-sharing schemes to allow private companies to enter the loop.[23]

The activities of private actors in the space sector have increased exponentially in recent years, thanks to the proliferation of start-ups active in the field, particularly in the USA, with strong relations with governmental actors.[24]

A case that needs to be mentioned is the American start-up Planet, formerly Planet Labs, whose mission is to image the entire Earth daily, to gather information on how it is changing. This goal is achieved through Earth-imaging satellites, known as "Doves", which together form a satellite constellation that provides a complete image of the Earth at 3–5 m optical resolution and open data access. The low-cost approach adopted by Planet is in line with the ideal involvement of private entities in sustainable development. Indeed, Planet's images have been used for climate monitoring, crop yield monitoring, urban planning and disaster response.[25]

The contribution of Planet to Earth observation is highly relevant to sustainable development. Planet is in a leading position to improve people's life on Earth, through the protection of the environment and its strong links with the UN. The images gathered by the "doves" can be a major component for achieving the majority of the 17 SDGs. An interesting example is the REDD+ (Reducing Emissions from Deforestation and Degradation) initiative: the aim is to help assessing changes

[21] United Nations Conference on Trade and Development (UNCTAD). "UNCTAD: Investing in Sustainable Development Goals. Action Plan for Private Investments in SDGs". UNITED NATIONS, 2015. http://unctad.org/en/PublicationsLibrary/osg2015d3_en.pdf

[22] Mohieldin, Mahmoud, and Svetlana Klimenko. "The Private Sector and the SDGs". Project Syndicate, 6 Feb 2017. https://www.project-syndicate.org/commentary/financial-markets-sustainable-development-by-mahmoud-mohieldin-and-svetlana-klimenko-1-2017-02?barrier=accessreg

[23] "Series of Dialogues on Means of Implementation of the Post-2015 Development Agenda Engaging with the Private Sector in the Post-2015 Agenda Consolidated Report on 2014 Consultations". UNIDO 23 Oct. 2017 https://www.unido.org/fileadmin/user_media_upgrade/Resources/Publications/Final_Consultation_Report_Engaging_with_the_Private_Sector.pdf

[24] National Aeronautics and Space Administration (NASA), "Emerging Space. The evolving landscape of twenty-first century American Spaceflight". NASA, 2015. https://www.nasa.gov/sites/default/files/files/Emerging_Space_Report.pdf

[25] "Welcome to the insights economy." Planet 23 Oct. 2017 https://www.planet.com/

in deforestation to accomplish the SDG 15, "protect, restore and promote sustainable use of terrestrial ecosystems, sustainably manage forests, combat desertification, halt and reverse land degradation and halt biodiversity loss".[26] Many other objectives are aligned with the work of Planet, which is continuously working with the UN, and especially with the Secretary-General's Global Pulse Initiative.[27]

Planet, compared to other private companies, formed its relations with institutions quite recently. When the start-up was established, its purposes were exclusively commercial. After the first year of activity, institutional actors started to be more interested in the company.[28]

As previously stated, Planet's mission is to take images of the whole Earth every day and make the global change visible, accessible and actionable. This kind of objective can have a strong impact on humanitarian challenges, such as the monitoring of ecosystems, managing of climate risks, enhancing food security, the building and development of resilient cities, poverty reduction and improved government.

Planet works closely with UN bodies, member states and partner organisations to help the advancement of the SDGs process and to ensure the right use of space technologies for development. It is worth to mention the participation of Planet's CEO, Will Marshall, in the United Nations Summit on SDGs on 27 September 2015. The participation of Planet in international fora is an important achievement that shows a new approach being adopted by the UN in opening a platform dialogue with private actors. Planet is part of this new process: indeed, collaboration with the UN is occurring in the frame of the Secretary-General's Global Pulse Initiative, which aims to explore innovative ways to connect space technologies, in particular, remote sensing, for inclusive sustainable development.[29]

Planet's commitment is to make satellite imagery open and accessible in support of the Global Goals. In this context it has developed a Global Partnership for Sustainable Development Data,[30] in which Planet confirmed its involvement in remote sensing through the donation of $60 million in imagery for selected regions,

[26] "Planet Monitoring for REDD+ MRV." Planet 23 Oct. 2017 https://www.planet.com/markets/redd/

[27] Marshall, Will. "A Commitment to Sustainability", 28 Sept. 2015. Planet 23 Oct. 2017 https://www.planet.com/pulse/globalgoals/

[28] Schingler, Robbie and Richard B. Leshner. "The Space renaissance: the government as an early adopter" 1 Mar. 2016, Space News 23 Oct 2017 http://spacenews.com/op-ed-the-space-renaissance-the-government-as-an-early-adopter/

[29] Zolli, Andrew. "Advancing the Sustainable Development Goals" 22 Sept. 2015, Planet 23 Oct. 2017 https://www.planet.com/pulse/unsdg/

[30] The Global Partnership for Sustainable Development Data is a multi-stakeholder network of more than 150 data champions harnessing the data revolution for sustainable development. Its members represent the full range of data producers and users, including governments, companies, civil society groups, international organisations, academic institutions, foundations, statistics agencies and data communities. The Global Partnership serves as an invaluable convener, connector and catalyst, building trust and encouraging collaboration among stakeholders to fill critical data gaps and ensure data is accessible and usable to end extreme poverty, address climate change and pave a road to dignity for all by 2030. "The Global Partnership for Sustainable Development Data." Data4SDGs 23 Oct. 2017 http://www.data4sdgs.org/

to assist the population to develop relevant tools and capabilities. These data are open and free, thanks to Planet's Open Region initiative whose principal example is Open California, a big example of available open data imagery archives.[31]

"Make global change visible, accessible and actionable" should be the principal goal for the entire international community. For this purpose it is highly desirable to enhance international partnership at any level. It is important to gather institutional actors, stakeholders and civil society together in this effort to establish new cooperative financial mechanisms. The existence of obstacles to sustainable development because of market access issues and the lack of a legal framework and rules is undeniable. Another obstacle is the absence of space-related local markets and tax policies that would favour the emergence of a space sector in the less developed countries.[32]

Also in this context, the SDGs are fundamental for putting a tile in the big mosaic of sustainable development. SDGs Goal 8, "Develop a global partnership for development", is the next challenge for private and public institutions.

It would be useful to set up Public-Private Partnership (PPP) schemes that will help to move to more sustainable business models, better regulatory frameworks, and public incentives. It is important to highlight the relevant role that private actors can play in different fields, among which is space, especially regarding Earth observation (EO) and connectivity.[33]

Acknowledgements A special mention goes to Edward Anderson, Senior Innovation and Technology Specialist at the World Bank, and to Andrew Zolli, Vice President of Global Impact Initiatives at Planet, who brought discussions and a fruitful exchange of ideas, thanks to their contributions to the 10th ESPI Autumn Conference. The examples brought to the attention by the World Bank and Planet showed the attention of both the institutional and private actors to sustainable development as well as the importance of the international space community to contribute to it.

[31] "Open California." Planet 23 Oct. 2017 https://www.planet.com/products/open-california/

[32] Zolli, Andrew (2015). *In* ibid.

[33] Ferretti, Stefano, Jean- Jacques Tortora, Elisabeth Veit, Alessandra Vernile, "Space for Sustainable Development. Outcome of the 10th ESPI Autumn Conference. 14–15 September 2016." European Space Policy Institute (ESPI), November 2016. http://www.espi.or.at/images/10th_autumn_conf/ESPI_Executive_Report_Outcome_10th_Autumn_Conference.pdf

Chapter 13
Satellite Connectivity for Development

Christine Leurquin and Alessandra Vernile

13.1 Introduction

Today, there is a strong consensus among top policy makers in Europe and Africa about the important role information and communications technology (ICT) plays when tackling development challenges. Digitalisation can provide new cost-effective solutions, drive inclusion by removing physical boundaries, and accelerate sustainable growth by bringing social and economic benefits through a variety of concrete ICT applications. Extending access connectivity in developing countries helps them to improve the social and economic development of their communities, as well as guarantee their access to basic needs.

Yet, the percentage of the African population that has Internet access today is only 16%.[1] This percentage is expected to grow by 2025 in line with demographic growth, reaching a peak of 50%.[2] Considering the demographic development of the next years in Africa, the 2025 prediction will not be enough to fill the connectivity gaps. If evaluations are correct, the impact of a 34% growth in connectivity on the economy will be significant. The sales on e-commerce, for instance, will reach over

[1] In total, only 167 million are Internet users, 67 million use smartphones around the whole continent, and more than 400 million have access to terrestrial fibre node. See "Offline and falling behind: Barriers to Internet adoption, in Technology, Media, and Telecom Practice" Aug. 2014, McKinsey & Company 26 Oct. 2017 http://www.mckinsey.com/~/media/McKinsey/dotcom/client_service/High%20Tech/PDFs/Offline_and_falling_behind_Barriers_to_Internet_adoption_Briefing.ashx

[2] The number of Internet users will increase to 600 million, and the number of smartphone users will be about 360 million. Cf. note 713.

C. Leurquin (✉)
Institutional Relations and Communications, SES, Brussels, Belgium
e-mail: christine.leurquin@ses.com

A. Vernile
Former Resident Fellow at European Space Policy Institute (ESPI), Vienna, Austria

C. Al-Ekabi, S. Ferretti (eds.), *Yearbook on Space Policy 2016*,
Yearbook on Space Policy, https://doi.org/10.1007/978-3-319-72465-2_13

$75 billion, and productivity in key sectors will amount to more than $300 billion, the same amount that would contribute to the African GDP.[3]

Yet today, in 2016, the African population with Internet access remains low. It is important to highlight that fibre nodes are along the coasts of Africa; on the inside it is harder to get connection, especially because of the huge distance and the existence of no related infrastructures. Today, research shows that on the African continent around 300 million people live at more than 50 km from a fibre or cable broadband connection and 400 million people have no Internet connectivity at all.[4] Knowing that the further away the connections are, the worse the broadband quality becomes demonstrates how important it is to increase connectivity access. The contribution that satellites can make to this goal can help accelerate the continent's development through the exploitation of local capabilities and the involvement of local authorities.[5]

The added value brought by a satellite is entirely related to the services they connect and the needs of the end consumer. Connectivity for the sake of connectivity will not in itself bring the solution. The assessment of the added value of satellite services can only be done by taking into consideration the impact it could bring while allowing for the deployment of new services and applications in rural areas. The relevance of connectivity is related to access to services and applications that people need or want and, in particular, the willingness of users to pay for a service.

Telecommunications operators should adopt satellite-based e-platforms to achieve their full business potential in less developed countries and rapidly answer to basic needs such as healthcare, education, and finance in rural areas or not well-connected parts of the world.

13.2 The SES Experience

An example can be gained from the commercial satellite operator, SES, which has fully committed to contributing to humanitarian public-private partnership projects in Africa, adding a value to the society also with the contribution of local authorities and NGOs. SES provides reliable and secure connectivity that fuels life-changing applications, to reduce isolation and the lack of terrestrial infrastructures among African communities. SES activities in development projects in Africa include the following: e-finance, e-health, e-learning, e-elections, and disaster response communications.

---

[3] Cf. Note 713.

[4] "Value of connectivity, Economic and Social benefits of expanding Internet access", Deloitte, February 2014.

[5] Ferretti S., Tortora J.J., Veit E. and Vernile A. "Engaging with Stakeholders in Preparation for UNISPACE +50". European Space Policy Institute (ESPI), November 2016.

13.2.1 E-Finance

The involvement of satellite links in e-finance is fundamental to connect bank offices and ATMs; adding to mobile networks might be congested and unable to offer continuous quality services. The added value of connecting financial institutions is fundamental for African partners because it reinforces corporate services, tailored to the needs of the financial institutions and society.

The beneficial contributions of satellite services are particularly visible for microfinance institutions, which are part of a fast-growing market in developing countries. The main issue is related to remote sites not well served by telecommunications means. The major challenge is to offer the same service both in remote areas and main cities. To make this happen, it is necessary to improve both IT security (i.e. to make transactions possible and safe) and the reliability of the communication network. Usually, financial institutions in these areas are too small to justify investments in VSAT links and terrestrial mobile services, so these institutions tend to remain behind the general conditions necessary to guarantee safe business transactions. In this context, the major constraint is the ability to guarantee an economical and sustainable service, correspondent to the expectations and the needs of a small rural community.

SatFinAfrica is an example of how satellites can be helpful for finance. This pilot project, launched in 2009 and completed in 2014, was a joint initiative between SES, the European Space Agency (ESA), SatDSL, and local Internet service providers to develop with African partners a reliable and secure communication system for African financial institutions. The main applications involved money transfer, ATMs, and VPN.[6]

13.2.2 E-Health

Remote areas of Africa suffer from a lack of healthcare services because of the absence of connectivity, inadequate IT infrastructure, lack of well-educated professionals, and the difficulties for people to access these services when they do exist. SES, working with public entities, has overcome these challenges through the development of two initiatives, one concluded and one still on-going: the sub-Saharan initiative for telemedicine in Sahel and the SATMED platform.

The first initiative, the sub-Saharan initiative for telemedicine, was launched in 2007 in Kenya and Senegal. The aim was to promote satellite-supported activity of e-health in Africa. The implementation of this initiative was a cornerstone for the development of a pan-African telemedicine network. The sub-Saharan initiative for telemedicine was supported and funded by the European Commission and ESA. This

[6] "SatFinAfrica – Reliable and secured financial services in remote areas". ESA Business Portal 23 Oct. 2017 https://business.esa.int/projects/satfinafrica

was the first step in the configuration of an integral system of telehealth (e-health) that includes services of telemedicine, epidemic control, and continuous training of health personnel, as basic pillars upon which other future services are structured.[7]

This pilot project found its application in three major areas:

- Medical e-learning that provided training for healthcare professionals in rural settings
- Clinical e-services that provided a link for dispensaries and isolated medical centres to clinical institutes that could help provide assistance for diagnostics and treatment of patients
- Computerised health management systems that helped to manage patient files and record and collect medical data, as well as identify pandemic threats and react to them[8]

Based on the experience of the Sahel project and the support of NGOs, the more recent initiative, SATMED, was established in April 2014. The purpose of SATMED is to bring together services for medical care, education, and health management through an open platform that can be accessed anywhere, thanks to cloud services for sharing and storing sensitive patient data, which is vital to facilitate communication between healthcare professionals around the world.

The deployment of SATMED does not need any additional IT infrastructure at the local site and can provide broadband Internet through satellite service in areas where there is no terrestrial connectivity. Pilot projects have been established in Benin, Eritrea, Niger, the Philippines, and Bangladesh. These projects were supported by medical NGOs, in particular Friendship from Luxembourg.

SATMED is designed to connect doctors and nurses based in remote areas of the world through satellite connectivity. This enables healthcare professionals to access the platform's medical applications and also to have medical consultations. SES provides connectivity through its satellite fleet and SATMED web applications, and the available backups are hosted and secured in Luxembourg. The deployment of the platform is delivered through a fully operational and managed service.

Since its launch in 2014, SES has contributed to the deployments of ten SATMED platforms across Africa and Asia. SES and the Ministry of Foreign and European Affairs of Luxembourg have recently decided to extend and reinforce the SATMED contract until 2020. This is a strong endorsement by local and international communities, since the innovative solutions offered are fundamental for remote areas in less developed countries. Thanks to these new deployments, the future evolution of

[7]"Sahel project: telemedicine for sub-Saharan Africa services". 24 May 2013. DigitalV Magazine 23 Oct. 2017 http://www.digitalavmagazine.com/en/2013/05/24/proyecto-sahel-servicios-de-telemedicina-para-el-africa-subsahariana/

[8]"Telemedicine Initiative for Sub-Saharan Africa". 23 June 2006. European Space Agency 23 Oct. 2017 http://www.esa.int/Our_Activities/Telecommunications_Integrated_Applications/Telemedicine_initiative_for_sub-Saharan_Africa

SATMED will be global and will do much towards fulfilling Sustainable Development Goal 3: "Good Health and Well-Being".[9]

One example that can be presented to show the utilisation of SATMED is the project to connect the maternity hospital "Ahozonnoude" in Allada, Benin, which was established between June 2015 and June 2016. The maternity clinic, built by the Fondation Follereau Luxembourg (FFL), used the SATMED platform for digital registration of patients through a simplified health record system to improve quality control of healthcare.

The added value of using SATMED in this framework is to improve medical control and to give women more opportunity to access quality healthcare.[10]

13.2.3 E-Learning

As an enabler for development, satellite links help in the improvement of socio-economic conditions in less developed countries. In the frame of e-learning activities, the fundamental challenge is to offer new generations the necessary tools to improve their lives. For this reason, SES has been involved in the Space4edu[11] project to support education in rural schools in South Africa in partnership with Openet, ESA, and Rally to Read. The project's aim is to connect school classes in Africa and Germany to deliver to students and teachers tools to enrich their educational experience through exchanging views and experiences between African and German students and teachers. In the Space4edu project, the SES satellite broadband services were used to give Internet access to schools in the KwaZulu-Natal province in South Africa.

As part of the project, the provision of equipment for 12 schools is planned. each of them will be equipped with satellite terminals, one solar panel, laptops, tablets, a video projector, a webcam, and an audio kit. This new equipment will allow teachers to access e-learning courses and have a means of communication between remote locations. Such a network provides a fast and easy way to install a permanent connection between rural schools and training programmes and also ensures that teachers have access to high-quality teaching resources, developing at the same time a new understanding of ICT.[12]

---

[9] "SES and Luxembourg Government Extend SATMED E-health Contract" 5 Apr. 2017 SES 23 Oct. 2017 https://www.ses.com/press-release/ses-and-luxembourg-government-extend-satmed-e-health-contract

[10] "SATMED Project in Benin" SATMED 23 Oct. 2017 https://satmed.com/project_benin.php

[11] The project, co-funded by the European Space Agency within the ARTES 3–4 Program (Satcom Applications) and proposed by OPENET (Prime) in partnership with SES, aims at implementing an ICT (Information Communication Technology) satellite network. See "SWAY4EDU". OpenNet 23 Oct.2017 http://www.openet.it/index.php/en/r-d-2/sway4edu

[12] "SES Techcom Services At IBC: Satellite-Based E-Platforms Bring Essential Services to Unconnected Areas" 12 Sept. 2015. Realwire 23 Oct. 2017 https://www.realwire.com/releases/SES-Techcom-Services-At-IBC-Satellite-Based-E-Platforms

One of the most fascinating SES success stories in e-learning has been the use of connectivity and solar panels in the Zaatari refugee camp in Jordan to provide education to children and women. SES together with SOLARKIOSK, a German clean energy provider, developed the "Connected Solar School" project, a combination of SES satellite connectivity empowering e-learning applications with the SOLARKIOSK's e-HUBB providing solar energy to enable connectivity and power the school's lights and computer.

13.2.4 E-Election

Free and transparent elections are the first step to ensuring the democratic and stable development of a country. The added value of a developed satellite voting system is to enable rapid and secure transmission of voting data from electoral booths to the central electoral commission to facilitate a legal and fair electoral process.

The SES VSAT link for e-elections was recently deployed in Burkina Faso during the presidential elections of November 2015 under the initiative of the Commission Électorale Nationale Indépendante (CENI), following from the successful support of SES to the previous municipal elections of 2012. The biggest challenge for the country was to coordinate polling stations and publish the results of the presidential elections in an efficient and transparent manner. For this reason, SES consulted directly with national institutions to ensure it could meet their expectations, in particular to guarantee a transparent and efficient electoral process.

The Burkina Faso electorate is composed of 5.5 million voters. In this context 368 polling stations were equipped at municipal level with satellite services and served as hubs for the collection and transmission of the votes from over 18,000 electoral offices across the whole country to the CENI collection centre in the capital, Ouagadougou. Each site was equipped with a VSAT station, making possible the rapid transmission of electoral data to the collection server located at the central CENI office. The data was then sent to an ad hoc server developed by local partner Unicom for processing.

The voting results were transmitted immediately via satellite communication technology. As a result, the CENI of Burkina Faso became the fastest electoral administration to deliver provisional electoral results in the African continent in a transparent way.

This activity was developed by Newtec and the local partners, Unicom, Satplay, and Accessat.

13.2.5 Disaster Response Communications

Timely disaster response communications are crucial to ensuring rapid coordination with local and international authorities to guarantee a fast reaction, increasing the chance of saving lives.

One of the latest SES disaster response operations was in Nepal after the 7.8 magnitude earthquake that hit Nepal on April 25, 2015. This rapid response intervention occurred under a public-private partnership, emergency.lu, set up between the government and three Luxembourg-based companies, namely, SES, Hitec Luxembourg, and Luxembourg Air Ambulance, that was first launched by the Luxembourg Ministry of Foreign Affairs shortly after the Haiti earthquake in 2010.

Emergency.lu offers benefits to the international humanitarian and disaster relief community due to the development of a rapid response solution for disaster relief and humanitarian operations, offering complementary solutions to the international humanitarian tool box through end-to-end services adapted to the requirements of the international community. These activities are managed in close coordination with the Emergency Telecommunications Cluster (ETC), the World Food Programme (WFP), and the European Union. Another relevant aspect is the coverage of the entire service chain including air transport, satellite infrastructure, terminals, and services, offered by the private companies involved in the PPP.[13]

In the case of the Nepal earthquake, telecommunications landlines and terrestrial wireless systems were destroyed and overloaded by people seeking help and information. Satellite infrastructures helped to quickly restore communications networks through a rollout of coordinated humanitarian assistance and aid, adapted to the mountainous region.

Emergency.lu provided a critical communication infrastructure in the aftermath of the disaster. Communication kits were deployed in cooperation with the UN Emergency Telecommunications Cluster.

13.3 Continuous Innovative Business Models

To meet the development goals, continuous innovative solutions have to be deployed. In July 2016, SES completed the acquisition of O3b Networks.[14] O3b deployed a next-generation satellite constellation, delivering cost-efficient and fast connectivity to the customer. The commercial activity of O3b Networks began on 1 September 2014. The characteristics of this satellite constellation can be summarised in the positioning in medium Earth orbit (MEO) of the satellite constellation, higher throughput, lower latency, and reduced costs.

Thanks to the innovation brought by this new technology, and the unique combination of GEO and MEO satellite solutions, SES has partnered with the Luxembourg and Burkinabe governments to bring fibre-like capacity to the 13 provinces of Burkina Faso and to deploy connectivity for e-government, education, and health applications at about 900 sites. The project is part of the development cooperation

[13] "Emergency.lu Missions" Emergency.lu 23 Oct. 2017 http://www.emergency.lu

[14] "SES EXERCISES CALL OPTION TO ACQUIRE 100% OF O3B NETWORKS" 4 July 2016. SES 23 Oct. 2017 https://www.ses.com/press-release/ses-exercises-call-option-acquire-100-o3b-networks

established for the period 2017–2021 between Luxembourg and Burkina Faso to support Burkina Faso e-governance policy.

That project is a perfect example of a new innovative way of working where SES will invest efforts and resources in deploying the connectivity network with a mix of satellite, fibre, and wireless technologies countrywide, while the Burkinabe government will operate the network and develop the right applications to improve the quality, reliability, and accessibility of IT and communication infrastructure throughout Burkina Faso and bring essential public services to the citizens and the youth.

13.4 Conclusion

As shown through the examples presented here, SES satellites bring connectivity to communities to overcome the digital divide, deliver humanitarian projects to guarantee the development of African society, and accelerate the achievement of the UN Sustainable Development Goals. SES e-inclusion activities are an important contribution to local authorities and societal needs.

From SATMED to e-elections, each of these projects enriches and contributes to saving lives. Public-private partnerships are key instruments to ensure inclusive connectivity and enable governments and institutions to meet the challenge of digital divide. PPPs are today considered as the best sustainable business model to support development in Africa. However, in order to encourage the private sector to invest more in PPP in Africa, African countries need to improve their business environment.

In some cases, the deployment of satellite services is indeed not easy to pursue because of legal uncertainty and a lack of predictability in the region, due to the absence of rules, or non-transparent practices that increase costs through additional taxes. This impedes the handling of licences, customs fees, or management of radio spectrum. In addition to this, for some countries in Africa, access to the market is still an obstacle to guarantee the sustainable development of the country. For instance, in Rwanda, Nigeria, or Zimbabwe, the governments have imposed the necessity of having a hub or gateway on their territory, which does not help in the fulfilment of market and societal needs. For this reason, it is important to ask for the exemption of extra taxes for the implementation of IT systems and satellite links in the fields of education and health.

The private sector also needs clear commitments from the EU, the AU, and the African beneficiary governments to ensure fast track, resilient, efficient, and long-term PPPs. A possible EU-AU Space for Sustainable Development programme should definitely include satellite telecommunications and PPPs, being effective tools that would help make the difference.

Chapter 14
At the Edges: Vulnerability, Prediction, and Resilience

Ashley Dara Dotz

> *With a focus on working with vulnerable populations, my goal is to empower people at the fringes – from outer space to the edge of disaster. One thing that you find when you move beyond the mainstream, is that the edges have unique characteristics of their own. They are exposed, fragile and unpredictable.*
>
> Ashley Dara Dotz

In space, pioneers risk their lives to push the boundaries of what we know and how we understand our place in the Universe. Humans aren't actually made for space. The Universe is a hostile place; it is almost as though space is literally trying to kill these pioneers at every turn. Further, the supply chain problems are immense. Getting the right tool or solution at the right time can be critical to survival. Currently, astronauts depend on an estimate/prediction of what they'll need. They do this through tremendous amounts of training, preparation, and redundancy. In order to get these supplies, the support team on Earth packs them into an extremely tiny nose cone on top of what is essentially a massive explosive. Some rockets don't make it. Meanwhile a never before seen, or experienced, life-threatening event could be occurring up above, and the astronauts don't have the actual item they need. Time is not on their side. Think Apollo 13. One false move, one bad day, and one leaking gasket can lead to disaster.

At the other extreme, in post-disaster communities here on Earth, people face challenges to stay alive and maintain their humanity every day. Getting the right thing to the right person at the right time can face incredible obstacles, from unpredictable landslides and floods to political or societal stumbling blocks, including graft and corruption. Even worse, instead of treating these precious lives as heroes, we treat them as victims who can't possibly be a part of the solution.

Technology, scenario planning, or good intentions alone can't solve the brittleness and unpredictability of these fragile systems. The complement to planning and prediction is resiliency. As more of our world becomes interdependent, I believe we will need to invest greater energy towards building resiliency into our systems. I

A. D. Dotz (✉)
Field Ready, San Francisco, CA, USA
e-mail: dara@fieldready.org

C. Al-Ekabi, S. Ferretti (eds.), *Yearbook on Space Policy 2016*,
Yearbook on Space Policy, https://doi.org/10.1007/978-3-319-72465-2_14

believe that the signals we gain at the extremes, the lessons we learn solving problems in the most challenging environments today, could be of incredible future value to the more established and mainstream parts of society. When we look at the biggest challenges identified within the Sustainable Development Goals (SDGs), we must wonder how might the efforts at one edge, space exploration, help us meet the goals at the other edge? In what ways could we consider those on the edges here on Earth in a different way, as pioneers and innovators capable of being a part of the solution, for not only spaceship Earth (as they solve challenges locally to meet the SDGs) but also for humanity's ability to thrive off-planet? I believe this will require a shift in mindset within the aid and space community. We can find heroes in the most unlikely places.

14.1 Moore's Law's Table Scraps

"I like to say I use Moore's law's table scraps…" using technology to aid those at the edges – from disaster survivors to astronauts. Moore's law refers to the exponential growth of technology that comes from the yearly doubling of computing power predicted by the co-founder of Intel, Gordon Moore. While technology is amazing, and can solve very complicated challenges, it can't solve everything alone. All too often technologists seem to believe that there is a fix with technology alone for any challenge. But even with all the technology, prediction, and planning of the greatest minds of our time, things still go wrong. There is still no way to accurately predict what will happen on long-distance space missions or in disasters – which, by definition, are chaotic. So how do we harness all the benefits that technology can bring in the service of resiliency and autonomy? How can we help people have agency and dignity and participate in their own futures? I believe that to reach the Sustainable Development Goals, we will need help from the edges of society to the edges of outer space. I wonder if there is some way that we can not only learn from both extremes but also weave them together in more powerful and fruitful ways.

How we can use technology from space to meet the Sustainable Development Goals, and what can crisis zones teach us about living off-planet?

14.2 Hyperlocal Manufacturing

14.2.1 Communities Living on the Fringes: Outer Space

We put the first human in orbit around the Earth in 1961. It is now 2016, why aren't we an interplanetary species yet? Stephen Hawking has noted that we really shouldn't have all of our eggs in one basket. He is one of the smartest men of our time, on this planet, and he's concerned that we are facing challenges even greater that those outlined in the SDGs.

Years ago, we believed that by now we'd be living in giant space habitats, growing food, and living life in orbit. But today, pioneers still risk their lives to push the boundaries of what we know and how we understand our place in the Universe. The challenges are far greater than anyone understood at the dawn of the space age. Two of the biggest issues that stop us from becoming an interplanetary species are cost and safety. It is incredibly dangerous, and costs fluctuate around $20,000 USD per pound to send anything to space, which means we can't bring everything up with us. We must find another way. We need to build "in situ" from what we find and have right there, in space.

A few years ago, I began working at a small space start-up, based at NASA Ames Research Park, called Made In Space. Our mission was to find a way to manufacture off-planet using an emerging disruptive technology called additive manufacturing, often referred to as 3D printing. After over 400 flight parabolas to test our approach to printing in zero gravity, we finally achieved success. We then successfully launched the first gravity independent 3D printer to the International Space Station (ISS) on September 21, 2014. We ran a series of test prints by sending 20 CAD (computer-aided design) files up with the printer itself. The final 21st print was to be (essentially) emailed. The printer was effectively operated from here on Earth. After the first 20 test prints, Butch, the mission commander, was overheard saying he wished he had a ratchet to fit one of our samples. We scrapped our final CAD file and custom designed him a ratchet that was then emailed to the ISS.

It was the fastest delivery ever to space. We were actually able to surprise him, which is pretty hard to do with as much training as astronauts go through. The file was created and sent to the ISS in a matter of days. The "all-in-one" design, optimised with moving parts built in, enabled it to be printed on station with no support material. That meant that Butch could take it out of the printer and use it immediately.

How can we take this sort of exponential technology and empower other communities at the edge to create their own solutions, on the fly, when they need it most?

14.2.2 Communities Living on the Fringes: Post-disaster

Remember Haiti? "Oh no, what now?" you might think, if you're paying attention to Haiti at all. That's the point. With our 24-h news cycle, once a disaster is out of the headlines, it fades from most people's attention. But locals still need to survive. When I went to Haiti 2 years after the quake, it had an eerie similarity to what the crew of the Apollo 13 faced, but Haitians couldn't call mission control.

In many cases of broken wells or complex machinery, all that was needed to fix the system was a missing screw, adapter, or washer. Money donated seemed to have been spent on feeling good about helping and sending relief, not so much on amplifying the resiliency of the Haitians or trusting that they were some of the more industrious people you would ever meet. I found countless examples of clever "hacks", from cardboard used to make impromptu splints to old rubber gloves used to hold makeshift oxygen splitters together. You see, often for lack of a missing screw, an entire, costly, donated piece of equipment or system became nothing more than another reminder that Haiti had been forgotten by the rest of the world.

In 2010 I lived in Nepal, but more recently, I had to return in tragic circumstances. I was on the ground within 36 h of the 2015 earthquake, paired with an amazing group of volunteers comprised of experienced geospatial analysts, veterans, and pilots. We were there to find survivors and assess damage.

In 2015 the ideal approach would have been to use helicopters in the mountains to find survivors and airlift them out, as well as to assess dangerous/isolated areas so that rescue and recovery teams could be sent out via car or on foot. In post-earthquake Nepal, however, this was a very different story with only three helicopters seemingly available at that time and the scale of devastation being so immense. We brought in an emerging exponential technology in the form of portable unmanned aerial vehicles (UAVs or drones). Although compared to helicopters, these UAVs are dramatically cheaper and easier to use; we still had huge challenges. We had to carry heavy equipment into range through and over volatile and dangerous terrain, with land giving way around almost every other bend.

To effectively perform the operation, we collaborated with local authorities and partners on the ground to cross-share and coordinate efforts as well as assess the changes in the territory. While in developed countries GPS (Global Positioning System) seems to work just about anywhere, it is definitely not dependable enough to safely fly a UAV through the Himalayas. These sorts of geolocation systems are often prone to gaps in coverage because of cloud cover or geological interference. While they can sometimes take advantage of Wi-Fi or cell tower networks to triangulate location and altitude with increased accuracy, with the Internet being either down or intermittently available at best, and with no robust cell or Wi-Fi infrastructure, we had to fly the UAVs by line-of-sight navigation. We transported equipment into the area by foot over treacherous and unstable terrain. We then flew the UAVs to capture footage from a "bird's eye view" and had trained human eyes and then analyse what we had discovered – by downloading the memory cards from the drones – to find potential survivors.

To get a better sense of the damage to infrastructure, we used another technology, in combination with UAVs, called photogrammetry, to stitch our images together and create a 3D model of the areas around us. This enabled us to assess the scene from a variety of angles and look for places or critical infrastructure like hospitals, for instance, which may not have been safe to enter, climb inside of, or even be near.

In the end, it was the combination of hi tech, low tech, and human resourcefulness that enabled the success of the mission.

A constant benefit of exponential technology is that the cost and access to this sort of capability consistently drop dramatically over time. A challenge can then be that this "easy access" can encourage dangerous misuse by well-meaning but often clueless "disaster hobbyists". In this type of scenario, there have been cases of amateur "drone pilots" recklessly flying UAVs inches away from children's faces or pilots being dismissive of local authorities, not sharing information that may have been urgent and valuable, and using UAVs to capture sensational footage to sell to news organisations.

It is useful to note that while technology that comes from investments in space, and pushing the frontiers of human capabilities, can give us amazing new reach and enhancements, those capabilities are a two-edged sword – they can be used appropriately to cocreate solutions, but they can also be used in dangerous ways that undermine efforts and can do more damage than good. It is essential to consider the context and problems on the ground at the edge and not to get overenthusiastic without an appreciation of risks and challenges.

14.3 Case Studies

14.3.1 Umbilical Cord Clamp

[SDG #3 Good Health and Well-Being/SDG #9 Industry, Innovation, and Infrastructure/SDG #12 Responsible Consumption and Production]

It turns out that, just like in space, getting supplies to Haiti post-disaster was incredibly expensive, time-consuming, and dangerous. People make do with what they have.

Nurse Maeve McGoldrick, while volunteering in the rural village of Arcahaie, Haiti, had to deliver a baby while completely out of medical supplies – all but her last pair of latex gloves. She still had to tie the umbilical cord of the infant. What was she to do? Her solution? Removing the fingers off of her gloves to create a sterile tie, ingeniously preventing neonatal sepsis. This blood infection commonly found in infants is frequently caused by umbilical cords being tied too tightly, too loosely, or simply tied with a contaminated material. The simplest solution, frequently used in the developing world, is a sterile premeasured umbilical cord clamp. And of course, that supply had run dry.

Maeve, a trained and knowledgeable professional, knew what she was doing, was familiar with her environment, and yet, still, even at the edges, had to improvise. While this was an excellent substitute for the infants, it was incredibly risky for her. As it turned out, she would deliver not one but five infants that night – with no gloves to protect her from the potentially HIV-positive mothers – all for lack of a simple clamp.

These small rural clinics can't buy at economies of scale, nor do they have the infrastructure to support this kind of "on-demand" logistics. It can take months or even years to gain access to costly supplies at outrageous rates. Further, graft and corruption can levy a toll as well. Getting the right thing to the right person at the right time can be met with incredible obstacles from unpredictable landslides and floods to political or societal stumbling blocks.

After hearing about what had happened to Maeve, I became irate. Enough was enough. Something had to be done! I thought – I bet those umbilical cord clamps could be 3D printed. Once there was power and Internet, I made a request on social media to see if anyone had an old 3D printer, and, before I knew it, I was learning to 3D print in a shipping container in Haiti.

It was great that I could 3D print – but what about Haitians printing their own supplies? It was then that I supported the first Maker Lab run by Haitians for Haitians. I found that rather than treating locals as victims, I could work with them to cocreate solutions that fitted their needs, along the way giving them a sense of agency and control. The best part is that we can have locals produce a variety of items all with one 3D printer. A 3D printer is, for some things, like a universal warehouse of supplies. Using one type of filament you can produce numerous kinds of products, on demand. For example, in 2013, Haitians could make clamps in minutes, as needed, and make them for less than $0.30, which was about 1/10 of the cost that small unsupported clinics would have to pay for mass-manufactured clamps. This means that it is now cheaper and quicker for a donor to hire a local to produce needed items while also creating paid work for people in the community.

It is possible to create an almost closed-loop system by recycling the material of the used clamps following their application. Initially, virgin polymer can be used for medical devices then, and when no longer needed, it can be sanitised and ground up back into plastic pellets and turned back into more filaments. This can then be used to make other needed items such as oxygen splitters, nebulizer adapters, or spigots.

This kind of work isn't about giving aid or having a set solution; it's about collaboration, iteration, codesign methods, and building muscle in the community to facilitate resiliency.

14.3.2 Baby Incubator

[SDG #3 Good Health and Well-Being/SDG #9 Industry, Innovation, and Infrastructure/SDG #12 Responsible Consumption and Production]

In Nepal after the earthquake, Field Ready noticed donated pieces of equipment that had been stored in the back of the building, unused. For example, there was a baby incubator with a broken corner bracket. There was an improvised metal replacement part in place of the bracket, but it had sharp edges and was dangerous. Our engineers analysed the missing part and, using computer-aided design (CAD) tools, modelled a replacement part that would be safer and stronger. They then 3D printed the replacement right there in field (Fig. 14.1). Once in place, the baby incubator was immediately put back into action, and we know of over 20 babies so far who have benefitted by having it back online.

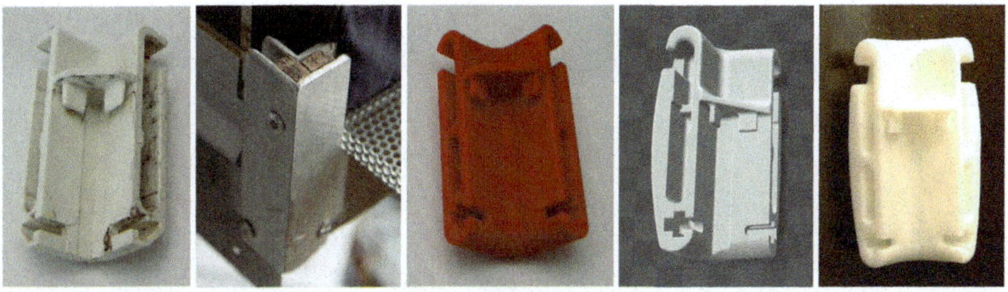

Fig. 14.1 3D-printed corner brackets printed by Field Ready

14.3.3 Ham Radio Antenna

[SDG #9 Industry, Innovation, and Infrastructure]

After the Nepal earthquake, most agencies were fighting over bandwidth on their satellite phones. Satellite phones are tremendously useful in crises and are yet another gift from space technology. However, what we found was that few phones were actually reliably working. One form of communication that did actually work was the few ham radios operating in the country. At Field Ready, we were able to work with local Nepali engineering students to create new and improved ham radio antennas using 3D printing to improve their ability to send and receive signals.

14.3.4 MakeFit App

[SDG #6 Clean Water and Sanitation]

One of the biggest challenges to using technologies such as 3D printing to meet unexpected needs in the field is learning CAD modelling to create new designs or modify existing ones. We wondered if we could remove that barrier in Nepal. We partnered with Imperial College of London students to create a new app that would automatically generate a 3D printable CAD file to connect two different-diameter water pipes together. This is a new kind of exponential technology called "generative design". It uses information embedded in algorithms so that the computer can dynamically combine various engineering restraints with user input goals to create bespoke solutions automatically. We called the software app "MakeFit", and it has dramatically simplified the creation of solutions in the field. All the person on the ground has to do is measure the diameter of each pipe and type in the two measurements. The software then generates a solid model of the needed fitting, which is then used to 3D print on demand. Broken pipes that are wasting or contaminating critical, life-saving water can be reconnected and flowing again in a matter of hours, instead of months.

Fig. 14.2 Hyper local digital manufacturing
This is what we call hyperlocal digital manufacturing – we power the printer off the car in an IDP camp – creating exactly what they need, where they need it, and when they need it. Here Mark Mellors is in Bahrabise Internally Displaced Person Camp demonstrating how to use a 3D printer to make water pipe fittings.

Note that by designing these sorts of solutions under the most severe constraints at the very edges of survival, you can often find approaches that could be incredibly useful in the developed world as well. With the rise of 3D printing in metals (called desktop metal additive manufacturing), your plumber could visit your home, take some measurements, and use a printer in his truck to automatically make the needed fitting, no warehousing of thousands of different fittings and no driving back and forth to the shop or ordering a custom part from a faraway warehouse (Fig. 14.2).

On the other edge, in space, this sort of generative design tool could be tremendously useful in long-distance space missions, such as those being considered to return to the Moon or put humans on Mars. Imagine that our explorers are on a planet such as Mars. Currently, it can take anywhere from 8 to 41 min to send and receive signals to the Curiosity rover on Mars. If engineers back on Earth are working to come up with a fix for some challenge or other, it could take hours or days. On top of that, context matters – back on Earth those engineers can't really know much about the actual on-the-ground situation. So, once they upload a given solution (and the pioneers wait another period of time for the file to make its way across the deepness of space), it may not work as intended, and the process repeats. Precious seconds, hours, or days could be lost as this back and forth plays out.

With generative design software such as MakeFit, the algorithms could learn from local conditions and the various contextual inputs the astronauts enter as they develop solutions. Within minutes, those solutions could be printed out on-site, tested, and refined. It is thus likely that the only way we'll have a robust interplanetary presence is by giving the tools of design and creation to the pioneers at the edges themselves rather than assuming all solutions will come from some centralised mission control.

14.3.5 Cook Stove Entrepreneur

[SDG #7 Affordable and Clean Energy/SDG #9 Industry, Innovation, and Infrastructure/SDG #12 Responsible Consumption and Production]

Sometimes the missing element is not a "thing" but a "way" – a way of looking at problems or getting over stumbling blocks. A lot of what we need to do to address the SDGs is about changing mindsets – not just of the locals on the edge but of the aid and space communities as well. In this example, one of our Field Ready engineers, Ram Chandra Thapa, was contacted by a local craftsman in Kathmandu about how to improve his cook stove burner. The entrepreneur had made multiple versions but could not get the angles right. Ram, who himself was a newly trained 3D modelling and printing professional at our lab in Nepal, demonstrated the rapid capability of creating various shapes for the design and showed the local entrepreneur what was possible. Ram could use the precision of 3D CAD and printing to create angles and designs that the entrepreneur could not create on his own. Building off local capacities – working with local metal workers – the mould was 3D printed and then casted using traditional techniques in metal. The combination of old tech with new tech, mixed with human ingenuity, resulted in an updated stove top burner that had 27% more burning efficiency. This approach prevented not only the waste of wood but produced less carbon pollution so that the stove was safer for children and yet still fitted within the culture. With no need for a tremendous product overhaul for adoption, it was created for local users by local craftsmen. With these new environmental and health efficiencies, 210,000 have been ordered already. Foundries in Nepal are now booked for the next 2 years, and numerous additional social businesses will likely spin up to distribute these burners.

14.4 Dream Bigger

Today only a handful of countries can dream about space. What if instead of a handful of countries, the entire world – all seven billion people – could dream about the next frontier? If we truly dream about establishing a permanent presence in outer space (and thus not put all our eggs in one basket), we need big ideas. Unfortunately, today, only a small number of privileged communities are even working on the problem.

Those scientists, engineers, and entrepreneurs are largely part of a "good ol' boys club" that consists primarily of men from spacefaring countries. While there has been some progress and there are wonderful female engineers and astronaut role models, most women, and the vast majority of the planet, have been left out of the equation. What if we could change the story before we even leave the Earth's atmosphere? What if we empowered the disenfranchised and encouraged all minds to get on deck? What if we gave girls permission to dream bigger and gave them opportunities to work locally on space challenges? What if families in parts of the world that have never had access to the Internet, let alone space, suddenly saw it as their birth right? Instead of a fraction of the world engaging in and dreaming about the challenges, we could have a global grassroots effort to build a more resilient future in space and right here on Earth too.

As we know, living in space is incredibly challenging and our heroes at the frontier make it all look easy, risking their lives every day. They dream of making space a home for humanity. What if the people of the world could collaborate with astronauts to help them make their lives better? Real challenges faced every day by real people?

I believe we can solve some of the most extreme challenges of sustainability under some of the most trying circumstances of all, both on- and off-planet, if we tie these two extremes together and learn from heroes at both edges.

As we know, the MDGs were designed in 2000, wrapped up in 2015, and only created for a few countries. Now with the SDGs, it is about the whole world taking action and being held responsible. It is my belief that combined efforts to solve the problems inherent in sustainable living in space will help us appreciate and change our own habits here on Earth.

The act of noticing and identifying both extremes will help create the right sense of identity, agency, and empathy that we need to get the global citizenry to truly understand the vital importance of the SDGs.

This is about inspiring the next generation, harnessing the power of crowds, and solving real, not "made-up" problems. This means that we need to collaboratively design and develop, side by side with those at the edges, using cutting-edge technologies and traditional ones, new skills and old. We need all minds on deck, particularly the best and brightest designers, engineers, businesspeople and policymakers, and those actually "on the ground"– from astronauts in space and on the Moon or Mars to farmers and emerging talents in Nepal.

This is about building a more resilient future through creating together, both for and with those at the edges.

The question is, are we ready to dream again?

Chapter 15
Emerging Approaches in Development Efforts: Chinese Perspective on Space and Sustainable Development

Yun Zhao

Abstract The issue of sustainable development came to the forefront in view of the increasingly serious concerns over space debris. However, this is only the vertical aspect of sustainable development in outer space; space sustainability needs to take into account the horizontal aspect of sustainable development, i.e. all the countries, irrespective of their economic, social and technological development levels, should be able to benefit from outer space and space activities. This paper aims to examine new approaches and perspectives in realizing space sustainability through international cooperation, with China as an example. China's efforts in promoting space cooperation through overseas assistance program exemplify the importance of financial and non-financial assistance efforts in the realization of the horizontal aspect of space sustainability for both space-faring and non-space-faring nations. Space sustainability cannot be achieved without taking into account the interests of developing countries. The China-Brazil cooperation presents an excellent example that space cooperation can take place between and/or among developing countries. While benefiting one state at one stage, space cooperation will bring benefits to cooperating countries in the long term; such benefits will not simply be restricted to these cooperating countries, with proper arrangement, other states can similarly benefit from such cooperation. The paper concludes that space sustainability, as an issue for both space-faring and non-space-faring nations, can only be achieved through international cooperation among nations, regardless of their level of economic and technological development.

Y. Zhao (✉)
Department of Law, The University of Hong Kong, Pok Fu Lam, Hong Kong
e-mail: zhaoy@hku.hk

C. Al-Ekabi, S. Ferretti (eds.), *Yearbook on Space Policy 2016*,
Yearbook on Space Policy, https://doi.org/10.1007/978-3-319-72465-2_15

15.1 Introduction

The number of space activities that are being conducted on a daily basis is continuing to grow; consequently, the issue of sustainable development has come to the forefront in view of the increasingly serious concerns over space debris.[1] Sustainable development is an important concept in the field of international law, in particular international environmental law. While acknowledging the important of economic development, this concept argues that such development shall not be at the sacrifice of environmental protection and conservation of natural resources.[2] The 1972 Stockholm declaration spelt out relevant rules calling for "the preservation and improvement of the human environment, for the benefit of all the people and for their posterity."[3] The 1992 Rio Declaration on Environment and Development took sustainable development as its main theme and clarified the scope and contents of this concept in the field of environmental law.[4] Within a short period of time, the concept of sustainable development had received widespread endorsement from international society. It has quickly extended to other fields, including outer space.

Nevertheless, the concept of sustainable development is a difficult one to define; its use in the space field is not limited to such traditional areas as environmental protection and space debris mitigation. Space sustainability is to be understood in a broad sense, covering a wide range of issues.[5] The final document entitled "The Space Millennium: Vienna Declaration on Space and Human Development", adopted at the Third United Nations (UN) Conference on the Exploration and Peaceful Uses of Outer Space in July 1999, pointed out the importance of space technologies and systems in realizing sustainable development on Earth and the Millennium Development Goals.[6] The concept of sustainable development is thus closely connected with outer space, though at this stage space technologies and systems are the only kind of tools conducive to environmental protection and resource conservation.

[1] Peter Martinez, Space Sustainability, in K.U. Schrogl et al. (eds.), Handbook of Space Security, 259–260 (New York: Springer, 2015).

[2] Tim Hillier, Principles of Public International Law 332 (London: Cavendish, 1999).

[3] "Declaration of the United Nations Conference on the Human Environment", 16 June 1972 United Nations Environment Programme, 21st plenary meeting http://www.unep.org/documents.multilingual/default.asp?documentid=97&articleid=1503

[4] "Rio Declaration on Environment and Development", United Nations Environmental Programme 23 Oct. 2017 http://www.unep.org/documents.multilingual/default.asp?documentid=78&articleid=1163

[5] The Secure World Foundation defines space sustainability as "ensuring that all humanity can continue to use outer space for peaceful purposes and socioeconomic benefits", Secure World Foundation, Space Sustainability: A Practical Guide, http://swfound.org/1808/space_sustainability_booklet.pdf

[6] Report of the Third United Nation Conference on the Exploration and Peaceful Uses of Outer Space, Vienna, 19–30 July 1999 (United Nations Publications, Sales No. E.00.I.3), chap. I, resolution 1.

With the rapid development of space technologies and activities, environmental protection and resource conservation are no longer issues that are only relevant to Earth. The outer space environment is facing serious challenges posed by space debris, congestion of the geostationary orbit and harmful interference of signals.

The 1967 Outer Space Treaty (OST) solemnly prescribes that outer space shall be the province of mankind and that space activities shall be carried out in the interests and for the benefit for all mankind.[7] The right of access to outer space serves as the basis for the other two types of freedoms identified in the OST, namely, the right to scientific experiment and the right to use and exploration.[8] Without the ability to access outer space, the majority of states will only be passive actors in the space field; this will defeat the ultimate goal that all states have equal access to and the ability to benefit from outer space as defined in the OST.[9] This is exactly the other side of the coin in understanding the concept of space sustainability.

As such, this paper intends to examine the other aspect of space sustainability. On the horizontal aspect, space sustainability emphasizes the importance of future generations in using outer space, which has been widely discussed in various platforms on environmental protection. The horizontal aspect of space sustainability has long been neglected; this aspect stresses the importance of developing countries' ability to access outer space, now.

From the start of the space era, international society has acknowledged the importance of international space cooperation in bringing benefits to human beings.[10] The principle of international cooperation remains the effective way to achieve space sustainability.[11] International cooperation must be carried out in a way that is mutually beneficial to the cooperating states, though such benefits are not necessarily exchanged at the same stage. Even if the benefits achieved at an earlier stage do not fully represent the efforts contributed by a state, such voluntary contributions will, in the long run, bring benefits to be enjoyed by international society as a whole, including the earlier contributing state.[12]

As elaborated in the United Nations General Assembly (UNGA) resolution in 1996, international cooperation can take a wide range of forms in a wide variety of areas in space activities.[13] China, as an advanced space-faring nation, has actively

[7] Outer Space Treaty, Article I.

[8] Ibid.

[9] Shouping Li and Yun Zhao, Introduction to the Law of Outer Space, 15 (Beijing: Guangming Daily Press, 2009).

[10] S N Hosenball, The United Nations Committee on the Peaceful Uses of Outer Space: Past Accomplishments and Future Challenges, 7 Journal of Space Law 95–106 (1979).

[11] Chukeat Noichim, International Cooperation for Sustainable Space Development, 31 Journal of Space Law 338 (2005).

[12] Susana Camargo Vieira, Brazil-South Africa: South-South Cooperation for Sustainable Development, 32 South African Yearbook of International Law 361–375 (2007).

[13] Declaration of Legal Principles Governing the Activities of States in the Exploration and Uses of Outer Space, UNGA Resolution 1962 (XVIII), 13 December 1963.

participated in space cooperation.[14] It provides an excellent example for the examination of financial and nonfinancial aspects of sustainable development through the provision of official development assistance (ODA) to other developing countries.[15] In turn, some of the recipient developing countries are now able to contribute to the theme of space sustainability.[16] Changing the role of former recipient countries in development strategies provides a vivid and persuasive case exemplifying the horizontal aspect of space sustainability, i.e. that international cooperation results in common development in the space field.

As a result, ODA to developing countries is a mutually beneficial mechanism for international space cooperation, which will lead to common development of international society. By sharing common resources and value, such a new approach in development efforts shall contribute to the realization of the ultimate goal of the peaceful uses of outer space, which is that it is in the common interests of all mankind. Accordingly, international society has realized the importance of examining space environment and application in a comprehensive manner.

This paper aims to examine new approaches and perspectives in realizing space sustainability, with China as an example. Part 2 offers a general overview of the concept of space sustainability and its development, to be followed by a thorough analysis of the theoretical and substantive aspects of space sustainability. The discussions in this part provide a theoretical basis for understanding the relationship between international cooperation and space sustainability. China's efforts in promoting space cooperation through overseas assistance programmes will be elaborated in Part 3, exemplifying the importance of financial and nonfinancial assistance efforts in the realization of the horizontal aspect of space sustainability for both space-faring and non-space-faring nations. The paper concludes that space sustain-

---

[14] "The Chinese government holds that each and every country in the world enjoys equal rights to freely explore, develop and utilize outer space and its celestial bodies, and that all countries' outer space activities should be beneficial to economic development, the social progress of nations, and to the security, survival and development of mankind". For more Chinese policies and events in space cooperation, see China's Space Activities in 2011, the Information Office of the State Council of the People's Republic of China, 29 December 2011, http://www.scio.gov.cn/ztk/dtzt/69/3/Document/1073810/1073810.htm

[15] "Official development assistance (ODA) is defined as government aid designed to promote the economic development and welfare of developing countries. Loans and credits for military purposes are excluded. Aid may be provided bilaterally, from donor to recipient, or channelled through a multilateral development agency such as the United Nations or the World Bank". For further information on the definition of ODA and China's ODA to other developing countries, see Net ODA, https://data.oecd.org/oda/net-oda.htm; Lean Alfred Santos, Building the Whole Picture of China's Growing ODA, 18 July 2014, https://www.devex.com/news/building-the-whole-picture-of-china-s-growing-oda-83916

[16] Such as the case of China assisting the Nigerian Space Program: "In June 2013 Nigeria's Minister of Science and Technology Ita Ewa announced at a press conference in Abuja that the government would prepare astronauts for space travel likely by 2015. In cooperation with the Chinese government Nigeria would build a spacecraft. A 20 kilometre launch site has already been established in Epe, Lagos State, and 12 engineers from the National Space Research and Development Agency are currently training in China". http://china.aiddata.org/projects/30988?iframe=y

ability, as an issue for both space-faring and non-space-faring nations, can only be achieved through international cooperation among nations, regardless of their level of economic and technological development.

15.2 International Cooperation and Space Sustainability

15.2.1 Initial Development of the Concept of Sustainable Development in Outer Space

Outer space is no longer monopolized by the two superpowers of the cold-war period. More and more countries have successfully joined the space club in the last one or two decades, developing indigenous launching capabilities. Outer space and space resources are increasingly open to international society. Satellite applications, such as telecommunications, remote sensing and television broadcasting, bring unprecedented convenience to our daily life.

Nevertheless, more frequent use of outer space at the same time brings serious challenges. The increased number of satellites being launched and manoeuvred in outer space directly leads to the first challenge: space debris. The congested slots in geostationary orbit also raised the issue of managing spectrum resources and potential harmful interference of signals. More recently, the possible use of ground-based missiles against space-based assets has caused alarms over space security. All these challenges prove that outer space is fragile; it is time for the international society to take the issue seriously and take some measures to keep outer space a safe and secure place for international society.

Earlier responses were rather sporadic, without systemic approaches in dealing with the above issue. It started with the study of possible measures to mitigate space debris. As early as 1986, the delegate from the former Soviet Union stated the view that the issue of space debris requires urgent attention.[17] Unfortunately, the legal subcommittee of the UNCOPUOS failed to reach any formal agreements after years of discussions. In the meantime, the Inter-Agency Space Debris Coordination Committee (IADC) provided a platform for member space agencies to exchange information and explore possible options for space debris mitigation. After years of work, the IADC was able to produce its Space Debris Mitigation Guidelines in 2002.[18] This document provides the basis for further discussions within the UNCOPUOS Scientific and Technical Subcommittee (STSC), which formally adopted the UN Space Debris Mitigation Guidelines in 2007.

---

[17] UN Press Release, Outer Space Committee Considers Agenda of Legal Sub-Committee, OS/1259, 11 June 1986.

[18] IADC Space Debris Mitigation Guidelines, issued by Steering Group and Working Group 4, IADC-02-01, Revision 1, September 2007, http://www.unoosa.org/documents/pdf/spacelaw/sd/IADC-2002-01-IADC-Space_Debris-Guidelines-Revision1.pdf

However, it was soon found that the issue of space debris is only one of many challenges arising in outer space. In 2006, the International Academy of Astronautics published a report titled "Cosmic Study on Space Traffic Management", placing the issue of space debris mitigation among possible first steps for improving space traffic.[19] Regulatory issues that were identified for further research included the prioritization of space activities, the notification/information system, the prevention of an arms race in outer space and the interests and expectations of private actors.[20]

As mentioned above, space sustainability is to be understood broadly. Apart from the traditional areas such as environmental protection and space debris mitigation, space sustainability should also include the horizontal aspect of sustainable development that all states have equal access to space and the ability to benefit from space.

15.2.2 The Realization of Space Sustainability Through International Cooperation

The UNCOPUOS has played an important role in dealing with the issue of space sustainability.[21] As early as 1994, the UNCOPUOS STSC included the issue of space debris in its agenda[22] and prioritized in the following year the consideration of space debris,[23] which led to the production of the nonbinding Space Debris Mitigation Guidelines in 2007.[24] However, more and more scholars believe that space debris is only part of the problem in outer space; a more systematic approach should be adopted to deal with the space environment. Under such circumstances, the United Nations Committee of Peaceful Uses of Outer Space (UNCOPUOS) decided in 2009 in its 52th session to include the item "long-term sustainability of outer space activities" on the agenda of its STSC for discussions starting from 2010.[25] In the same meeting, the UNCOPUOS agreed to set up a working group to study the item.

---

[19] "Cosmic Study on Space Traffic Management", 2006, International Academy of Astronautics 23 Oct. 2017 https://iaaweb.org/iaa/Studies/spacetraffic.pdf

[20] Ibid., 16.

[21] Nandasiri Jasentuliyana, Space Debris and International Law, 26 Journal of Space Law 146 (1998).

[22] "The STS agreed that consideration of space debris was important, and that international cooperation was needed to evolve appropriate and affordable strategies to minimize the potential impact of space debris on future space missions." UNCOPUOS STS, Report on its 31st Session, 21 February-3 March 1994, P 64, U.N.Doc. A/AC.105/571 (10 March 1994).

[23] UNCOPUOS STS, Report on its 32d Session, 6-16 February 1995, P 95, U.N.Doc. A/AC.105/605 (24 February 1995).

[24] UNCOPUOS STS, Report on its 44th Session, 12-23 February 2007, Annex IV P 1, U.N.Doc. A/AC.105/890 (6 March 2007).

[25] Official records of the General Assembly, Sixty-fourth Session, Supplement No. 20 (A/64/20), para. 161.

The STS established a Working Group (WG) on the Long-term Sustainability of Outer Space Activities in 2010 to study the issue.[26] The consensus of the UNCOPUOS to extend the concept of "sustainable development" in environmental law to outer space, culminated in the adoption of the Terms of Reference in 2011 to examine the issue of long-term sustainability in a wider context of sustainable development on Earth.[27]

The WG stressed the importance of international cooperation in realizing space sustainability, taking into account the concerns and interests of all countries, in particular those of developing countries.[28] It further defines that consideration should be given to "acceptable and reasonable financial and other connotations" to ensure that all countries are able to have equitable access to outer space and resources and benefits associated with it.[29] The process of international cooperation is mutually beneficial; voluntary contributions from the cooperating parties shall lead to the end result benefiting international society as a whole.

The WG was tasked with coming up with a set of voluntary, recommended guidelines for space activities.[30] After 6 years of hard work, the UNCOPUOS adopted the first set of guidelines in June 2016. It is expected that more guidelines will be adopted in the near future when consensus can be reached.

Space sustainability is not an issue peculiar only to space-faring nations; all members of international society have a stake in this important issue. Developing countries, including China, have played an active role during this process. China submitted its position paper shortly before the adoption of the first set of guidelines.[31] Brazil has similarly submitted a proposal together with other countries.[32] The adoption of guidelines has important implications for the future development of space activities. As a major space-faring nation and the largest developing country, China has a role to play in the further development of guidelines on space sustainability. It would thus be important to study the Chinese perspective on the issue of space sustainability and its impact on the future development of guidelines on the matter.

The principle of international cooperation, believed to be a means to realize peaceful uses of outer space, has been considered a fundamental principle for

[26] Report of the Scientific and Technical Subcommittee, forty-seventh session, Vienna, 8 to 19 February 2010, U.N.Doc. A/AC.105/958.

[27] Report of the Committee on the Peaceful Uses of Outer Space, fifty-fourth session, 1–10 June 2011, UN General Assembly document A/66/20, Annex II.

[28] Terms of Reference and Methods of Work of the WG on the LTS of Outer Space Activities of the Scientific and Technical Subcommittee, UN General Assembly document A/66/20.

[29] Ibid.

[30] UNCOPUOS, Fifty-third session, Vienna, 9–18 June 2010, Item 8 of the provisional agenda, Report of the STS on its forty-seventh session, 8 June 2010, A/AC.105/L.277.

[31] China's Position Paper on the Issues of Long-Term Sustainability of Outer Space Activities, Committee on the Peaceful Uses of Outer Space Scientific and Technical Subcommittee, Fifty-third session, Vienna, 15–26 February 2016, A/AC.105/C.1/2016/CRP.13, 16 February 2016.

[32] Long-term sustainability of outer space activities: proposal to adopt a first set of guidelines together with a renewed work plan for the Working Group on the Long-term Sustainability of Outer Space Activities of the Scientific and Technical Subcommittee, Committee on the Peaceful Uses of Outer Space, Fifty-ninth session, Vienna, 8–17 June 2016, A/AC.105/L.305, 28 April 2016.

space activities from the very beginning of the space era.[33] This principle was laid down in UNGA Resolution in 1961 and later adopted formally in the 1967 Outer Space Treaty.[34]

The 1996 UNGA Resolution on space cooperation further develops this principle by elaborating on its actual uses in space activities and encouraging various forms of cooperation in a wide range of areas of space activities.[35] It defines the formal and substantive requirements for space cooperation.[36] Party autonomy continues to guide space cooperation with states free to decide on all aspects of space cooperation.[37]

From the formal aspect, states are free to decide on the effectiveness and appropriateness of specific modes of cooperation, which include governmental and non-governmental cooperation, commercial and non-commercial cooperation and global, multilateral, regional and bilateral cooperation.[38] Space cooperation can also be conducted between countries at all levels of development, including space-faring and non-space-faring nations and developing and developed countries.[39]

From the substantive aspect, the UNGA resolution, while emphasizing the importance of party autonomy, provides some kind of minimum standards to protect the rights and legitimate interests of relevant states. More specifically, space cooperation should be carried out on a mutually acceptable and equitable basis, showing the real intention of the states. The term and conditions for cooperation shall be reached through negotiation; such terms and conditions should be fair and reasonable, not only to the two cooperating parties, but also to international society as a whole.[40] Furthermore, space cooperation shall take into account the special needs, interests and limitations of developing countries.[41]

---

[33] International Cooperation in the Peaceful Uses of Outer Space, UNGA Resolution 1721 A and B (XVI), 20 December 1961.

[34] Articles 1 (3), 3, and 9 of the OST.

[35] Declaration on International Cooperation in the Exploration and Use of Outer Space for the Benefit and in the Interest of All States, Taking into Particular Account the Needs of Developing Countries (The 1996 UNGA Resolution), UNGA Resolution 51/122, 13 December 1996.

[36] For further discussions on the requirements for space cooperation, see Yun Zhao, The Role of Bilateral and Multilateral Agreements in International Space Cooperation, 35 Space Policy 12–18 (2016).

[37] "States are free to determine all aspects of their participation in international cooperation in the exploration and use of outer space on an equitable and mutually acceptable basis. Contractual terms in such cooperative ventures should be fair and reasonable and they should be in full compliance with the legitimate rights and interests of the parties concerned, as, for example, with intellectual property rights". Paragraph 2 of the Annex of the 1996 UNGA Resolution.

[38] Paragraph 4 of the Annex of the 1996 UNGA Resolution.

[39] Ibid.

[40] See Paragraph 2 of the Annex of the 1996 UNGA Resolution.

[41] United Nations Committee on the Peaceful Uses of Outer Space, 598th Meeting, Wednesday 3 June 2009, Vienna, COPUOS/T.598, http://www.unoosa.org/pdf/transcripts/copuos/COPUOS_T598E.pdf

15.3 China's Overseas Assistance Programme in Realizing Space Sustainability

China firmly believes that it is important to help developing countries with their space capacity building.[42] Long-term sustainability of space activities is an issue that was raised during the development of space industry; correspondingly, such issues arising from the development process should be resolved by means of development.[43] Thus, international society needs to further achieve consensus, attain synergetic development and build common capacities to tackle the common threats it faces.[44] For example, it would be of utmost importance to have concerted actions from international society to monitor and forecast space weather as well as prevent space weather hazards.[45]

China has been, and is, developing practical collaboration with states and advocates enhancing mutual trust and understanding through collaboration.[46] It is China's firm position that international cooperation should be carried out on the basis of equality and mutual benefits and that China is willing to provide technical and financial assistance to achieve common development.[47]

15.3.1 China's Official Development Assistance Scheme

As the world's largest developing country, China provides assistance to other developing countries within the framework of South-South Cooperation through various channels, with the aim of promoting economic and social development in

[42] Tomasz Nowakowski, China's Agreement with United Nations to Help Developing Countries Get Access to Space, 28 July 2016, http://www.spaceflightinsider.com/organizations/china-national-space-administration/china-agreement-with-united-nations-to-help-developing-countries-get-access-to-space/#dcF1u0Fjz1jZIRO4.99

[43] China's Position Paper on the Issues of Long-Term Sustainability of Outer Space Activities, Working Paper by China, Committee on the Peaceful Uses of Outer Space, Scientific and Technical Subcommittee, Fifty-third session, Vienna, 15–26 February 2016, Item 14 of the provision agenda: Long-term sustainability of outer space activities, 16 February 2016, A/AC.105/C.1/2016/CRP.13.

[44] Chukeat Noichim, International Cooperation for Sustainable Space Development, 31 Journal of Space Law 338 (2005).

[45] "IGG and the South Korea Space Weather Center Sign a Cooperation" 2016, Memorandum of Space Weather Observation Research 23 Oct. 2017, http://english.igg.cas.cn/NC/News2015/2016/201609/t20160913_167638.html

[46] Such as cooperation with Russia, "Sino-Ukrainian Space Cooperation Program", "Status Quo of China-Europe Space Cooperation and the Cooperation Plan Protocol", China-Brazil Earth resources satellites bilateral cooperation, etc.

[47] "China's Space Activities in 2011", the Information Office of the State Council of the People's Republic of China, 29 Dec. 2011, http://www.scio.gov.cn/ztk/dtzt/69/3/Document/1073810/1073810.htm

developing countries.[48] Such cooperation is carried out at different levels—bilateral, multilateral, or regional.[49] China's foreign aid policies follow the principles of mutual respect, equality, keeping promises, mutual benefits and achieving win-win for both parties.[50]

The budget for China's foreign aid, called official development assistance (ODA), is under the unified management of the Ministry of Finance.[51] Normally there are three major types of financial resources for foreign assistance: grants (aid gratis), interest-free loans and concessional loans.[52] Concessional loans are to be raised by the Export-Import Bank of China on the market as a kind of financial subsidy at a low interest fixed rate.[53] China has sometimes granted debt relief as one way to provide financial assistance to developing countries.[54] Other official flows (OOFs) include non-concessional loans, such as preferential export credits, market-rate export buyers' credits and commercial loans from Chinese banks.[55]

Currently, major forms of assistance include completion of projects and provision of goods and materials, technical cooperation and human resources development cooperation.[56] First of all, China helps developing countries to improve infrastructure and promote the development of information-based societies. For example, China assisted Turkmenistan, Togo and Eritrea in building IT-related projects, including optical cable telecom networks, e-government websites and radio and television frequency modulation transmitters.[57] China also launched a

[48] China's Foreign Aid (2014), the State Council of the People's Republic of China, 10 July 2014, http://english.gov.cn/archive/white_paper/2014/08/23/content_281474982986592.htm

[49] Ibid.

[50] Shixue Jiang, "China's Principles in Foreign Aid", 29 November 2011, http://www.china.org.cn/opinion/2011-11/29/content_24030234.htm

[51] Rong Xiang, "Regulation of Foreign Aid: China", 6 September 2015, https://www.loc.gov/law/help/foreign-aid/china.php

[52] Tomasz Nowakowski, China's Agreement with United Nations to Help Developing Countries Get Access to Space, 28 July 2016, http://www.spaceflightinsider.com/organizations/china-national-space-administration/china-agreement-with-united-nations-to-help-developing-countries-get-access-to-space/#dcF1u0Fjz1jZIRO4.99

[53] Ibid.

[54] China maintains a strong commitment to debt relief. By the end of 2009, China had signed debt relief protocols with fifty countries throughout Africa, Asia, Latin America, the Caribbean and Oceania, writing off 380 matured debts amounting to RMB25.58 billion (about US$ 4.08 billion).

[55] Deborah Brautigam, Chinese Development Aid in Africa: What, Where, Why, and How Much?, in Jane Golley and Ligang Song (eds.), Rising China: Global Challenges and Opportunities 205 (Canberra: ANU E Press, 2011).

[56] Tomasz Nowakowski, China's Agreement with United Nations to Help Developing Countries Get Access to Space, 28 July 2016, http://www.spaceflightinsider.com/organizations/china-national-space-administration/china-agreement-with-united-nations-to-help-developing-countries-get-access-to-space/#dcF1u0Fjz1jZIRO4.99

[57] Ibid.

telecommunications satellite for Venezuela on 3 October 2008[58] and provided broadcasting and telecommunications facilities to African countries in 2015.[59]

Second, China helps to strengthen capacity building in developing countries in the space field. In the area of human resources cooperation, China provides training sessions for technical personnel in the area of telecommunications services and disaster relief and prevention. For example, China provided the maintenance of data processing and a broadcasting system and training sessions for Pakistan Observatory Centre in 2014.[60] Training courses have also been provided to Caribbean countries in the establishment of earthquake and tsunami early warning and monitoring systems.[61] In the area of technical cooperation, China sends experts to transfer applicable techniques and help improve technical management capacities in radio and television.[62]

Third, China helps strengthen environmental protection through assisting in construction projects in the area of meteorological information services in relevant developing countries[63]; China further provides assistance in capacity building through the provision of training sessions in early warning of meteorological disasters.[64]

China further provides foreign assistance under regional cooperation mechanisms and platforms, such as the Forum on China-Africa Cooperation (FOCAC) and the China-ASEAN Summit.[65] For example, under the FOCAC, China has helped African countries to improve their ability to cope with climate change and strengthen cooperation in meteorological satellite monitoring.[66] Further, for African countries, China has built meteorological stations and high-altitude observation radar stations and provided personnel training and exchanges.[67]

---

[58] "Symbol of China's strength: China's Space Assistance Arousing excitement in Venezuela", 24 September 2009, http://junshi.xilu.com/2009/0924/news_8_347534_1.htm

[59] "The Forum on China-Africa Cooperation Johannesburg Action Plan (2016-2018)", 10 December 2015, http://www.fmprc.gov.cn/mfa_eng/zxxx_662805/t1323159.shtml

[60] "Chinese Meteorological Experts Providing Technical Assistance in Pakistan", 2 May 2014, China Meteorology Press, 23 Oct. 2017 http://www.cma.gov.cn/2011xwzx/2011xqxxw/2011xqxyw/201405/t20140502_244961.html

[61] "Foreign Assistance under Regional Cooperation Mechanism", 10 July 2014, http://www.china.org.cn/government/whitepaper/2014-07/10/content_32924010.htm

[62] "The Interpretations of the Johannesburg Summit of the FOCAC and the Sixth Ministerial Conference on the 10 Major China-Africa Cooperation Plans in Economic and Trade Domains", 16 December 2015, http://english.mofcom.gov.cn/article/policyrelease/Cocoon/201512/20151201219036.shtml

[63] "China's Foreign Aid Helps Developing Countries", 11 July 2014, China Daily USA 23 Oct. 2017 http://usa.chinadaily.com.cn/epaper/2014-07/11/content_17729005.htm

[64] Ibid.

[65] Tomasz Nowakowski, China's Agreement with United Nations to Help Developing Countries Get Access to Space, 28 July 2016, http://www.spaceflightinsider.com/organizations/china-national-space-administration/china-agreement-with-united-nations-to-help-developing-countries-get-access-to-space/#dcF1u0Fjz1jZIRO4.99

[66] Ibid.

[67] "China Promotes New Strategic Partnership with Africa: White Paper", 10 July 2014, http://www.focac.org/eng/zxxx/t1173107.htm

15.3.2 Case Study: China-Brazil Space Cooperation

China-Brazil space cooperation provides an excellent example showing that South-South Cooperation can be carried out leading to the realization of mutual benefits in the end.[68] The China-Brazil Earth Resource Satellite (CBERS) project was set up in 1986 with the signing of the Protocol on Research and Production of the Earth Resource Satellite in 1988,[69] leading to the launch of the first satellite (CBERS-1) in 1999.[70] Several protocols were signed afterwards for the launch of more satellites.[71] Apart from the satellite launch, cooperation between the two countries covers a wide range of areas, including facilitating the entry and exit of equipment and materials required for the implementation of the project, as well as for the provision of appropriate documentation for citizens to enter, exit and reside in order to carry out relevant activities.[72]

The protocols reached between the two parties took full account of the large differences in the level of space development and agreed at the earlier stage of the cooperation that China would take up 70% of the total cost. The two countries reached a new protocol in 2002 prescribing an equal sharing of the total investment.[73]

The first three CBERS satellites were launched from China; according to the 2002 protocol, the CBERS-4 was scheduled to be launched from the Alcantara Launch Centre in Maranhão, Brazil.[74] However, different from originally planned, the CBERS-4 was launched from Taiyuan Satellite Launch Centre in China on 7 December 2014.[75] The original launch arrangement could have served as a test of the viability of the Brazilian launch centre as a future commercial launching site. Nevertheless, even though this did not happen as scheduled, Brazil has made rapid developments in its indigenous launching capability; as such, space cooperation

[68] Yun Zhao, The 2002 Space Cooperation Protocol Between China and Brazil: An Excellent Example of South-South Cooperation, 21 Space Policy 213–219 (2005).

[69] Zi Yuan, "CBERS (China-Brazil Earth Resources Satellite)", 21 July 2011, Globalsecurity 23 Oct. 2017 http://www.globalsecurity.org/space/world/china/zy-1.htm

[70] "CBERS: A Chinese-Brazilian Collaboration", 20 May 2011, Earthzine IEEE, 23 Oct. 2017, https://earthzine.org/2011/05/20/cbers-a-chinese-brazilian-collaboration/

[71] Jose Monserrat Filho and Alvaro Fabricio dos Santos, Chinese-Brazilian Protocol on Distribution of CBERS Products, 31 Journal of Space Law 271 (2005).

[72] "The Interpretations of the Johannesburg Summit of the FOCAC and the Sixth Ministerial Conference on the 10 Major China-Africa Cooperation Plans in Economic and Trade Domains", 16 December 2015, http://english.mofcom.gov.cn/article/policyrelease/Cocoon/201512/20151201219036.shtml

[73] The 2002 Space Cooperation Protocol Between China and Brazil, Article 11.

[74] "CBERS-4", 20 March 2015, Wikipedia, 23 Oct 2017, https://en.wikipedia.org/wiki/CBERS-4

[75] Barbosa, Rui C. "200th Long March rocket launches CBERS-4 for Brazil", 6 December 2014, Nasaspaceflight.com 23 Oct. 2017, https://www.nasaspaceflight.com/2014/12/200th-long-march-launches-cbers-4-brazil/

between China and Brazil has had far-reaching effects on satellite launches among developing countries.[76]

The above space cooperation does not simply benefit Brazil in developing its space launching capability; both parties benefit equally from the satellites and their remote sensing data.[77] When it comes to domestic use of CBERS data in China, the former Commission for Science, Technology and Industry for National Defense (COSTIND) declared a free online distribution policy in 2006.[78] One year later, COSTIND released a document providing that the data is owned by the state and that the Centre for Resources Satellite Data and Applications (CRESDA) is authorized to process, archive and distribute the data.[79] In the same year, the COSTIND enacted Administrative Rules on CBERS-01/02/02B Domestic Data (Trial Version), which provide detailed rules on acquisition, storage, distribution and use of CBERS data.[80]

It is to be noted that the impact of space cooperation between China and Brazil goes beyond these two states, benefiting other countries. As early as 2004, China and Brazil reached consensus on data policy for the use of CBERS data outside both territories.[81] While the on-board data recorder shall be exclusively operated by the two countries, the downlink data are accessible to all countries and international organizations.[82] The annual fee of using the data shall be determined by the conditions of ground stations.[83] Both countries also reached an agreement on an international price list for images and equally shared the revenues from distribution of data.[84]

Four years later, both countries reached yet another agreement that all Latin American countries and some African countries could obtain images for free.[85] Consensus was further reached in 2010 that CBERS data should be freely and openly accessible to all developing countries.[86]

[76] "The Interpretations of the Johannesburg Summit of the FOCAC and the Sixth Ministerial Conference on the 10 Major China-Africa Cooperation Plans in Economic and Trade Domains", 16 December 2015, http://english.mofcom.gov.cn/article/policyrelease/Cocoon/201512/20151201219036.shtml

[77] Yun Zhao, National Space Law in China: An Overview of the Current Situation and Outlook for the Future, 92–93 (Leiden: Brill&Nijhoff, 2015).

[78] "The State Council, The COSTIND Announces the Free Distribution Policy for CBERS-2 Data", Mar 2006, gov.cn, 23 Oct. 2016 http://www1.www.gov.cn/gzdt/2006-03/20/content_231785.htm

[79] Article 1, Several Opinions on Encouraging Domestic Users to Use CBERS Data, the COSTIND, Ke Gong Yi Si [2007], No. 1191, 29 October 2007.

[80] Administrative Rules on CBERS-01/02/02B Domestic Data (Trial Version), Ke Gong Yi Si [2007], No. 1417, 15 November 2007.

[81] "CBERS Data Policy", June 2004, APPL-07-2004, INPE 23 Oct. 2016 http://mtc-m16c.sid.inpe.br/col/dpi.inpe.br/banon/2006/08.03.19.25/doc/appl_07_2004.pdf

[82] CBERS Data Policy (2004), Article 2.

[83] "CBERS Data Policy (2004), Article 6(b).

[84] "CBERS Data Policy (2004), Articles 2 and 3.

[85] "CBERS-2B Completes a Full Year in Orbit", 19 September 2008, CBERS 20 Sep. 2016 http://www.cbers.inpe.br/en_noticias/index.php?cod=not101

[86] "Statement by Brazil GEO – The Beijing Ministerial Summit", 5 November 2010, INPE 23 Oct 2017, http://www.inpe.br/noticias/arquivos/pdf/GEO_china.pdf

15.3.3 *Implications of China-Brazil Space Cooperation*

China-Brazil space cooperation proves to be the most successful South-South Cooperation in the space field. It demonstrates vividly to the international community that space cooperation can be carried out between and among countries at different levels of development. We should not simply look at cooperation in its formal aspect; more importantly, substantive equality in cooperation is more important to the realization of the goal that outer space can benefit the cooperating states. China-Brazil cooperation also shows that such cooperation can not only benefit the cooperating states but also other members of international society.

15.4 Conclusion

It is believed that space sustainability can be achieved through international cooperation. The two concepts, i.e. space sustainability and international cooperation, are interdependent.[87] Space debris mitigation and the promotion of peaceful uses of outer space have been key sustainability concerns for a long time.[88] However, in this new era, space sustainability has gained another momentum, which goes to the essence of equal access to outer space for all countries. As widely accepted, spectrum and orbit are limited natural resources, but space is also a limited commodity[89]; sustainable development in outer space cannot simply be understood as something only relevant to developed or space-faring nations.[90] The ability to use outer space or the freedom of action in outer space is something we should strive for; it is something fundamental to the understanding of space sustainability, through which developing countries can realistically benefit as enshrined in the Outer Space Treaty.[91] It has thus been advocated that "more established stakeholders [should] engage emerging space actors on space sustainability efforts and facilitate multi-stakeholder collaboration in promoting responsible behaviour in space".[92]

---

[87] "Declaration of the United Nations Conference on the Human Environment", 16 June 1972 United Nations Environment Programme, 21st plenary meeting http://www.unep.org/documents.multilingual/default.asp?documentid=97&articleid=1503

[88] See generally Surya Gablin Gunasekara, Mutually Assured Destruction: Space Weapons, Orbital Debris, and the Deterrence Theory for Environmental Sustainability, 37 Air & Space Law 141–164 (2012).

[89] Rajeswari Pillai Rajagopalan, The International Code of Conduct and Space Sustainability, in C. Al-Ekabi et al. (eds.) Yearbook on Space Policy 2014, 241 (Wien: Springer-Verlag, 2016).

[90] "Declaration of the United Nations Conference on the Human Environment", 16 June 1972 United Nations Environment Programme, 21st plenary meeting http://www.unep.org/documents.multilingual/default.asp?documentid=97&articleid=1503

[91] Outer Space Treaty, Article I.

[92] Laura Delgado Lopez, Space Sustainability Approaches of Emerging Space Nations: Brazil, Colombia, and Mexico, 30 Space Policy 1 (2015).

Accordingly, the concept of sustainable development in the field of environmental law has received momentum for future expansion and concretization in the field of space law. Space sustainability should not be limited only to the vertical aspect of sustainable development, i.e. environmental protection and space debris mitigation. At this critical stage, we should not neglect the horizontal aspect of sustainable development, i.e. all countries, irrespective of their economic, social and technological development levels, should be able to benefit from outer space and space activities.

International cooperation is one major mechanism that we can rely upon to realize the above goal of equal access to outer space. As elaborated in the 1996 United Nations General Assembly (UNGA) Resolution, international cooperation can be carried out in variety of forms, at different levels and in a wide range of areas of space activities.[93] Beyond the space field, the official development assistance (ODA) has been used frequently as the major form of financial and nonfinancial assistance to developing countries. While acknowledging each state's unique priorities and interests, one needs to be wary of the justifications and accountability required of the ODA from both the donors and the developing countries.[94] Donors need to persuade relevant domestic authorities to approve relevant assistance to certain countries, while the receiving states need to make sure that such assistance does not in any way violate international rules on sovereignty. This article specifically examines Chinese approaches in dealing with space sustainability and how it affects the future development of relevant guidelines for space sustainability.

As stated earlier, international cooperation can be carried out in various ways, including those that are at the bilateral, multilateral and regional levels. While bilateral and multilateral cooperation is relatively easier, we cannot disregard the importance of regional engagement. Regional cooperation can take into account regional perspectives and solidarities, strengthen space development of the region as a whole and put forward a consistent regional voice in the international arena. This is particularly important for developing countries in space affairs.

Space sustainability cannot be achieved without taking into account the interests of developing countries. The China-Brazil cooperation model presents an excellent example that space cooperation can take place between and/or among developing countries, which exemplifies the horizontal aspect of space sustainability. While benefiting one state at one stage, space cooperation will bring benefits to cooperating countries in the long term; such benefits will not simply be restricted to these cooperating countries—with proper arrangement, other states can similarly benefit

[93] Long-term sustainability of outer space activities: proposal to adopt a first set of guidelines together with a renewed workplan for the Working Group on the Long-Term Sustainability of Outer Space Activities of the Scientific and Technical Subcommittee, Committee on the Peaceful Uses of Outer Space, Fifty-ninth session, Vienna, 8–17 June 2016, A/AC.105/L.305, 28 April 2016.

[94] Mitsuaki Furukawa, "Management of the International Development Aid System and the Creation of Political Space for China: The Case of Tanzania", JICA Research Institute Working Paper no. 82, October 2014, at 41, https://www.jica.go.jp/jica-ri/publication/workingpaper/jrft3q00000025tq-att/JICA-RI_WP_No.82.pdf

from such cooperation. This is exactly what we strive for, “the exploration and use of outer space [...] shall be carried out for the benefit and in the interests of all countries, irrespective of their degree of economic or scientific development”.[95]

Consequently, sustainable development of outer space is an issue common to all countries. International cooperation is vital to the realization of space sustainability, not only for countries to reach consensus on relevant guidelines, but also for all countries to benefit from space activities.

[95] Outer Space Treaty, Article I.

Part III
Facts and Figures

Chapter 16
Chronology: 2016

Cenan Al-Ekabi

16.1 Access to Space[1]

All launch dates are calculated using Greenwich Mean Time (GMT); hence the date at the launch site may differ from the date listed here by 1 day.

Europe	Other countries
Launch log	
January 2016	
27 Ariane 5 ECA – Intelsat 29e (C)*	15 Long March 3B – Belintersat 1 (C)
	17 Falcon 9 v1.1 – Jason 3 (S)
	20 PSLV XL – IRNSS 1E (N)
	29 Proton M Briz-M (Ph.3) – Eutelsat 9B / EDRS A (C)
February 2016	
	01 Long March 3C – BD-3 M3-S (N)
	05 Atlas 5 (401) – GPS-2F 12 (N)
	07 Soyuz 2.1b Fregat-M – Cosmos 2514 (N)
	07 Unha-3 – Kwangmyŏngsŏng 4 (D) (launch failure: successfully launched in to SSO; however no signals have been detected, and the spacecraft appears to be tumbling)
	10 Delta 4 Medium+ (5,2) (upgrade) – Topaz 4 (EW)
	16 Rokot KM – Sentinel 3A (S)
	17 H-IIA 202 – Astro H (S), Horyu 4 (D), ChubuSat 2 (D), ChubuSat 3 (D)

(continued)

[1] Federal Aviation Administration. The Annual Compendium of Commercial Space Transportation: 2017. Washington, DC: FAA, Jan 2017: 95–99.

C. Al-Ekabi (✉)
European Space Policy Institute (ESPI), Vienna, Austria
e-mail: Cenan.Al-Ekabi@espi.or.at

C. Al-Ekabi, S. Ferretti (eds.), *Yearbook on Space Policy 2016*,
Yearbook on Space Policy, https://doi.org/10.1007/978-3-319-72465-2_16

Europe	Other countries
March 2016	
09 Ariane 5 ECA – Eutelsat 65 West A (C)*	04 Falcon 9 FT – SES 9 (C)*
	10 PSLV XL – IRNSS 1F (N)
	13 Soyuz 2.1b – Resurs-P 3 (R)
	14 Proton M Briz-M (Ph.3) – TGO (ExoMars 2016) (S), EDM (Schiaparelli) (S)
	18 Soyuz-FG – Soyuz TMA-20M (MF)
	23 Atlas 5 (401) – Cygnus CRS-6 (ISS)*, Diwata 1 (R), Flock-2e′ 1 (R)*, Flock-2e′ 2 (R)*, Flock-2e′ 3 (R)*, Flock-2e′ 4 (R)*, Flock-2e′ 5 (R)*, Flock-2e′ 6 (R)*, Flock-2e′ 7 (R)*, Flock-2e′ 8 (R)*, Flock-2e′ 9 (R)*, Flock-2e′ 10 (R)*, Flock-2e′ 11 (R)*, Flock-2e′ 12 (R)*, Flock-2e′ 13 (R)*, Flock-2e′ 14 (R)*, Flock-2e′ 15 (R)*, Flock-2e′ 16 (R)*, Flock-2e′ 17 (R)*, Flock-2e′ 18 (R)*, Flock-2e′ 19 (R)*, Flock-2e′ 20 (R)*, Lemur-2 5 (R)*, Lemur-2 6 (R)*, Lemur-2 7 (R)*, Lemur-2 8 (R)*, Lemur-2 9 (R)*, Lemur-2 10 (R)*, Lemur-2 11 (R)*, Lemur-2 12 (R)*, Lemur-2 13 (R)*
	24 Soyuz 2.1a Kosmos 2515 (EW)
	29 Long March 3A BD-2 I6 (N)
	31 Soyuz 2.1a Progress MS-2 (ISS), Tomsk-TPU 120 (D)
April 2016	
25 Soyuz-STB Fregat-M – Sentinel 1B (R), MICROSCOPE (S), AAUSAT 4 (D), e-st@r 2 (D), OUFTI 1 (D)	05 Long March 2D – Shijian 10 (S)
	08 Falcon 9 v1.2 – Dragon CRS-8 (ISS)*, BEAM (D)*
	28 Soyuz 2.1a Volga – MVL-300 (S), Aist 2D (D), SamSat-218/D (D)
	28 PSLV XL – IRNSS 1G (N)
May 2016	
24 Soyuz-STB Fregat-MT – Galileo-FOC FM10 (N), Galileo-FOC FM11 (N)	06 Falcon 9 FT – JCSat 14 (C)*
	15 Long March 2D – Yaogan 30 (EW)
	27 Falcon 9 FT – Thaicom 8 (C)*
	29 Soyuz 2.1b Fregat-M – Glonass M 753 (N)
	30 Long March 4B – ZY-3 02 (R), ÑuSat 1 (R), ÑuSat 2 (R)
June 2016	
18 Ariane 5 ECA – EchoStar 18 (C)*, BRIsat (C)*	04 Rokot KM – Kosmos 2517 (R)
	09 Proton M Briz-M (Ph.4) – Intelsat 31 (C)*
	11 Delta 4 Heavy (upgrade) – Orion 9 (EW)
	12 Long March 3C – BD-2 G7 (N)
	15 Falcon 9 FT – Eutelsat 117 West B (C)*, ABS 2A (C)*

(continued)

Europe	Other countries
	22 PSLV XL – Cartosat-2C (R), SkySat 3 (R), BIROS (R), M3MSat (C), LAPAN A3 (R), GHGSat-D (R)*, Flock-2p 1 (R)*, Flock-2p 2 (R)*, Flock-2p 3 (R)*, Flock-2p 4 (R)*, Flock-2p 5 (R)*, Flock-2p 6 (R)*, Flock-2p 7 (R)*, Flock-2p 8 (R)*, Flock-2p 9 (R)*, Flock-2p 10 (R)*, Flock-2p 11 (R)*, Flock-2p 12 (R)*, SathyabamaSat (R), Swayam (D), BeeSat 4 (D) 24 Atlas 5 (551) – MUOS 5 (C) 25 Long March 7 – DFFC (D), Aoxiang Zhixing (D), Tiange Feixingqi 1 (D), Tiange Feixingqi 2 (D), Aolong 1 (D), ZGZ Shiyan Zhuangzhi (D) 29 Long March 4B – Shijian 16 (EW)
July 2016	
	07 Soyuz-FG – Soyuz MS 1 (MF) 16 Soyuz U – Progress MS-3 (ISS) 18 Falcon 9 FT – Dragon CRS-9 (ISS)*, IDA 2 (ISS) 28 Atlas 5 (421) – Quasar 20 (C)
August 2016	
24 Ariane 5 ECA – Intelsat 33e (C)*, Intelsat 36 (C)*	05 Long March 3B – Tiantong-1 (C) 09 Long March 4C – Gaofen 3 (R) 14 Falcon 9 FT – JCSat 16 (C)* 15 Long March 2D – QSS (D), Lixing 1 (D), 3Cat 2 (D) 19 Delta 4 Medium+ (4,2) – GSSAP 3 (EW), GSSAP 4 (EW) 31 Long March 4C – Gaofen 10 (R) (launch failure: failure due to malfunction in its third stage)
September 2016	
16 Vega – PerúSat 1 (EW), SkySat 4 (R)*, SkySat 5(R)*, SkySat 6 (R)*, SkySat 7 (R)*	01 Falcon 9 v1.2 AMOS 6 (C)* (launch failure: prelaunch failure during testing) 08 GSLV Mk.2 – Insat 3DR (M) 08 Atlas 5 (411) – OSIRIS-REx (S) 13 Shavit 2 – Ofeq 11 (EW) 15 Long March 2F – Tiangong 2 (TSS), Banxing 2 (D) 26 PSLV G(3) – SCATSAT 1 (M), AlSat 1B (R), AlSat 2B (R), BlackSky Pathfinder 1 (R)*, Pratham (D), PISat (R), AlSat-Nano (R), CanX 7 (D)

(continued)

Europe	Other countries
October 2016	
05 Ariane 5 ECA – NBN-Co 1B (C)*, GSat 18 (C)	16 Long March 2F – Shenzhou 11 (MF) 18 Antares 230 – Cygnus CRS-5 (ISS)*, Lemur-2 14 (R)*, Lemur-2 15 (R)*, Lemur-2 16 (R)*, Lemur-2 17 (R)* 19 Soyuz-FG – Soyuz MS 2 (MF)
November 2016	
17 Ariane 5 ES – Galileo-FOC FM7 (N), Galileo-FOC FM12 (N), Galileo-FOC FM13 (N), Galileo-FOC FM14 (N)	02 H-IIA-202 – Himawari 9 (M) 03 Long March 5 – Shijian 17 (D) 09 Long March 11 – XPNAV 1 (S), Xiaoxiang 1 (D), Lishui 1-01 (D), Pina-2 01 (D), Pina-2 02 (D), CAS 2T and KS 1Q (D) 11 Atlas 5 (401) – WorldView 4 (R), RAVAN (D), OptiCube 4 (D), Aerocube 8C (D), Aerocube 8D (D), Prometheus 2.1 (D), Prometheus 2.2 (D), CELTEE 1 (D) 11 Long March 2D – Yunhai-1 01 (R) 17 Soyuz-FG – Soyuz MS 3 (MF) 19 Atlas 5 (541) – GOES R (M) 22 Long March 3C – Tian Lian 1D (C)
December 2016	
05 Vega – Göktürk 1A (EW) 21 Ariane 5 ECA – Star One D1 (C)*, JCSat 15 (C)*	01 Soyuz U – Progress MS-4 (ISS) (launch failure: failure due to a malfunction in the third stage engine) 07 PSLV XL – Resourcesat 2A (R) 07 Delta 4 Medium +(5,4) (Upgrade) – WGS 8 (C) 09 H-IIB-304 – HTV 6 (ISS), Lemur-2 18 (R)*, Lemur-2 19 (R)*, Lemur-2 20 (R)*, Lemur-2 21 (R)*, TechEdSat 5 (D), EGG (D), TuPOD (D), AOBA-VELOX 3 (D), STARS C (D), FREEDOM (D), ITF 2 (D), Waseda-SAT 3 (D), OSNSAT (D), Tancredo 1 (D) 10 Long March 3B – FY 4A (M) 15 Pegasus XL – CYGNSS A (S), CYGNSS B (S), CYGNSS C (S), CYGNSS D (S), CYGNSS E (S), CYGNSS F (S), CYGNSS G (S), CYGNSS H (S) 18 Atlas 5(431) – EchoStar 19 (C)* 20 Epsilon 2 – ERG (S) 21 Long March 2D – TanSat (R), Spark 01 (R), Spark 02 (R), Yijian (R) 28 Long March 2D – GaoJing-1 01 (R), GaoJing-1 02 (R), BY70-1 (D)

C communications, *D* development, *I* intelligence, *ISS* International Space Station, *M* meteorological, *MF* manned flight, *N* navigation, *R* remote sensing, *S* scientific, *TSS* Tiangong Space Station, *EW* early warning system, * commercial

16.2 Space Science and Exploration

Europe	Other countries
Earth sciences	
25 April – Launch of MICROSCOPE (CNES, France) scientific mission to test the equivalence principle (EP), regarding the universality of free fall, with unprecedented accuracy	*17 January* – Launch of the Jason 3(NOAA, USA) satellite for continuity of high-precision ocean topography measurements *05 April* – Launch of Shijian 10 (CAST, China) recoverable satellite to conduct microgravity science and space life science *15 December* – Launch of CYGNSS A-to-H (NASA, USA) to probe the rapidly changing air-sea interaction processes that take place near the inner core of storms and hurricanes
Astronomy	
14 September – ESA published the first catalogue based on data collected Gaia (ESA, Europe) during its first 14 months of its 5-year scientific mission to measure the positions and motions of a billion stars of the roughly 100 billion stars in our galaxy to create the most accurate map yet of the Milky Way[a]	*17 February* – Launch of the Astro H (JAXA, Japan) next-generation X-ray astronomy satellite by JAXA providing imaging spectroscopy in the hard X-ray band above 10 keV *28 April* – Launch of the MVL-300 (Moscow State University, Russia) satellite to observe ultrahigh energy cosmic rays and study transient phenomena in the Earth's upper atmosphere *29 September* – The first scientific results of ISRO's Astrosat mission (ISRO, India) providing optical, ultraviolet and X-ray images of black holes and other related phenomena were presented at the Inter-University Centre for Astronomy and Astrophysics (IUCAA), Pune, India[b] *20 December* – Launch of the ERG (JAXA, Japan) satellite to observe the dynamics of space storms and the relativistic electron acceleration mechanism in the context of the cross-energy coupling via wave-particle interactions
Exploration	
14 March – Launch of ExoMars Trace Gas Orbiter and Schiaparelli Landing Demonstrator Module to Mars[c] *30 September* – Rosetta spacecraft (ESA, Europe) concluded its mission by colliding with the Ma'at region of the Comet 67P/Churyumov-Gerasimenko's small lobe[d]	*08 September* – Launch of OSIRIS-REx (NASA, USA) sample-return mission to an asteroid *09 November* – Launch of the XPNAV 1 (Maichong Xing Shiyan Weixing) small satellite to test autonomous spacecraft navigation and a more precise deep-space navigation by using the signals of X-ray pulsars

(continued)

Europe	Other countries
Manned spaceflight and cargo transfers	
15 December (2015)–18 June (2016) – Tim Peake (UK) participated in ISS Expeditions 46 and 47 as a flight engineer for ESA's Principia mission[e] *19 November* – Thomas Pesquet (France) participated in ISS Expeditions 50 and 51 as a flight engineer for ESA's Proxima mission[f]	*18 March* – Launch of Soyuz TMA-20M (Roscosmos, Russia) with three ISS Expedition 47/48 crews[c] *23 March* – Launch of Cygnus CRS-6* (Orbital ATK, USA) commercial resupply services mission to ISS[c] *31 March* – Launch of Progress MS 2 (Roscosmos, Russia) cargo transfer vehicle to the ISS[c] *08 April* – Launch of Dragon CRS-8* (SpaceX, USA) commercial resupply services mission to ISS[c] *07 July* – Launch of Soyuz MS 1 (Roscosmos, Russia) with three ISS Expedition 48/49 crew[c] *16 July* – Launch of Progress MS-3 (Roscosmos, Russia) cargo transfer vehicle to the ISS[c] *18 July* – Launch of Dragon CRS-9* (SpaceX, USA) commercial resupply services mission to ISS[c] *15 September* – Launch of the Tiangong 2 (CNSA, China) Space Station *16 October* – Launch of the Shenzhou 11 (CNSA, China) crewed mission to the Tiangong 2 Space Station *18 October* – *Cygnus CRS-5** (Orbital ATK, USA) commercial resupply services mission to ISS[c] *19 October* – Launch of Soyuz MS 2 (Roscosmos, Russia) with three ISS Expedition 49/50 crew[c] *17 November* – Launch of Soyuz MS 3 (Roscosmos, Russia) with three ISS Expedition 50/51 crew[c] *01 December* – *(failed launch)* launch of *Progress MS 4* (Roscosmos, Russia) cargo transfer vehicle to the ISS[c] *09 December* – *HTV 6 (JAXA, Japan)* cargo transfer vehicle to the ISS[c]

[a]Gaia's Billion-Star Map Hints at Treasures to Come." 14 Sept. 2016. ESA 22 June 2017 http://www.esa.int/Our_Activities/Space_Science/Gaia/Gaia_s_billion-star_map_hints_at_treasures_to_come
[b]A science meet to commemorate one year of AstroSat in orbit." 29 Sept. 2016. ISRO 27 June 2017 http://www.isro.gov.in/update/29-sep-2016/science-meet-to-commemorate-one-year-of-astrosat-orbit
[c]"Value of connectivity, Economic and Social benefits of expanding Internet access", Deloitte, February 2014
[d]"Mission Complete: Rosetta's Journey Ends in Daring Descent to Comet." 30 Sept. 2016. ESA 20 June 2017 http://www.esa.int/Our_Activities/Space_Science/Rosetta/Mission_complete_Rosetta_s_journey_ends_in_daring_descent_to_comet

[e]"Tim Peake returns to Earth." 18 June 2016. ESA 8 May 2017 http://m.esa.int/Our_Activities/Human_Spaceflight/Principia/Tim_Peake_returns_to_Earth
[f]"ESA ASTRONAUT THOMAS PESQUET ARRIVES AT THE INTERNATIONAL SPACE STATION." 19 Nov. 2016. ESA 8 May 2017 http://www.esa.int/Our_Activities/Human_Spaceflight/Proxima/ESA_astronaut_Thomas_Pesquet_arrives_at_the_International_Space_Station

16.3 Applications

Europe	Other countries
Earth observation	
16 February – Launch of Sentinel 3A (ESA, Europe) oceanography and land-vegetation monitoring satellite contributing to the EU Copernicus Earth Observation flagship programme[a] *25 April* – Launch of Sentinel 1B (ESA, Europe) radar imaging satellite contributing to the EU Copernicus Earth Observation flagship programme[a] *2 June* – Launch of BIROS (DLR, Germany)[a]	*13 March* – Launch of Resurs-P 3 (Roscosmos, Russia)[a] *23 March* – Launch of Flock-2e′ 1 to 20* (Planet, USA), Lemur-2 5 to 13* (Spire, USA) and Diwata 1 (DOST, Philippines) remote sensing cubesats[a] *30 May* – Launch of the ÑuSat 1 and 2 (Satellogic S.A., Argentina) and ZY-3 02 (CNSA, China) remote sensing cubesats[a] *04 June* – Launch of Kosmos 2517 (Russian Aerospace Forces, Russia)[a] *22 June* – Launch of the Cartosat-2C (ISRO, India), SkySat-2 1 (Terra Bella, USA) and LAPAN A3 (LAPAN, Indonesia) small satellites and Flock-2p 1 to 12* (Planet, USA), GHGSat-D* (GHGSat Inc., Canada) and SathyabamaSat (Sathyabama University, India) remote sensing cubesats[a] *09 August* – Launch of the Gaofen 3 (CNSA, China) remote sensing satellite[a] *31 August* – Launch of the Gaofen 10 (CNSA, China) remote sensing satellite (failed launch)[a] *08 September* – Launch of the Insat 3DR (Insat, India) meteorology satellite[a] *16 September* – Launch of the SkySat-2 2-to-5* (Terra Bella, USA) small satellites[a] *26 September* – Launch of the SCATSAT 1 (ISRO, India) meteorology small satellite, AlSat 1B and 2B (Agence Spatiale Algérienne, Algeria) remote sensing small satellites and PISat (PESIT, India), BlackSky Pathfinder 1* (BlackSky Global, USA) and AlSat-Nano (Agence Spatiale Algérienne, Algeria) remote sensing cubesats[a]

(continued)

Europe	Other countries
	18 October – Launch of Lemur-2 14 to 17* (Spire, USA) remote sensing cubesats[a]
	02 November – Launch of the Himawari 9 (Japanese Ministry of Transport Civil Aviation Bureau & Meteorological Agency, Japan) meteorology satellite[a]
	11 November – Launch of WorldView 4* (DigitalGlobe, USA)[a]
	11 November – Launch of Yunhai-1 01 (SAST, China)[a]
	19 November – Launch of GOES R (NOAA, USA) meteorology satellite[a]
	07 December – Launch of Resources at 2A (ISRO, India)[a]
	09 December – Launch of Lemur-2 18 to 21* (Spire, USA) remote sensing cubesats[a]
	10 December – Launch of the FY 4A (China Meteorological Administration, China) meteorology satellite[a]
	21 December – Launch of the TanSat (CAS, China) remote sensing small satellite and Spark 01 and 02 (CAS, China) and Yijian (CAS, China) remote sensing cubesats[a]
	28 December – Launch of the GaoJing-1 01 and 02 (Beijing Space View Tech Co Ltd, China) remote sensing small satellites[a]
Intelligence and early warning	
	10 February – Launch of the Topaz 4 (NRO, USA) radar reconnaissance satellite[a]
	24 March – Launch of the Kosmos 2515 (VKO, Russia) optical reconnaissance satellite[a]
	15 May – Launch of the Yaogan 30 (CNSA, China) optical reconnaissance satellite[a]
	11 June – Launch of the Orion 9 (NRO, USA) signals intelligence satellite[a]
	29 June – Launch of the Shijian 16 (PLA, China) signals intelligence satellite[a]
	19 August – Launch of GSSAP 3 and 4 (USAF, USA) space surveillance satellites[a]
	13 September – Launch of the Ofeq 11 (Tsahal, Israel) optical reconnaissance satellite[a]
	16 September – Launch of the PerúSat 1 (Peruvian armed forces, Peru) optical reconnaissance satellite[a]
	05 December – Launch of the Göktürk 1A (Turkish Ministry of Defence, Turkey) optical reconnaissance satellite[a]

(continued)

Europe	Other countries
Navigation	
24 May – Launch of Galileo-FOC 10 and FOC 11 (ESA, Europe) satellites *17 November* – Launch of Galileo-FOC 7, FOC 12, FOC 13 and FOC 14 (ESA, Europe) satellites	*20 January* – Launch of the IRNSS 1E (ISRO, India) satellite[a] *01 February* – Launch of the BD-3 M3-S (CNSA, China) satellite[a] *05 February* – Launch of the Navstar GPS-2F 12 (USAF, USA) satellite[a] *07 February* – Launch of the Uragan-M #44 (Kosmos 2514) (Russian Aerospace Forces, Russia) satellite[a] *10 March* – Launch of the IRNSS 1F (ISRO, India) satellite[a] *29 March* – Launch of the BD-2 I6 (CNSA, China) satellite[a] *28 April* – Launch of the IRNSS 1G (ISRO, India) satellite[a] *29 May* – Launch of the Uragan-M #45 (Kosmos 2516) (Russian Aerospace Forces, Russia) satellite[a] *12 June* – Launch of the BD-2 G7 (CNSA, China) satellite[a]
Telecommunications/broadcasting	
27 January – Launch of Intelsat 29e* (Intelsat, Luxembourg)[a] *29 January* – Launch of Eutelsat 9B* (Eutelsat, France)[a] *04 March* – Launch of SES 9* (SES, Luxembourg)[a] *09 March* – Launch of Eutelsat 65 West A* (Eutelsat, France)[a] *09 June* – Launch of Intelsat 31* (Intelsat, Luxembourg)[a] *24 August* – Launch of Intelsat 33e* (Intelsat, Luxembourg) and Intelsat 36* (Intelsat, Luxembourg)[a]	*15 January* – Launch of Belintersat 1 (Belintersat, Belarus)[a] *06 May* – Launch of JCSat 14* (JSAT Corporation, Japan)[a] *27 May* Launch of Thaicom 8* (Thaicom, Thailand)[a] *15 June* – Launch of Eutelsat 117 West B* (Satellites Mexicanos S.A. de C.V., Mexico) and ABS 2A* (ABS, China)[a] *18 June* – Launch of EchoStar 18* (EchoStar, USA), BRIsat* (Bank Rakyat Indonesia, Indonesia)[a] *22 June* – Launch of M3MSat (CSA, Canada)[a] 24 *June* – Launch of MUOS 5 (US Navy, USA)[a] *28 July* – Launch of Quasar 20 (NRO, USA)[a] *05 August* – Launch of Tiantong-1 01 (CAST, China)[a] *14 August* – Launch of JCSat 16* (JSAT Corporation, Japan)[a] *01 September* – Launch of AMOS 6* (Spacecom Ltd., Israel) (prelaunch failure)[a] *05 October* – Launch of GSat 18 (ISRO, India) and NBN-Co 1B* (NBN-Co Limited, Australia)[a]

(continued)

Europe	Other countries
	22 November – Launch of Tian Lian 1D (CAST, China)[a] *07 December* – Launch of WGS 8 (USAF,USA)[a] *18 December* – Launch of EchoStar 19* (EchoStar, USA)[a] *21 December* – Launch of Star One D1* (Star One, Brazil) and JCSat 15* (JSAT Corporation, Japan)[a]
Technology development	
25 April – Launch of AAUSAT 4 (Aalborg University Cubesat, Denmark), e-st@r 2 (Politecnico di Torino, Italy) and OUFTI 1 (Université de Liège, Belgium) development cubesats[a] *22 June* – Launch of BeeSat 4 (TU Berlin, Germany) development cubesat[a] *15 August* – Launch of 3Cat 2 (Universidad Politécnica de Cataluña (UPC), Spain) development cubesat[a] *09 December* – TuPOD (Gauss Srl., Italy) development cubesat[a]	*07 February* – Launch of Kwangmyŏngsŏng 4 (NADA, North Korea) small satellite (failed launch)[a] *17 February* – Launch of ChubuSat 2 and 3 (Nagoya University/Daido University, Japan) and Horyu 4 (Kyushu Institute of Technology, Japan) development cubesats[a] *31 March* – Launch of the Tomsk-TPU 120 (Tomsk-TPU, Russia) cubesat[a] *08 April* – Launch of the BEAM* (Bigelow Aerospace, USA) technology demonstration[a] *28 April* – Launch of Aist 2D (Samara Aerospace University, Russia) small satellite and SamSat-218/D (Samara Aerospace University, Russia) cubesat[a] *22 June* – Launch of the Swayam (College of Engineering, Pune (COEP), India) development cubesat[a] *25 June* – Launch of the DFFC (CNSA, China), Tiange Feixingqi 1 and 2 (Unknown, China), Aolong 1 (CALT, China), ZGZ Shiyan Zhuangzhi (Unknown, China) satellites and Aoxiang Zhixing (Shaanxi Engineering Laboratory (SELM)/Northwestern Polytechnical University (NPU), China) development cubesat[a] *15 August* – Launch of the QSS (CAS, China) and Lixing 1 (CAS, China) small satellites[a] *15 September* – Launch of the Banxing 2 (SAST, China) cubesat[a] *26 September* – Launch of the Pratham (IIT Bombay, India) and CanX 7 (UTIAS, Canada) development cubesats[a] *03 November* – Launch of the Shijian 17 (CAST, China) technology demonstrator[a]

(continued)

Europe	Other countries
	09 November – Launch of the Xiaoxiang 1 (Changsha Gaoxinqu Tianyi Research Institute, China), Lishui 1-01 (Unknown, China), Pina-2 01 and Pina-2 02 (Unknown, China) and CAS 2T and KS 1Q (CAMSAT, China) development cubesats[a] *11 November* – Launch of the RAVAN (Johns Hopkins APL, USA), OptiCube 4 (NASA, USA), Aerocube 8C and 8D (Aerospace Corporation, USA), Prometheus 2.1 and 2.2 (Los Alamos National Laboratory, USA) and CELTEE 1 (M42 Technologies, USA) development cubesats[a] *09 December* – Launch of the TechEdSat 5 (SJSU, USA), EGG (University of Tokyo, Japan), AOBA-VELOX 3 (Nanyang Technological University, Singapore; Kyushu Institute of Technology (Kyutech), Japan), STARS C (Kagawa University, Japan), FREEDOM (Tohoku University, Japan), ITF 2 (University of Tsukuba, Japan), Waseda-SAT 3 (Waseda University, Japan), OSNSAT (Open Space Network, USA) and Tancredo 1 (Escola Municipal Presidente Tancredo de Almeida Neves, Brazil) development cubesats[a] *28 December* – Launch of the BY70-1 (CAST, China) development cubesat[a]
Business	
12 July British government awarded feasibility study contracts for horizontal or vertical launch proposals[b] *30 November* – CNES announced that it would sell its stake in Arianespace will increase to 74%, and the three directors representing CNES will be replaced by three directors appointed by Airbus Safran Launchers, while CNES will join ESA as statutory censors of Arianespace's Board of Directors[c]	*19 April* – Roscosmos provided a 20 billion rouble cash infusion to the space-hardware builder Khrunichev Space Centre to repay its suppliers[d] *28 April* – Russia's Vostochny Cosmodrome conducted its first launch using a Soyuz 2.1a[e] *23 May* – Telecom Regulatory Authority of India (TRAI) published a "preconsultation paper" highlighting the savings that satellite television broadcasters could realise if they transmitted popular satellite television programmes using common transponder space shared by multiple DTH providers[f] *18 August* – Boeing and its Russian and Ukrainian partners in the Sea Launch joint venture reached a preliminary framework agreement to settle a lawsuit initiated by Boeing in February 2013[g]

(continued)

Europe	Other countries
	21 October – the Russian government sent a formal warning to the French government wanting a resolution of the Arianespace's payment to Roscosmos by March 2017; the payment was blocked by an ongoing dispute by former shareholders of Russia's Yukos energy company. Left unresolved, Russia will take France to court for violating a 1989 bilateral treaty, and other Euro-Russian space projects would be at risk if the stalemate continued[h]

[a]"Value of connectivity, Economic and Social benefits of expanding Internet access", Deloitte, February 2014
[b]De Selding, Peter B. "Britain selects U.S., French, British teams to study spaceport feasibility." 13 July 2016. SpaceNews 10 Nov. 2016 http://spacenews.com/britain-selects-u-s-french-british-teams-to-study-spaceport-feasibility/
[c]Messier, Doug. "Airbus Safran Launchers Buys CNES Shares of Arianespace." 3 Dec. 2016. Parabolic Arc 21 Aug. 2017 http://www.parabolicarc.com/2016/12/03/airbus-safran-launchers-buys-cnes-shares-arianespace/
[d]De Selding, Peter B. "Russia bails out debt-ridden Khrunichev." 20 Apr. 2016. SpaceNews 10 Nov. 2016 http://spacenews.com/russian-government-provides-debt-ridden-khrunichev-with-cash-and-loans/
[e]Haines, Lester. "First rocket finally departs Russia's Vostochny cosmodrome." 28 Apr. 2016. The Register 15 Aug. 2017 https://www.theregister.co.uk/2016/04/28/vostochny_launch/
[f]De Selding, Peter B. "Satellite operators give negative reviews of Indian regulator's satellite-TV proposal." 8 June 2016. SpaceNews 10 Nov. 2016 http://spacenews.com/satellite-operators-give-negative-reviews-of-indian-regulators-satellite-tv-proposal/
[g]Foust, Jeff. "Boeing and Energia negotiating Sea Launch settlement." 19 Aug. 2016. SpaceNews 18 Aug. 2017 http://spacenews.com/boeing-and-energia-negotiating-sea-launch-settlement/
[h]De Selding, Peter B. "Eutelsat, freed by Paris court ruling, makes overdue payment on $424 million RSCC contract." 28 Nov. 2016. SpaceNews 4 Jan. 2017 http://spacenews.com/eutelsat-freed-by-paris-court-ruling-pays-russias-rscc-long-due-424-million/

16.4 Policy and International Cooperation

Europe	Other countries
General policy	
12–13 January – Eighth high-level conference on EU space policy held in Brussels, Belgium[a] *7 April* – UK Space Agency published its Corporate Plan for the period 2016/2017, laying out an overview of the Agency's mandate, strategy and key targets and milestones[b]	*22 November* – Canada's Innovation, Science and Economic Development Canada (ISED) Minister Navdeep Bains announced that a new Canadian space strategy would be released in June of 2017[i] *23 December* – US President Obama signed the National Defense Authorization Act for 2017 (NDAA-17) which partly repealed the provision permitting the use of the Russian-built RD-180 engine that was enacted in the previous year's NDAA[j]

(continued)

Europe	Other countries
23 June – UK citizens voted to end the UK's membership of the EU; the Brexit process will trigger Article 50 of the Treaty on European Union by the end of March 2017[c] *28–29 June* – EUMETSAT Council endorsed its new strategy "Challenge 2025" which establishes the framework for its activities in the next decade[d] *26 October* – European Commission released its Space Strategy for Europe[e] *26 October* – ESA Director General Jan Woerner and European Commissioner Elżbieta Bieńkowska signed a 'Joint Statement on Shared Vision and Goals for the Future of European Space'[f] 30 November – European Commission released its European Defence Action Plan[g] *14 December* – ASI released its Strategic Vision Document for the period 2016–2025[h]	
General cooperation	
19 January – DLR's Project Management Department and China's National Centre for Science and Technology Evaluations (NCSTE) enter a 3-year agreement for cooperation in the development of research and innovation[k] *25 January* – CNES and ISRO signed three cooperation agreements during French President François Hollande's state visit to India[l] 25 January – ASI and the United Arab Emirates (UAE) Space Agency sign a MoU providing a broad framework agreement to carry out cooperative space activities between the two entities[m] *25 February* – DLR and JAXA sign an 'Interagency Arrangement for Strategic Partnership' to achieve common strategic goals[n] *18 May* – ASI and the Argentinian Space Agency (CONAE) signed a Letter of Intent to extend and strengthen collaboration in the field of space activities[o] *02 June* – DLR and CNES renewed their 2002 framework agreement for bilateral cooperation[p] *17 June* – ASI and Roscosmos signed an MoU to cooperate in the field of remote sensing and Earth observation during the Saint Petersburg Economic Forum	*15–26 February* – UN COPUOS 53rd Scientific and Technical Subcommittee meeting took place in Vienna, Austria *04–15 April* – UN COPUOS 55th Legal Subcommittee meeting took place in Vienna, Austria *08–17 June* – UN COPUOS 59th Plenary Meeting took place in Vienna, Austria *26–30 September* – 67th International Astronautical Congress (IAC) held in Guadalajara, Mexico *26 December* – Russia and Kazakhstan signed a cooperation agreement extending Russia's use of Kazakhstan's Baikonur Cosmodrome, including an 8-year road map up to the year 2025[u]

(continued)

Europe	Other countries
6 September – CNES signed an agreement with the Vietnamese Academy of Science and Technology (VAST) to reinforce cooperation between France and Vietnam[q]	
14 September – DLR and CNES sign agreement for the design, construction and operation phases of the Methane Remote Sensing LIDAR Mission (MERLIN) environmental satellite[r]	
27 October – ASI and the government of Australia signed a collaborative partnership agreement to pursue space activities, promoting future joint research and development, academic exchange and industry cooperation[s]	
7 December – ESA and NASA agreed to extend their collaboration in human space exploration, wherein ESA will provide a second service module to support NASA's first crewed Orion mission, expected to launch on its SLS as early as 2021[t]	

[a]"8th Annual Conference on EU Space Policy." Cosmos Space – NCP Network 26 Sept. 2017 http://ncp-space.net/8th-annual-conference-on-eu-space-policy/

[b]"UK Space Agency Corporate Plan 2016 – 2017." 7 Apr. 2016. GOV.UK 27 July 2017 https://www.gov.uk/government/publications/uk-space-agency-corporate-plan-2016-2017

[c]"Brexit: Theresa May to trigger Article 50 by end of March." 2 Oct. 2016. BBC News 17 Aug. 2017 http://www.bbc.com/news/uk-politics-37532364

[d]"EUMETSAT Strategy Challenge 2025 Summary." 15 June 2016. EUMETSAT 31 July 2017 https://www.eumetsat.int/website/wcm/idc/idcplg?IdcService=GET_FILE&dDocName=PDF_BR_COR02SUM_EN&RevisionSelectionMethod=LatestReleased&Rendition=Web

[e]Commission of the European Communities. Communication from the Commission to the European Parliament, the Council, the European Economic and Social Committee and the Committee of the Regions. Space Strategy for Europe. COM (2016) 705 final of 26 October 2016. Brussels: European Union https://ec.europa.eu/transparency/regdoc/rep/1/2016/EN/COM-2016-705-F1-EN-MAIN.PDF

[f]"Shared vision and goals for the future of Europe in space." 26 Oct. 2016. ESA 31 July 2017 http://m.esa.int/About_Us/Welcome_to_ESA/Shared_vision_and_goals_for_the_future_of_Europe_in_space

[g]"European Defence Action Plan: Towards a European Defence Fund." 30 Nov. 2016. European Commission Press Release Database 21 Aug. 2017 http://europa.eu/rapid/press-release_IP-16-4088_en.htm

[h]Battiston, Roberto. "Documento di Visione Strategica 2016-2025." 14 Dec. 2016. ASI 2 Aug. 2017 http://www.asi.it/it/news/documento-di-visione-strategica-2016-2015

[i]Boucher, Marc. "New Canadian Space Strategy Coming in June 2017." 22 Nov. 2016. SpaceRef Canada 21 Aug. 2017 http://spaceref.ca/news/new-canadian-space-strategy-coming-in-june-2017.html

[j]"Statement by the President on Signing the National Defense Authorization Act for Fiscal Year 2017." 23 Dec. 2016. The White House President Barack Obama 15 Aug. 2017 https://obamawhitehouse.archives.gov/the-press-office/2016/12/23/statement-president-signing-national-defense-authorization-act-fiscal

[k]"Germany and China broaden their collaboration in science management." 22 Jan. 2017. DLR 3 Aug. 2017 http://www.dlr.de/dlr/presse/en/desktopdefault.aspx/tabid-10172/213_read-16485/#/gallery/21731

[l]"List of Agreements/MOUs signed during the State Visit of President Francois Hollande of the French Republic to India (January 25. 2016)." 25 Jan. 2016. Ministry of External Affairs – Government of India 20 Aug. 2017 http://www.mea.gov.in/bilateral-documents.htm?dtl/26294/List+of+AgreementsMOUs+signed+during+the+State+Visit+of+President+Francois+Hollande+of+the+French+Republic+to+India+January+25+2016

[m]"ESA TO SUPPLY SERVICE MODULE FOR FIRST CREWED ORION MISSION." 7 Dec. 2016. ESA 8 May 2017 http://www.esa.int/Our_Activities/Human_Spaceflight/Orion/ESA_to_supply_Service_Module_for_first_crewed_Orion_mission

[n]"DLR is expanding cooperation with Japan and South Korea." 29 Feb. 2016. DLR 3 Aug. 2017 http://www.dlr.de/dlr/en/desktopdefault.aspx/tabid-10857/1527_read-16946/#/gallery/22220

[o]"Italy - Argentina means space collaboration." 18 May 2016. ASI 1 Aug. 2017 http://www.asi.it/en/news/italy-argentina-means-space-collaboration

[p]"DLR and CNES renew their framework agreement for bilateral cooperation." 2 June 2016. DLR 3 Aug. 2017 http://www.dlr.de/dlr/en/desktopdefault.aspx/tabid-10081/151_read-18064/year-2016/151_page-4/#/gallery/23231

[q]"France/Vietnam : la visite de François Hollande relance la coopération économique." 7 Sept. 2016. Le Moci 19 Aug. 2017 http://www.lemoci.com/actualites/pays-marches/france-vietnam-la-visite-de-francois-hollande-relance-la-cooperation-economique/

[r]"Climate protection – DLR and CNES sign an agreement for the construction and operation of the MERLIN environmental satellite." 14 Sept. 2016. DLR 3 Aug. 2017 http://www.dlr.de/dlr/en/desktopdefault.aspx/tabid-10081/151_read-19317/year-2016/151_page-3/#/gallery/24323

[s]"Australia, Italy sign contract for a collaborative partnership in space." 27 Oct. 2016. Geospatial World 1 Aug. 2017 https://www.geospatialworld.net/news/australia-italy-sign-contract-collaborative-partnership-space/

[t]"ESA TO SUPPLY SERVICE MODULE FOR FIRST CREWED ORION MISSION." 7 Dec. 2016. ESA 8 May 2017 http://www.esa.int/Our_Activities/Human_Spaceflight/Orion/ESA_to_supply_Service_Module_for_first_crewed_Orion_mission

[u]"Why Russia Won't Be Leaving Kazakhstan's Baikonur Cosmodrome Anytime Soon." 29 Dec. 2016. Sputniknews.com 16 Aug. 2017 https://sputniknews.com/politics/201612291049119821-russia-kazakhstan-baikonur-cooperation/

16.5 Country Profiles 2016

Austria	
Population[a]	8.690 million
GDP[b]	€349.34 billion
Responsibility[c]	The Austrian Space Programme is financed by the Federal Ministry for Transport, Innovation and Technology (BMVIT) and managed by the Aeronautics and Space Agency (ALR), which is integrated into the Austrian Research Promotion Agency (FFG), the central organisation for fostering science and technology activities in Austria, and serves as a docking station to the international aerospace world for Austrian business and science
Activities[d]	ALR of FFG implements national aerospace policy and represents Austria in numerous European and international aerospace institutions – from ESA and EUMETSAT through EU committees and other international and national organisations responsible for aerospace policy. Austrian representation in the European Space Agency is of special strategic importance in this context. In addition to ESA programmes, the Austrian Space Applications Programme (ASAP) and the space programmes under the 7th EU Framework Programme are of special relevance for space research projects
Budget[e]	€73.2 million (ESA contribution, €47.6 million; National Civil Space Budget, €16.6 million; and EUMETSAT, €9.0 million)
Staff[f]	12 (ALR)
Direct employment in space manufacturing industry[g]	422

[a]All population figures are from Eurostat: http://appsso.eurostat.ec.europa.eu/nui/show.do?dataset=demo_gind&lang=en. Accessed 1 August 2017

[b]All GDP figures are from Eurostat: http://appsso.eurostat.ec.europa.eu/nui/submitViewTableAction.do. Accessed 29 July 2017

[c]All information on the ESA Member States “Responsibilities” was obtained from direct email exchange with the different agencies and organisations responsible for space in each country and updated with the ESTMP 2016

[d]All information on the ESA Member States “Activities” was obtained from direct email exchange with the different agencies and organisations responsible for space in each country and updated with the ESTMP 2016

[e]All contributions to 2016 ESA Budget figures are from ESA: http://www.esa.int/spaceinimages/Images/2016/01/ESA_budget_2016. Accessed 9 August 2017. All Contributions to 2016 EUMETSAT Budget figures are best estimates for the year. All budget figures on National expenditures, both civil and military, originates from EUROCONSULT

[f]All information on the ESA Member States “Staff” in their national public organisations and agencies for space was obtained from direct email exchange with the different agencies and organisations responsible for space in each country

[g]All “direct employment in space manufacturing industry” figures are taken from ASD-Eurospace. “Facts and Figures – The European Space Industry in 2016”

Belgium	
Population	11.311 million
GDP	€421.61 billion
Responsibility	Two-level competence system
	Federal: space in the frame of international agreements
	Regions and communities: basic and applied research
	The Belgian Federal Science Policy Office (BELSPO) with its Department of Space Research and Applications manages the scientific aspects of the space policy. The Belgian High Representation for Space Policy (BHRS), acting under the auspices of the federal minister in charge of Science Policy, defends Belgian interests in the international institutions dealing with space (ESA and EU levels). Other federal ministries involved with space are the Ministries of Defence and Foreign Affairs. On the regional level, the Ministries for Economic Development, Transport and Environment have stakes in space applications and industrial development
Activities	Strategy and operational goals
	To create and/or strengthen expertise and industrial capabilities in high-growth areas of the space sector
	To increase the return on investment of the public money invested in space at the ESA level, in the frame of the EU, as well as in bilateral cooperation projects
	To support industry to be able to tackle new markets and positive impact on economy
	To give to the public authority the space means necessary to define and to implement its policies
	To represent Belgium's interests in the definition and implementation of the European space policy
	Cooperation with ESA, EUMETSAT, EC and other aerospace-related agencies
	Within ESA, Belgium traditionally has major interests in launchers, telecommunications, integrated applications and technology support programmes, earth observation
Budget	€208.9 million (ESA contribution, €188.9 million; National Civil Space Budget, €9.0 million; and EUMETSAT, €11.0 million)
Staff	BELSPO Department of Space Research and Applications: 20
Direct employment in space manufacturing industry	1752

Czech Republic	
Population	10.554 million
GDP	174.41 billion Euro

(continued)

Responsibility	The Ministry of Transport of the Czech Republic is the coordinator of all space activities in the Czech Republic. For the coordination purposes, it established the Coordination Council for Space Activities under its leadership, which includes all relevant ministries and institutions. The Ministry of Transport is also directly responsible for the major space activities in the Czech Republic, including participation in ESA and EU space policy and international cooperation in space. The Ministry of Transport is responsible for the implementation of the Galileo programme in the Czech Republic, and it bears the responsibility for satellite systems' follow-up applications (e.g. intelligent transport systems)
Activities	The Czech Republic is subscribed in ESA Optional Programmes such as MTG, MetOp-SG, EOEP, GSC, ARTES, EGEP, GSTP, MREP, ELIPS, SSA, FLPP and PRODEX
Budget	€20.8 million (ESA contribution, €15.6 million; National Civil Space Budget, €1.0 million; and EUMETSAT, €4.2 million)
Direct employment in space manufacturing industry	187

Denmark	Ministry of Higher Education and Science — Danish Agency for Science, Technology and Innovation
Population	5.707 million
GDP	€276.81 billion
Responsibility	In June 2015, the Ministry of Higher Education and Science was formally assigned responsible for matters relating to the regulation of Danish space activities and for coordinating space activities across ministries and public institutions. In 2016 a new national space strategy was published, and an Inter-Ministerial Space Committee, consisting of representatives from the ministries which have responsibilities relating to the space sector, was given the important task of coordinating and following up on the strategy and the initiatives it includes
Activities	Denmark supports space technology research and development through ESA's ELIPS, GSTP, PRODEX, ARTES, FLPP and GNSS Evolution Programmes
Budget	€38.0 million (ESA contribution, €29.5 million; National Defence Space Budget, €0.9 million; and EUMETSAT, €7.6 million)
Staff	29 (DTU)
Direct employment in space manufacturing industry	232

Estonia	
Population	1.316 million
GDP	€20.92 billion
Responsibility	Space activities in Estonia are coordinated by the Space Affairs Council (SAC), established in 2010 by the Ministry of Economic Affairs and Communications. The main task of SAC is initiation and governance of space-related programmes and activities on national and international level and coordination of the utilisation of resources. SAC is supported by secretariat, containing representatives of Enterprise Estonia and Estonian Research Council. Since 2006 Estonian public business development and support agency Enterprise Estonia (www.eas.ee) is assigned as implementing body (ESA delegation) for the ESA-Estonian agreements. Enterprise Estonia also carries out the tasks of Estonian Space Office, being intermediary between Estonian companies/ institutions and international space community. Main tasks concluded mapping and auditing space-related competences and capabilities, organising trainings for entrepreneurs and public sector officials, taking companies to expositions and study trips, encouraging the technology transfer, etc. Enterprise Estonia is a member of EURISY since 2008 and member of IAF since 2009
	Space science activities in Estonia are carried out by Tartu Observatory (www.to.ee), having long tradition in astronomy, cosmology and Earth observation. Nowadays Tartu Observatory serves as Estonian space technology development and testing centre
Activities	Estonian space focus of space activities is on earth observation, navigation, technology, especially related to ICT and science. Estonian space activities are mainly related to Horizon 2020 projects and cooperation with ESA and PECS programme, where Estonia has 27 projects. Estonia became the 21st ESA member state in September 2015. Estonia is actively preparing for Galileo and Copernicus programme implementation in public and private sectors
Budget	€3.6 million (ESA contribution, €0.9 million; National Space Budget, €2.3 million; and EUMETSAT, €0.4 million)
Staff	10 (related to space in Enterprise Estonia)
	80 (at Tartu Observatory)
Direct employment in space manufacturing industry	39

Finland	Tekes
Population	5.487 million
GDP	€214.06 billion
Responsibility	Finland's public sector space activities are funded by the Ministry of Employment and Economy (ESA membership), Tekes (ESA and national programmes), Academy of Finland (EISCAT, ESO, NOT) under the Ministry of Education and Culture and Finnish Meteorological Institute (EUMETSAT) under the Ministry of Transport and Communications. National coordination is provided by the Finnish Space Committee under Ministry of Economy
Activities	In addition to cooperation with ESA and EC (e.g., space programme under the Horizon 2020 EU Framework Programme), Tekes funds space technologies and applications nationally. The Finnish focus of space activities is on space science, Earth observation, satellite telecommunications, satellite navigation and space technologies. Tekes also funds scientific instruments for spacecraft
Budget	€44.3 million (ESA contribution, €21.6 million; National Civil Space Budget, €16.9 million; and EUMETSAT, €5.8 million)
Staff	6 (Tekes)
Direct employment in space manufacturing industry	191

France	cnes
Population	66.760 million
GDP	€2,228.86 billion
Responsibility	Space activities in France are under parent Ministries of Higher Education and Research and of Defence. Among different actors, France relies on its national space agency CNES (Centre National d'Etudes Spatiales) for space policy proposal and application
Activities	France inscribes its space activities in the frame of European space activities and intents to play there a leading role. Activities are conducted in strong interface with EC, ESA, EUMETSAT and other member states or at international level. National orientations include five acting domains, namely, Ariane (autonomous European access to space), Observation, Sciences, Telecommunications and Defence
Budget	€2524.3 million (ESA contribution, €844.5 million; National Space Budget, €1616.7 million; and EUMETSAT, €63.1 million)
Staff	2400 (CNES)
Direct employment in space manufacturing industry	14,140

Germany	
Population	82.176 million
GDP	€3,134.07 billion
Responsibility	Acting on behalf of the federal government, the Space Administration designs and implements Germany's space programme, which integrates all German space activities on the national and European plane. These activities include Germany's national space programme, DLR's "Space" research and development programme and Germany's contributions to the European Space Administration (ESA) as well as the European Organisation for the Exploitation of Meteorological Satellites (EUMETSAT). In addition, the administration designs and supervises space and security research projects under the sixth and seventh EU research framework program
	The Space Administration's principal client is the Federal Ministry for Economics and Technology. However, the Space Administration also works for other ministries, mainly in application related fields such as Earth observation, navigation and satellite communication. The Federal Government's Space Programme provides both business and science with a reliable political framework for independent planning and action, thus ensuring that public funds are used efficiently. The national programme is implemented by the Space Administration. Its major function is to promote and support Germany's strategic goals in the European programmes of ESA and the EU by purposefully equipping Germany's industry and scientific institutions for competition within the EU as well as for their tasks within the ESA framework
Activities	Germany has a national civil programme, which includes bi- or trilateral cooperations, a participation in the ESA programmes and a defence programme (e.g. SAR-Lupe, MUSIS). The major areas of interest for Germany's domestic programmes are Earth observation, navigation, telecommunications, space transportation systems, space infrastructures, microgravity research, space science, space technologies and space exploration
Budget	€1793.8 million (ESA contribution, €872.6 million; National Space Budget, €839.2 million; and EUMETSAT, €82.0 million)
Staff	7900 (DLR)
Direct employment in space manufacturing industry	7830

Greece	
Population	10.784 million

(continued)

GDP	€175.89 billion
Responsibility	The General Secretariat for Research and Technology (GSRT), under the responsibility of the Ministry of Education, Research and Religious Affairs, is responsible for Greek space activities
Activities	The technological activities of interest to Greece are mostly pursued through ESA EOEP, ARTES (focus on integrated applications) and GSTP
Budget	€25.3 million (ESA contribution, €11.9 million; National Civil Space Budget, €7.4 million; and EUMETSAT, €6.0 million)
Direct employment in space manufacturing industry	34 (GSRT)

Hungary	Magyar Űrkutatási Iroda
Population	9.830 million
GDP	€112.40 billion
Responsibility	The Hungarian Space Office (HSO), integrated into the body of the Ministry of National Development, manages, coordinates and represents Hungarian space activities. Hungary's space activity is supervised by the Minister, who makes his decisions based on the advices of the Scientific Council on Space Research in scientific issues and the Hungarian Space Board in strategic questions
Activities	Participation in microgravity, Earth observation, life and material sciences and GTSP programmes of ESA
Budget	€8.1 million (ESA contribution, €5.0 million; National Civil Space Budget, €0.4 million; and EUMETSAT, €2.7 million)
Direct employment in space manufacturing industry	97

Ireland	ENTERPRISE IRELAND where innovation means business
Population	4.725 million
GDP	€265.84 billion
Responsibility	Enterprise Ireland (EI) is the Irish Government's business development agency. EI manages industrial and technology development programmes relating to space activity on behalf of the Department of Jobs, Enterprise and Innovation (DJEI). EI promotes ESA and EU space programmes in Ireland, primarily to the industrial sector, as well as to the academic community and end-user organisations
Activities	ESA programmes (telecommunications (ARTES), earth observation, launchers, navigation, PRODEX, ELIPS and Technology (GSTP)); Horizon 2020 Space; EUMETSAT

(continued)

Budget	€27.3 million (ESA contribution, €23.3 million; and EUMETSAT, €4.0 million)
Staff	740 (Enterprise Ireland)
Direct employment in space manufacturing industry	61

Italy	
Population	60.666 million
GDP	1,672.44 billion Euro
Responsibility	The Italian Space Agency, Agenzia Spaziale Italiana (ASI), defines, coordinates and manages national space programmes and the Italian participation to European and international space projects on behalf of the Ministry of Education and University and Research (MIUR) and in coordination with the Ministry for the International Affairs. Other ministries, such as the Ministry of Economic Development, the Ministry of Defence and the Ministry of Transport and Infrastructures on navigation, support the space sector in agreement with ASI
Activities	ASI proposes its national strategy within its Strategic Vision Document and coordinates the realisation of competitive programmes and infrastructures in order to sustain intellectual and industrial growth through three main programmatic lines: science in space, technology for space and services from space. Its main investments, at national and international level, are made in Earth observation, launchers, science/exploration of the Universe and telecommunication fields. Another important aspect for Italy in space is education and outreach. For this reason ASI is involved in communication activities and spread of knowledge both in schools and universities
Budget	€854.4 million (ESA contribution, €512.0 million; National Space Budget, €294.6 million; and EUMETSAT, €47.8 million)
Staff	250 (ASI)
Direct employment in space manufacturing industry	4963

Luxembourg	
Population	576.25 thousand
GDP	54.20 billion Euro
Responsibility	The Ministry of the Economy, Directorate of ICT & Space Affairs, is in charge of space affairs in Luxembourg. Luxinnovation, the National Agency for Innovation and Research, manages the interface between industry, the public research sector and ESA. And the meteorological service, department of the Ministry for Sustainable Development and Infrastructure, represents Luxembourg in the EUMETSAT Council

(continued)

Activities	Technology R&D activities in Luxembourg are mainly focused on future telecommunication applications. Until now, these activities have been funded through the ARTES program. The adhesion of Luxembourg to ESA will offer new opportunities for supporting technology R&D activities. Luxembourg is active in telecommunications, Earth observation, navigation, security and space technologies
Budget	€70.5 million (ESA contribution, €22.0 million; National Space Budget, €47.6 million; and EUMETSAT, €0.9 million)
Staff	5 (3 Ministry of the Economy + 2 Luxinnovation)
Direct employment in space manufacturing industry	30

The Netherlands	SRON Netherlands Institute for Space Research / Netherlands Space Office
Population	16.979 million
GDP	697.22 billion Euro
Responsibility	The Netherlands Space Office (NSO) acts as the Dutch agency for space affairs. The NSO was established by the Dutch government in order to develop the Netherlands' space programme and to bring that programme to action. The NSO is the front office of the Dutch space community for international space organisations like ESA, NASA and JAXA as well as the central point of contact for the space community within the Netherlands. The NSO also works to innovatively and openly bring the story of spaceflight science, usage and exploration to teachers, students and the general public. The director of the NSO reports to the steering committee of the Ministry of Economic Affairs; Ministry of Education, Culture and Science; Ministry of Transport, Public Works and Water Management; and the Netherlands Organisation for Scientific Research (NWO)
Activities	In the ESA context, the Netherlands participate in all major programmes with a special interest in earth observation and environmental measurements and in robotics. The Netherlands Institute for Space Research (SRON) develops and uses innovative technology for ground-breaking research in space, focusing on astrophysical research, Earth science and planetary research. In addition to this, SRON has a line of research into new and more sensitive sensors for X-rays and infrared radiation. Space policy in the Netherlands is primarily focused on international cooperation in European contexts within the European Space Agency, the European Organisation for the Exploitation of Meteorological Satellites (EUMETSAT) and the European Union. The Netherlands also has a national programme guided by the ministries of Economic Affairs; of Education, Culture and Science; and of Traffic, Public Works and Water Management
Budget	€143.8 million (ESA contribution, €102.6 million; National Space Budget, €22.7 million; and EUMETSAT, €18.5 million)
Staff	26 (NSO)
	Approx. 200 (SRON)
Direct employment in space manufacturing industry	965

Norway	Norsk Romsenter NORWEGIAN SPACE CENTRE
Population	5.214 million
GDP	€334.94 billion
Responsibility	The Norwegian Space Centre (NSC), under the Ministry of Trade, Industry and Fisheries, is responsible for organising Norwegian space activities, particularly with respect to ESA and the EU, and for coordinating national space activities
Activities	In addition to ESA programmes (in particular Earth observation, telecommunications and launchers), Norway has national support programmes and commercial activities (Telenor). Moreover, Norway operates the Andøya rocket range and the Svalbard and Antarctica ground stations. Norway has also a bilateral agreement with Canada on the use of Radarsat 2 data
Budget	€96.8 million (ESA contribution, €59.6 million; National Space Budget, €26.1 million; and EUMETSAT, €11.1 million)
Staff	NSC: 40
Direct employment in space manufacturing industry	364

Poland	POLISH SPACE AGENCY
Population	37.967 million
GDP	€424.27 billion
Responsibility	Polish space activities are under the leading responsibility of the Ministry of Economic Development, in cooperation with the Ministry of Science and Higher Education, the Ministry of National Defence, the Ministry of Foreign Affairs, the Ministry of Environment and other appropriate ministries. The Ministry of Development represents Poland in the European Space Agency and in the European Union institutions dealing with issues related to space sector
Activities	The recently established Polish Space Agency (POLSA) is the implementing entity for Polish space activities under the responsibility of the Ministry of Economy, Department of Industry. Prior to its establishment, research and development on space technology in Poland has been carried out mainly by the Space Research Centre (SRC) of the Polish Academy of Sciences (PAS). The five main SRC research areas are physics of the Sun, study of planets and small solar system bodies, interplanetary space physics and astrophysics, plasma physics, and planetary geodesy and geodynamics. The SRC is the body cooperating with ESA
Budget	€47.0 million (ESA contribution, €29.9 million; National Civil Space Budget, €6.7 million; and EUMETSAT, €10.4 million)
Direct employment in space manufacturing industry	213

Portugal	
Population	10.341 million
GDP	€184.93 billion
Responsibility	The management and coordination of space activities in Portugal is carried out by the Foundation for Science and Technology (FCT), mainly through its Space Office. The FCT is the national funding agency for scientific and technological projects acting under the responsibility of the Portuguese Ministry of Science, Technology and Higher Education. The Ministry of Science is coordinating the Portuguese membership to ESA and is also providing the funding to ESA's mandatory activities and several optional programmes
Activities	Technology research and development activities are conducted in the frame of ESA TRP, Exploration, ARTES, EOEP and GSTP programmes
Budget	€21.1 million (ESA contribution, €16.0 million and EUMETSAT, €5.1 million)
Direct employment in space manufacturing industry	175

Romania	
Population	19.760 million
GDP	€169.58 billion
Responsibility	The Romanian Space Agency (ROSA) is the coordinator of Romania's national and international space activities. ROSA is a public institution entirely self-funded, operating under the decisions of the Ministry of Education and Research – National Authority for Scientific Research and Innovation (ANCSI). As a coordinator of national space research and application programme, ROSA designs and coordinates the implementation of the National Space Programme. Following its objectives, the agency is authorised to establish research and development centres
Activities	Main areas of interest are space exploration, space applications, technology and security. On behalf of the government, ROSA is the national representative in the cooperative agreements with international organisations, such as the European Space Agency (ESA) and Committee on Space Research (COSPAR), as well as bilateral governmental agreements. Together with the Ministry of Foreign Affairs, ROSA is representing Romania in the sessions of the United Nations Committee on the Peaceful Uses of Outer Space (COPUOS) and its subcommittees
Budget	€44.3 million (ESA contribution, €26.1 million; National Civil Space Budget, €14.4 million; and EUMETSAT, €3.8 million)
Direct employment in space manufacturing industry	64

Spain	
Population	46.440 million
GDP	€1,113.85 billion
Responsibility	The Centre for the Development of Industrial Technology (CDTI), under the Ministry of Economy and Competitiveness, channels the funding and supports applications for national and international R&D&i projects of Spanish companies, including the Spanish space activities in coordination with the Ministry of Industry, Energy and Tourism
	The National Institute of Aerospace Technology (INTA) is the other important Spanish actor in the space field. INTA is a public research organisation specialised in aerospace research and technology development reporting to the Ministry of Defence. The INTA has four main action lines related to design, development, integration, verification and testing of small- and medium-sized platforms and payloads, ground segment and satellite tracking and operations
Activities	In addition to ESA and EUMETSAT programmes, Spain has several national space programmes in the field of Earth observation (SEOSAT/INGENIO), communication satellites (Hispasat, Amazonas), defence space systems (SPAINSAT, XTAR-EUR, HELIOS, SECOMSAT, Pleiades and SEOSAR/PAZ), small satellites, space exploration (MSL-REMS) and ground control stations
Budget	€264.9 million (ESA contribution, €152.0 million; National Space Budget, €82.5 million; and EUMETSAT, €30.4 million)
Staff	319 (CDTI)
Direct employment in space manufacturing industry	3359

Sweden	
Population	9.851 million
GDP	€462.06 billion
Responsibility	The Swedish National Space Board (SNSB), a central governmental agency under the Ministry of Higher Education and Research, is responsible for national and international activities relating to space and remote sensing, primarily research and development
	The Swedish space programme is carried out by means of extensive international cooperation, in particular through Sweden's membership of the European Space Agency (ESA). SNSB's responsibility for international activities includes the Swedish involvement in ESA as well as bilateral cooperation within space. Most of the activities funded by the Swedish National Space Board are carried out in cooperation with other countries

(continued)

Activities	The Swedish space programme is mostly carried out through international cooperation. Sweden has a high level of activities in the technological ESA programmes GSTP and ARTES. And nationally, special focus is put on on-board computers, telecom equipment, MEMS technology, separation systems and adapters, turbines, AOCS and EP systems, green propellant, formation-flying platforms and the Ariane nozzle
Budget	€110.3 million (ESA contribution, €73.9 million; National Civil Space Budget, €24.3 million; and EUMETSAT, €12.1 million)
Staff	19
Direct employment in space manufacturing industry	855

Switzerland	Schweizerische Eidgenossenschaft Confédération suisse Confederazione Svizzera Confederaziun svizra
Population	8.327 million
GDP	€596.13 billion
Responsibility	The Swiss Space Office (SSO) under the authority of the State Secretariat for Education, Research and Innovation (SERI) of the Federal Department of Economic Affairs, Education and Research (EAER) is the administrative body in charge of planning and implementing the Swiss space policy. The SSO chairs the Interdepartmental Committee for Space Affairs (IKAR) and acts as the executive secretariat for the Federal Commission for Space Affairs (CFAS), which advises the Federal Council on matters relating to a coherent and forward-looking space policy by taking into consideration the European and global development and the national interests (among others of political, application-oriented, scientific, technological and industrial nature). It also issues recommendations to the Federal Council regarding the implementation of this policy and advises the Federal Council in terms of the evaluation of proposals for projects with important scientific or technological relevance
Activities	The Swiss participation to ESA's PRODEX programme is an important bridge to experimental space research and has led to a remarkable improvement in the country's space research capabilities. Other ESA programmes include general budget, science programme and CSG, Earth observation, launchers, human spaceflight and exploration, technology and telecommunication and navigation
Budget	€169.0 million Euro (ESA contribution, €146.4 million; National Civil Space Budget, €7.8 million; and EUMETSAT, €14.8 million)
Direct employment in space manufacturing industry	821

United Kingdom	
Population	655.383 million
GDP	€2,366.91 billion
Responsibility	The UK Space Agency, an executive agency of the Department for Business, Innovation and Skills (BIS), and reporting to the Minister of State for Universities and Science, is responsible for the strategic decisions on all UK space activities. The UK Space Agency also interfaces with other departments on security and military programmes. Although it does not manage these programmes directly, the UK Space Agency is kept informed and involved in decisions relating to the programmes as necessary
Activities	The focus of UK space activities is on innovative technologies that have the potential to maximise public and commercial mission objectives. The UK is largely supporting space science, Earth observation, ARTES, Galileo MREP and the ELIPS programme and increasing participation on the ISS. Technologies are developed both by means of national activities funded through UK industry such as NSTP and by ESA's TRP, GSTP, CTP, ARTES and ETP programmes
Budget	€671.7 million (ESA contribution, €324.8 million; National Space Budget, €290.5 million; and EUMETSAT, €56.4 million)
Staff	50 (UKSA)
Direct employment in space manufacturing industry	3722

Chapter 17
Bibliography of Space Policy Publications: 2016

Cenan Al-Ekabi

17.1 Monographs

Barentine JC (2016) The Lost Constellations. A History of Obsolete, Extinct, or Forgotten Star Lore. Series: Springer Praxis Books. Springer, Cham
Barentine JC (2016) Uncharted Constellations. Asterisms, Single-Source and Rebrands. Series: Springer Praxis Books. Springer, Cham
Bernardi G (2016) The Unforgotten Sisters. Female Astronomers and Scientists before Caroline Herschel. Series: Springer Praxis Books. Springer, Cham
Biesbroek R (2016) Lunar and Interplanetary Trajectories. Series: Springer Praxis Books. Springer, Cham
Bignami G, Sommariva A (2016) The Future of Human Space Exploration. Palgrave Macmillan, United Kingdom
Burgess C (2016) Aurora 7. The Mercury Space Flight of M. Scott Carpenter. Series: Springer Praxis Books. Springer, Cham
Burgess C (2016) Faith 7. L. Gordon Cooper, Jr., and the Final Mercury Mission. Series: Springer Praxis Books. Springer, Cham
Burgess C (2016) Sigma 7. The Six Mercury Orbits of Walter M. Schirra, Jr. Series: Springer Praxis Books. Springer, Cham
Burgess C, Vis B (2016) Interkosmos. The Eastern Bloc's Early Space Program. Series: Springer Praxis Books. Springer, Cham
Cockell CS (ed) (2016) Dissent, Revolution and Liberty Beyond Earth. Springer, New York
Conway E (2016) Exploration and Engineering: The Jet Propulsion Laboratory and the Quest for Mars. Johns Hopkins University Press, Maryland
David L (2016) Mars. Our Future on the Red Planet. National Geographic, Washington D.C.

C. Al-Ekabi (✉)
European Space Policy Institute (ESPI), Vienna, Austria
e-mail: Cenan.Al-Ekabi@espi.or.at

C. Al-Ekabi, S. Ferretti (eds.), *Yearbook on Space Policy 2016*,
Yearbook on Space Policy, https://doi.org/10.1007/978-3-319-72465-2_17

Davies JK (2016) The Life Story of an Infrared Telescope. Series: Springer Praxis Books. Springer, Cham
Dunn BD (2016) Materials and Processes for Spacecraft and High Reliability Applications. Series: Springer Praxis Books. Springer, Cham
Dyson MJ (2016) A Passion for Space. Adventures of a Pioneering Female NASA Flight Controller. Series: Springer Praxis Books. Springer, Cham
Ellery A (2016) Planetary Rovers. Robotic Exploration of the Solar System. Series: Springer Praxis Books. Springer, Cham
Guthrie J (2016) How to Make a Spaceship: A Band of Renegades, an Epic Race, and the Birth of Private Spaceflight. Penguin Press, United Kingdom
Häuplik-Meusburger S, Bannova O (2016) Space Architecture Education for Engineers and Architects. Springer, New York
Kokhanovsky A (2016) Light Scattering Reviews 10. Light Scattering and Radiative Transfer. Series: Springer Praxis Books. Springer, Cham
Kokhanovsky A (2016) Light Scattering Reviews, Volume 11. Light Scattering and Radiative Transfer. Series: Springer Praxis Books. Springer, Cham
Maciel WJ (2016) Introduction to Stellar Structure. Series: Springer Praxis Books. Springer, Cham
Massimino M (2016) Spaceman: An Astronaut's Unlikely Journey to Unlock the Secrets of the Universe. Crown Archetype, Massachusetts
Miller B (2016) The Aliens are Coming! The Extraordinary Science Behind Our Search for Life in the Universe. The Experiment Publishing, New York
Murdin P (2016) Rock Legends. The Asteroids and Their Discoverers. Series: Springer Praxis Books. Springer, Cham
Onstott, T (2016) Deep Life: The Hunt for the Hidden Biology of Earth, Mars, and Beyond. Princeton University Press, New Jersey
O'Sullivan J (2016) In the Footsteps of Columbus. European Missions to the International Space Station. Series: Springer Praxis Books. Springer, Cham
Pellegrino M, Stang G (2016) Space security for Europe. Institute for Security Studies, France
Pirard T, Cosyn P (2016) Aspiring Space Nations. Small Programs with Large Goals. Praxis, United Kingdom
Preston L (2016) Goldilocks and the Water Bears: The Search for Life in the Universe. Bloomsbury Sigma, New York
Rapp D (2016) Human Missions to Mars. Enabling Technologies for Exploring the Red Planet. Series: Springer Praxis Books. Springer, Cham
Schwartz JSJ, Milligan T (eds) (2016) The Ethics of Space Exploration. Springer, New York
Seedhouse E (2016) Mars via the Moon. The Next Giant Leap. Series: Springer Praxis Books. Springer, Cham
Seedhouse E (2016) SpaceX's Dragon: America's Next Generation Spacecraft. Series: Springer Praxis Books. Springer, Cham
Seedhouse E (2016) XCOR, Developing the Next Generation Spaceplane. Series: Springer Praxis Books. Springer, Cham

Shayler DJ, Harland DM (2016) Enhancing Hubble's Vision. Service Missions That Expanded Our View of the Universe. Series: Springer Praxis Books. Springer, Cham

Shayler DJ, Harland DM (2016) The Hubble Space Telescope. From Concept to Success. Series: Springer Praxis Books. Springer, Cham

Shetterly ML (2016) Hidden Figures: The American Dream and the Untold Story of the Black Women Mathematicians Who Helped Win the Space Race. Morrow, New York

Sobel D (2016) The Glass Universe. How the Ladies of the Harvard Observatory Took the Measure of the Stars. Viking Press, New York

Squeri L (2016) Waiting for Contact: The Search for Extraterrestrial Intelligence. University of Florida, Florida

Sterns PM, Tennen LI (eds) (2016) Private Law, Public Law, Metalaw and Public Policy in Space. Springer, New York

Thompson M (2016) A Space Traveler's Guide to the Solar System. Pegasus, United Kingdom

Tyson NdG, Simons J, Liu C (eds) (2016) StarTalk. Everything You Ever Need to Know About Space Travel, Sci-Fi, the Human Race, the Universe, and Beyond. National Geographic, Washington D.C.

Tyson NdG, Strauss M, Gott JR (2016) Welcome to the Universe: An Astrophysical Tour. Princeton University Press, New Jersey

von Ehrenfried D (2016) The Birth of NASA. The Work of the Space Task Group, America's First True Space Pioneers. Series: Springer Praxis Books. Springer, Cham

Willis J (2016) All These Worlds Are Yours: The Scientific Search for Alien Life. Yale University Press, Connecticut

Wohlforth C, Hendrix A (2016) Beyond Earth: Our Path to a New Home in the Planets. Pantheon Books, New York

17.2 Articles

Adamo DR, Logan JS (2016) Aquarius, a reusable water-based interplanetary human spaceflight transport. Acta Astronautica 128:160–179

Aganaba-Jeanty T (2016) Introducing the Cosmopolitan Approaches to International Law (CAIL) lens to analyze governance issues as they affect emerging and aspirant space actors. Space Policy 37:3–11

Aganaba-Jeanty T (2016) Overcoming the danger of a single story of space actors: Introducing the Cosmopolitan Approaches to International Law (CAIL) Lens to Analyze Global Space Governance. Space Policy 35:15–23

Aleina SC et al. (2016) Reusable space tug concept and mission. Acta Astronautica 128:21–32

Antoni N (2016) Book Review: Small Satellites – Regulatory Challenges and Chances, by Irmgrad Marboe (Brill | Nijhoff 2016). Air & Space Law 41:554–558

Balint TS, Stevens J (2016) Wicked problems in space technology development at NASA. Acta Astronautica 118:96–108

Baruah R (2016) WTO and Space Activities: A Legal Assessment on Liberalization of the Space Industry. German Journal of Air and Space Law 65:375–385

Blamont J (2016) We the people: Consequences of the revolution in the management of space applications. Space Policy 37:120–126

Bombardelli C et al. (2016) Mission analysis for the ion beam deflection of fictitious asteroid 2015 PDC. Acta Astronautica 118:296–307

Borowitz M, Battat J (2016) Multidisciplinary evaluation of next steps for human space exploration: Technical and strategic analysis of options. Space Policy 35:33–42

Brown F (2016) Space – Obstacles and opportunities: The 2015 Canada–UK colloquium. Space Policy 35:24–26

Bruhns S, Haqq-Misra J (2016) A pragmatic approach to sovereignty on Mars. Space Policy 38:57–63

Carpenter J et al. (2016) Establishing lunar resource viability. Space Policy 37:52–57

Chen Y (2016) China's space policy-a historical review. Space Policy 37:171–178

Crawford IA (2016) Introduction to the Special Issue on using extraterrestrial resources to facilitate space science and exploration. Space Policy 37:51

Crawford IA (2016) The long-term scientific benefits of a space economy. Space Policy 37:58–61

Danilenko GM (2016) International law-making for outer space. Space Policy 37:179–183

de Magalhães JP (2016) A direct communication proposal to test the Zoo Hypothesis. Space Policy 38:22–26

Dede G, Akçay M (2016) Technology foresight in practice: A proposal for Turkish space vision. Space Policy 35:1–5

Delgado LM (2016) When inspiration fails to inspire: A change of strategy for the US space program. Space Policy 37:190–194

Denis G (2016) The evolution of Earth Observation satellites in Europe and its impact on the performance of emergency response services. Acta Astronautica 127:619–633

Dennerley JA (2016) Emerging space nations and the development of international regulatory regimes. Space Policy 35:27–32

Di Pippo S (2016) Registration of Space Objects with the United Nations Secretary-General. German Journal of Air and Space Law 65:364–374

Drube L et al. (2016) The NEOTωIST mission (Near-Earth Object Transfer of angular momentum spin test). Acta Astronautica 127:103–111

Eerme T (2016) Indirect industrial effects from space investments. Space Policy 38:12–21

Elvis M (2016) What can space resources do for astronomy and planetary science? Space Policy 37:65–76

Elvis M et al. (2016) The peaks of eternal light: A near-term property issue on the moon. Space Policy 38:30–38

Eubanks TM, Radley CF (2016) Scientific return of a lunar elevator. Space Policy 37:97–102

Fan Y (2016) Latecomer's strategy: An assessment of BDS industrialization policy. Space Policy 38:1–11

Forshaw JL et al. (2016) RemoveDEBRIS: An in-orbit active debris removal demonstration mission. Acta Astronautica 127:448–463

Goemaere S (2016) Gaining deeper insight into the psychological challenges of human spaceflight: The role of motivational dynamics. Acta Astronautica 121:130–143

Gonzales AA, Stoker CR (2016) An efficient approach for Mars Sample Return using emerging commercial capabilities. Acta Astronautica 123:16–25

Guerra AGC (2016) On small satellites for oceanography: A survey. Acta Astronautica 127:404–423

Hall JL (2016) Columbia and Challenger: Organizational failure at NASA. Space Policy 37:127–133

Hameed H, Stefoudi D (2016) Report on the Symposium on Legal Aspects of Space Resource Utilization: IIASL, Leiden University, 17 April 2016. Air & Space Law 41:387–393

Harding RC (2016) Conclusions: Space policy in developing countries. Space Policy 37:48–49

Harding RC (2016) Introduction: Space policy in developing countries. Space Policy 37:1–2

Hastings DE (2016) When will on-orbit servicing be part of the space enterprise? Acta Astronautica 127:655–666

Haya-Ramos R et al. (2016) The design and realisation of the IXV Mission Analysis and Flight Mechanics. Acta Astronautica 124:39–52

Hempsell M et al. (2016) A Business Analysis of a SKYLON-based European Launch Service Operator. Acta Astronautica 121:1–12

Henri Y (2016) ITU World Radiocommunication Conference (WRC-15) Allocates Spectrum for Future Innovation. Air & Space Law 41:119–128

Hilborne MP (2016) China. Space Policy 37:39–45

Ho K et al. (2016) Campaign-level dynamic network modelling for spaceflight logistics for the flexible path concept. Acta Astronautica 123:51–61

Hobe S (2016) Space Traffic Management – Some Conceptual Ideas –. German Journal of Air and Space Law 65:3–21

Hobe S (2016) The International Institute of Space Law adopts Position Paper on Space Resource Mining. German Journal of Air and Space Law 65:204–209

Hulsroj P, Pecujlic AN (2016) New in the Nest: The Danish Space Act. Air & Space Law 41:503–510

Hussein A et al. (2016) From detection to deflection: Mitigation techniques for hidden global threats of natural space objects with short warning time. Acta Astronautica 126:488–496
Kansakar P, Hossain F (2016) A review of applications of satellite earth observation data for global societal benefit and stewardship of planet earth. Space Policy 36:46–54
Kazansky Y et al. (2016) The current and potential role of satellite remote sensing in the campaign against malaria. Acta Astronautica 121:292–305
Kelso A (2016) UK space policy and the politics of parliamentary debate. Space Policy 35:43–46
Kempf S et al. (2016) Simplified spacecraft vulnerability assessments at component level in early design phase at the European Space Agency's Concurrent Design Facility. Acta Astronautica 129: 291–298
Ketcham C (2016) Towards an ethics of life. Space Policy 38:48–56
Kitts C, Rasay M (2016) A university-based distributed satellite mission control network for operating professional space missions. Acta Astronautica 120:229–238
Kojima A (2016) To ignite the passion in children's hearts – Role and effect of space education, issues and consideration. Acta Astronautica 127:614–618
Lefeber R (2016) Relaunching the Moon Agreement. Air & Space Law 41:41–48
Lehnert C et al. (2016) The common objectives of the European Nordic countries and the role of space. Acta Astronautica 128:640–649
Liu H, Tronchetti F (2016) United Nations Resolution 69/32 on the 'No first placement of weapons in space': A step forward in the prevention of an arms race in outer space? Space Policy 38:64–67
Long J (2016) China's space station project and international cooperation: Potential models of jurisdiction and selected legal issues. Space Policy 36:28–37
López LD (2016) Space sustainability approaches of emerging space nations: Brazil, Colombia, and Mexico. Space Policy 37:24–29
Martinez P (2016) The development and initial implementation of South Africa's national space policy. Space Policy 37:30–38
Mayer R (2016) Beyond the Blue Marble: Artistic research on space and ecology. Acta Astronautica 128:573–579
McInnes CR (2016) Near Earth asteroid resource utilisation for large in-orbit reflectors. Space Policy 37:62–64
McKenna-Lawlor S et al. (2016) Performance of the mission critical Electrical Support System (ESS) which handled communications and data transfer between the Rosetta Orbiter and its Lander Philae while en route to and at comet 67P/Churyumov-Gerasimenko. Acta Astronautica 125:118–136
Metzger PT (2016) Space development and space science together, an historic opportunity. Space Policy 37:77–91
Michielsen P (2016) The Belgian Space Act: An Innovative Legal Safeguard to Boost the Space Industry. Air & Space Law 41:89–117
Miller TF (2016) Combustion-based power source for Venus surface missions. Acta Astronautica 127:197–208

Monzon A. et al. (2016) Methods for promoting knowledge exchange and networking among young professionals in the aerospace sector—IAF's IPMC workshop 2013 insights. Acta Astronautica 118:123–129

Morozova E (2016) All-Electric Satellites Seek Equal Rights in Space. Air & Space Law 41:193–199

Murray B (2016) Can space exploration survive the end of the Cold War? Space Policy 37:184–189

Myers PL, Spencer DB (2016) Application of a multi-objective evolutionary algorithm to the spacecraft stationkeeping problem. Acta Astronautica 127:76–86

Nagendra NP (2016) Diversification of the Indian space programme in the past decade: Perspectives on implications and challenges. Space Policy 36:38–45

Nagendra NP, Basu P (2016) Demystifying space business in India and issues for the development of a globally competitive private space industry. Space Policy 36:1–11

Nair KK (2016) Near Earth Asteroid Defence – Examining the Need, Organisation and Legal Options. German Journal of Air and Space Law 65:210–225

Ndiritu M et al. (2016) Kiswahili translation on the scientific and space-related terminology. Acta Astronautica 128:330–334

Pace S (2016) Space cooperation among order-building powers. Space Policy 36:24–27

Paikowsky D et al. (2016) Trends in space activities in 2014: The significance of the space activities of governments. Acta Astronautica 118:187–198

Payson D et al. (2016) The Mishin Diaries, a new significant primary source of space history information. Acta Astronautica 123:192–199

Perakis N (2016) Project Dragonfly: A feasibility study of interstellar travel using laser-powered light sail propulsion. Acta Astronautica 129:316–324

Peter N (2016) The changing geopolitics of space activities. Space Policy 37:145–153

Peters S et al. (2016) Mission concept and autonomy considerations for active Debris removal. Acta Astronautica 129:410–418

Petroni G, Bianchi DG (2016) New patterns of space policy in the post-Cold War world. Space Policy 37:12–19

Polansky JL, Cho M (2016) A university-based model for space-related capacity building in emerging countries. Space Policy 36:19–23

Pyne SJ (2016) Space: a Third Great Age of Discovery. Space Policy 37:113–119

Robinson J (2016) Transparency and confidence-building measures for space security. Space Policy 37:134–144

Rosher P, Shaw A (2016) Micro Satellites: The Smaller the Satellites, the Bigger the Challenges? Air & Space Law 41:311–328

Rovetto RJ (2016) Defending spaceflight – The echoes of Apollo. Space Policy 38:68–78

Rumpf C (2016) On the influence of impact effect modelling for global asteroid impact risk distribution. Acta Astronautica 123:165–170

Schrogl K-U, Summerer L (2016) Climate engineering and space. Acta Astronautica 129:121–129

Secara T, Bruston J (2016) Current barriers and factors of success in the diffusion of satellite services in Europe. Space Policy 37:154–161
Sheehan M (2016) Viewpoint: Space security and developing nations. Space Policy 37:20–23
Slann PA (2016) Anticipating uncertainty: The security of European critical outer space infrastructures. Space Policy 35:6–14
Smart J, Benaroya H (2016) An examination of non-linear and passive technology transfer in the space sector: Consideration of the Contingent Effectiveness Model as a basis for formal modelling. Space Policy 38:39–47
Smirnov NN (2016) Space Flight Safety – Discussing perspectives. Acta Astronautica 126:497–499
Smith CM (2016) An adaptive paradigm for human space settlement, Acta Astronautica 119:207–217
Smith LJ, Doldirina C (2016) Remote sensing: A case for moving space data towards the public good. Space Policy 37:162–170
Snelgrove KB, Saleh JH (2016) Toward a new spacecraft optimal design lifetime? Impact of marginal cost of durability and reduced launch price. Acta Astronautica 127:271–282
Soucek A (2016) Legal and Practical Questions in Applying Articles II and IV of the Registration Convention. German Journal of Air and Space Law 65:22–43
Sowers GF (2016) A cislunar transportation system fueled by lunar resources. Space Policy 37:103–109
Srivastava V et al. (2016) Microwave processing of lunar soil for supporting longer-term surface exploration on the Moon. Space Policy 37:92–96
Succa M et al. (2016) IXV avionics architecture: Design, qualification and mission results. Acta Astronautica 124:67–78
Supancana IBR (2016) The Legal Challenges of Implementing National Space Legislation – The Case of Indonesia. German Journal of Air and Space Law 65:226–238
Szocik K et al. (2016) Political and legal challenges in a Mars colony, Space Policy 38:27–29
Trimarchi A (2016) International Space Station – A Focus on Intellectual Property Rights – Main Emphasis on the ESA Perspective. German Journal of Air and Space Law 65:533–550
Tronchetti F (2016) Title IV – Space Resource Exploration and Utilization of the US Commercial Space Launch Competitiveness Act: A Legal and Political Assessment. Air & Space Law 41:143–156
Tsuda Y (2016) Watanabe, Flight status of robotic asteroid sample return mission Hayabusa2. Acta Astronautica 127:702–709
Tumino G, Mancuso S (2016) The success of the ESA Intermediate eXperimental Vehicle program. Acta Astronautica 124:1
Uhlig T et al. (2016) ISS emergency scenarios and a virtual training simulator for Flight Controllers. Acta Astronautica 128:513–520

van Wynsberghe E, Turak A (2016) Station-keeping of a high-altitude balloon with electric propulsion and wireless power transmission: A concept study. Acta Astronautica 128:616–627

Vigneron AC et al. (2016) Nonlinear filtering for autonomous navigation of spacecraft in highly elliptical orbit. Acta Astronautica 126:138–149

Ward ED et al. (2016) A method to evaluate utility for architectural comparisons for a campaign to explore the surface of Mars. Acta Astronautica 128:237–242

Weeden B (2016) The Iranian space endeavor: Ambitions and reality. Space Policy 37:46–47

Wheeler J (2016) The Consequences Post Referendum for the UK Satellite and Space Industry. Air & Space Law 41:445–457

Winter FH (2016) Did the Germans learn from Goddard? An examination of whether the rocketry of R.H. Goddard influenced German Pre-World-War II missile development. Acta Astronautica 127:514–525

Zhao Y (2016) The Role of bilateral and multilateral agreements in international space cooperation. Space Policy 36:12–18

About the Editors and Contributors

Editors

Cenan Al-Ekabi is a resident fellow at the European Space Policy Institute (ESPI) in Vienna, Austria. He joined ESPI in 2011, after completing two advanced studies LL.M. degrees in Air and Space Law and in European and International Business Law from Leiden University in the Netherlands; and he recently participated in the ISU Space Studies Program in 2016. He is a specialist in current and historical evolutions in the civil, military, industry and commercial space domains and has conducted macroeconomic and microeconomic assessments for ESPI's stakeholders in the European space sector. He is responsible for the management and production of ESPI's Yearbook on Space Policy book series and its Space Policies, Issues and Trends report series, including their related databases. His other projects at ESPI have addressed the stakes for European commercial spacecraft industry competitiveness, challenges for Europe to maintain autonomous access to space and measuring the future benefits of space exploration.

Stefano Ferretti is an ESA space policy officer seconded to the European Space Policy Institute (ESPI), as resident fellow since 2015, where he deals with governance, innovation and future space-based services. He is in charge of "Space for sustainable development" and the ESPI activities towards UNISPACE+50 while also addressing high-level policy across several areas with the European Union. Previously he worked as energy and infrastructure manager at ESA/ESRIN, and as International Space Station project manager at ESA/ESTEC, coordinating with international space agencies. At Thales Alenia Space, he was manager on ISS Node3 development activities. He carried out academic and industrial research at NASA, and during ESA parabolic flight campaigns, for which he received the International Astronautical Federation Napolitano Award in 2002. He holds a PhD, with a dissertation on Innovative Technologies for Space Habitats, a master in mechanical

C. Al-Ekabi, S. Ferretti (eds.), *Yearbook on Space Policy 2016*,
Yearbook on Space Policy, https://doi.org/10.1007/978-3-319-72465-2

engineering from the University of Bologna and a master of space studies from the International Space University. He attended executive programmes in space policy and law, innovation and entrepreneurship and leadership, at George Washington University and MIT.

Contributors

Josef Aschbacher is the ESA director of Earth observation programmes and head of ESRIN, ESA's centre for Earth observation, located in Frascati, Rome. Born in Austria, he studied at the University of Innsbruck, graduating with a master's and a doctoral degree in natural sciences. His professional career in ESA began in 1990 as a young graduate at ESRIN. From 1991 to 1993, he was seconded as ESA representative to Southeast Asia to the Asian Institute of Technology in Bangkok, Thailand. From 1994 to 2001, he was the scientific assistant to the director of the Space Applications Institute at the European Commission Joint Research Centre, Ispra, Italy. He returned to ESA HQ (Paris) in 2001 as programme coordinator, primarily responsible for advancing Copernicus activities within ESA. In 2006, he was nominated head of the Copernicus Space Office, leading all activities for Copernicus within the agency and with external partners, in particular the European Commission. In 2014, he was promoted to head of Programme Planning and Coordination, where he was responsible for planning ESA's Earth observation programmes and for formulating and implementing programmatic and strategic decisions across the directorate.

Leonora Barclay is communications coordinator at SOS Children's Villages International where she has worked since 2014. Previously she was publications PR and media assistant at the Cardiovascular and Interventional Radiological Society of Europe. She holds a master of science degree in gender and international relations and a bachelor of arts in philosophy and German degree from the University of Bristol.

Bhupendra Singh Bhatia is the programme director (space education), at the Vikram A Sarabhai Community Science Centre, Ahmedabad. He is also the former director of Development and Educational Communication Unit (DECU), Indian Space Research Organisation (ISRO). Born in 1944, Sri Bhupendra Singh Bhatia held numerous key positions in the Indian Space Research Organisation between 1971 and 2007. This included working as director of the DECU and as project director for EDUSAT, the first dedicated "educational satellite" that provided satellite-based two-way communication to classrooms for delivering educational materials in India. He has also been a consistent and passionate supporter of telemedicine projects in India, which have helped to connect over a hundred rural medical centres to urban specialty centres.

Clio Biondi Santi is a former trainee at the European Space Agency (ESA), Paris, under the supervision of Simonetta Cheli in the Coordination Office of Earth observation directorate. She has been an intern at DB (Deutsche Bank) in Geneva in the sector of private wealth management, and afterwards she worked as an international relations assistant for the deputy-secretary of State of the Ministry of Justice in Rome (Italy). She is now doing a master's degree in corporate finance (MCP) at Bocconi University.

Petra Dannecker is professor and head of the Department of Development Studies at the University of Vienna. She did her PhD in sociology (Bielefeld University, Germany) where, until 2007, she was an assistant professor and lecturer. Then she joined the German Development Institute in Bonn as a senior research fellow responsible for coordinating research and knowledge transfer between the Federal Ministry for Economic Cooperation and Development (BMZ) and scientific communities focusing on development research in Germany. In October 2008, she became a visiting professor for global studies and development sociology at the Department of Sociology at the University of Vienna and in 2011 a full professor for development sociology. She is working on issues related to development sociology, development politics, globalization and migration processes and gender, focusing regionally on South and Southeast Asia.

Simonetta Di Pippo is the director of the United Nations Office for Outer Space Affairs (UNOOSA), which is mandated to enhance international cooperation in space activities to promote their use for humanity. Prior to joining UNOOSA in 2014, she was the head of the European Space Policy Observatory at Agenzia Spaziale Italiana (ASI) in Brussels. Ms. Di Pippo also served as director of Human Spaceflight of the European Space Agency from 2008 to 2011 and director of the Observation of the Universe at ASI from 2002 to 2008, where she started her career in 1986. She has been an academician of the International Academy of Astronautics (IAA) since 2013 and since 2016 a member of the IAA Board of Trustees. Ms. Di Pippo holds a master's degree in astrophysics and space physics from "La Sapienza" University in Rome and an honoris causa degree in environmental studies from the St. John University in Vinovo. Ms. Di Pippo was knighted by the president of the Italian Republic in 2006. In 2008, the International Astronomical Union named asteroid 21887 "dipippo" in honour of her contribution to this field.

Ashley Dara Dotz holds a BS in industrial design with a minor in international business from MSCD. She combines her design strategy experience and rapid ethnographic assessment to empower others through technology. With a focus on disrupting supply chains and "making" in austere environments, she has cofounded Field Ready, focusing in on-demand manufacturing in post-disaster zones. Recently in Haiti, Field Ready brought 3D printers to teach locals how to make needed medical supplies and replacement parts for rural clinics. In the Bay Area, Ms. Dotz also works as human factors lead for Made In Space, Inc., a company that built and operates the first gravity-independent 3D printer on the International Space Station. In her free time, she flies drones and is attempting to break the space-time continuum. She believes that 3D printing in space is just the beginning.

Jennifer Frankel-Reed is a senior climate change specialist at the US Agency for International Development, where she provides technical assistance on climate change adaptation to USAID's regional and country missions and coordinates the USAID and NASA SERVIR programme. She has been a climate change consultant for leading international organizations including IRG, GIZ, GEF and the United Nations Development Programme in the United States and abroad. Jennifer holds an MEM in global change science and policy from Yale University and a BS in environmental science from Willamette University.

Petra Füreder is holding a master's degree in applied geoinformatics and is working as researcher at the Department of Geoinformatics (Z_GIS), University of Salzburg, since 2008. She has been involved in a range of national and EU-funded research projects in the field of human security, environmental monitoring and emergency response. Since 2011 her work is mainly concentrating on providing EO-based information products of refugee and IDP camps to Médecins Sans Frontières (MSF). Her research interests include development of automated image classification methods using object-based image analysis (OBIA), cartography and visualization.

Daniel E. Irwin is a research scientist at the NASA Marshall Space Flight Center and the NASA coordinator for the SERVIR programme. He has over 20 years of experience in satellite remote sensing applications and geographic information systems (GIS) in the developing world. He has also served as a lead remote sensing specialist in support of NASA's space archaeology program, with a focus on detecting spectral signatures of structures of the ancient Maya. Prior to joining NASA, Daniel lived and worked in Guatemala, developing leading GIS laboratories for conservation organizations and the Guatemalan government. Daniel holds an MS in environmental science from Miami University.

Stefan Lang (PhD) is the head of the Z_GIS Division Integrated Spatial Analysis (ISA) and research coordinator at Z_GIS. His research focuses on object-based image analysis (OBIA), systemic and hierarchical scale concepts, multisource data integration, multidimensional and multi-scale analysis, validation and accuracy. In several bilateral R&D agreements, Dr. Lang and his team have developed a humanitarian information service using latest geospatial technologies, specifically supporting NGOs active in humanitarian assistance and disaster response, including Doctors without Borders (MSF), SOS Children's Villages and the Red Cross Movement. He has organized a number of workshops and special sessions on the use of Earth observation (EO) technology in specific domains (e.g. EO4Hum, EO4Hab) and launched the series of GEOBIA conferences in 2006. Next to authoring or co-authoring numerous scientific articles, he also edited special issues in international journals and published a textbook on landscape analysis available in several languages. Dr. Lang is co-founder of the Z_GIS spin-off Spatial Services GmbH and responsible for IPR management between Z_GIS and the company.

Christine Leurquin is the vice president of Institutional Relations and Communications, at SES S.A., positioning the company within the European Union and other European institutions, such as ESA and OECD. She reviews regulations, provides political support and engages in setting up research and development and institutional projects. She follows the European institutions' political issues and debates of interest to her company, especially those concerning crisis management, satellite broadband and the mobile multimedia sector. She is a steering board member of Networld2020 and the 5G IA Association. She is also the key person in the company as regards the GNSS projects such as Galileo and Egnos. She has 30 years of experience in the satellite telecommunications field in Europe. In her previous position, she was the CEO of a company that she started which introduced new satellite services in Belgium. She turned it into a successful and profitable business. With her ability to speak several languages, Mrs. Leurquin has been able to market in areas where services are not yet established, and she provides SES with that extra incentive to attract clients in those regions.

Justin Loiseau is a senior policy associate at JPAL, where he supports J-PAL's Crime and Finance sectors and is the lead regional liaison to the Africa office. As a member of the policy group, Mr. Loiseau works with NGOs, foundations, governments and others to cultivate research partnerships and ensure that programme and policy decision-making is informed by scientific evidence. He is the author of J-PAL publications on microcredit and early childhood stimulation. Justin holds an MA in economics from the University of Auckland and dual BAs in economics and environmental studies from the University of North Carolina at Chapel Hill.

Andreas Papp is international director of Emergency Response at SOS Children's Villages International (SOS-CVI) since March 2015. His professional engagement for humanitarian aid began with Médecins Sans Frontières (MSF) in 2004, where he worked as logistics administrator and later coordinated missions in complex emergency settings. He then served as MSF-Austria's delegate and programme director, where his tasks included the initiation and implementation of space technology projects to support field operations. In his current position at SOS-CVI, Andreas Papp oversees emergency response activities in more than two dozen countries, fostering emergency response capacity building through the use of new technologies.

Wolfgang Rathgeber is an electrical engineer by education. He started his career as a doctoral student at the German Aerospace Center (DLR), and he obtained his PhD degree at the University of Karlsruhe with a thesis on remote sensing. He went on several secondments from DLR, including the European Space Policy Institute (ESPI) in Vienna and European Space Agency (ESA) Headquarters in Paris. He then joined ESA as a programme coordinator in the directorate of Earth observation programmes in Frascati, Italy.

Edith Rogenhofer started to work in the humanitarian field in 1991 and since 1998 has been with MSF. She has mainly worked in projects in Africa, with a few stints in Afghanistan, Pakistan, Bangladesh and Myanmar. Most of the projects have been carried out during complex emergencies, focusing on water supply and sanitation in camp settings and rural areas. She holds an MSc in water management as well as a diploma in drilling. She has been contracted since 2011 by MSF-Austria as WatSan consultant in the Delegate and Programme Department with the specific task of developing and integrating GIS applications in MSFs work and liaising with the projects. Within the EO4HumEn project, she coordinates work related to user requirements and user validation.

Alfredo Roma is an economist, now consultant for the aerospace industry, and he is a former member of the Advisory Council of ESPI in Vienna. He has been member of the RPAS steering committee, created by the European Commission for the integration of civil drones into the common airspace, and coordinator of the Group of Experts on social impact. He has been national coordinator for the Galileo Project at the Italian prime minister's cabinet and Italian delegate at the European Space Agency. From September 1998 to July 2003, he was chairman of ENAC, the Italian Civil Aviation Authority and, from June 2000 to July 2003, also president of ECAC (European Civil Aviation Conference). He was a member of the High Level Group formed by Loyola De Palacio, vice president of the European Commission, for the Single Sky. From 1992 to 1998, he was CEO of ANSA. From 1975 to 1992, he was an untenured professor of Business Finance, Faculty of Economics, University of Modena.

Alessandra Vernile is a former resident fellow at European Space Policy Institute (ESPI), with a secondment from the Italian Space Agency (ASI) and the Italian Society for International Organization. Previously she worked as intern at the Strategy Department at the European Space Agency HQ in Paris and as event manager at the NATO Defence College Foundation in Rome. She has an MA in international relations from LUMSA University, in Rome. In 2014 she attended a master of advanced studies in economic security, geopolitics and intelligence at SIOI, and in 2015, as a fellowship student, she received a master in institutions and space policy at SIOI, in Rome.

Yun Zhao is a professor and the head of the Department of Law at the University of Hong Kong, PhD (Erasmus University Rotterdam), LLM (Leiden University), LLM and LLB (China University of Political Science and Law). He is also the director of the Centre for Chinese Law. Prof. Zhao is Chen An Chair professor in International Law at Xiamen University (2015) and Siyuan scholar chair professor at Shanghai University of Foreign Trade (2012–2014). He is listed as an arbitrator in several international arbitration commissions.

The manufacturer's authorised representative in the EU is Springer Nature Customer Service Centre GmbH, Europaplatz 3, 69115 Heidelberg, Germany. If you have any concerns regarding our products, please contact ProductSafety@springernature.com

Printed and bound by CPI Group (UK) Ltd, Croydon, CR0 4YY
01/10/2026
02225959-0004